Aviation as a Business

CENTENNIAL OF FLIGHT SERIES

Roger D. Launius, *General Editor*

Aviation as a Business

CLEMENT M. KEYS AND THE FORMATION OF THE AMERICAN AVIATION INDUSTRY

EDWARD M. YOUNG

Foreword by F. Robert van der Linden

TEXAS A&M UNIVERSITY PRESS

COLLEGE STATION

♾ This paper meets the requirements of ANSI/NISO Z39.48–1992
(Permanence of Paper).
Binding materials have been chosen for durability.
Manufactured in the United States of America.

Library of Congress Cataloging-in-Publication Data

Names: Young, Edward M., 1948- author | Van der Linden, F. Robert writer of foreword
Title: Aviation as a business: Clement M. Keys and the formation of the American
    aviation industry / Edward M. Young; foreword by F. Robert van der Linden.
Other titles: Centennial of flight series
Description: First edition. | College Station: Texas A&M University Press, [2026] |
    Series: Centennial of flight series | Includes bibliographical references and index.
Identifiers: LCCN 2025050139 (print) | LCCN 2025050140 (ebook) | ISBN
    9781648433641 hardcover | ISBN 9781648433658 ebook
Subjects: LCSH: Keys, Clement M., 1876-1952 | Curtiss-Wright Corporation—History
    | Capitalists and financiers—Biography | Aeronautics, Commercial—United
    States—Finance—History—20th century | Aircraft industry—United States—
    Finance—History—20th century | BISAC: BIOGRAPHY & AUTOBIOGRAPHY
    / Aviation & Nautical | HISTORY / Military / Vehicles / Air | LCGFT: Biographies
Classification: LCC HD9711.U62 Y68 2025 (print) | LCC HD9711.U62 (ebook)
LC record available at https://lccn.loc.gov/2025050139
LC ebook record available at https://lccn.loc.gov/2025050140

This book complies with the EU General Product Safety Regulation (GPSR) (Regulation (EU) 2023/988).
For regulatory inquiries and product safety matters within the EU, contact our authorized representative:

Mare Nostrum Group B.V. | Doelen 72 | 4831 GR Breda | The Netherlands

Email: gpsr@mare-nostrum.co.uk | Website: https://mngbookshop.co.uk/

This publication has undergone a risk assessment and meets applicable safety standards.
Traceability identifiers: ISBN, batch number, and publisher contact details are provided for compliance.
No hazardous materials or components are included in this product.

Published by Texas A&M University Press
John H. Lindsey Building, Lewis Street | College Station, TX 77843 | www.tamupress.com

Dedicated to my wife,
Candis Elaine Litsey

CONTENTS

FOREWORD

F. ROBERT VAN DER LINDEN

Curator of Air Transportation
Smithsonian National Air and Space Museum

In the pantheon of aviation greats, Clement Melville Keys is perhaps the greatest name no one has ever heard of. At a time when Wilbur and Orville Wright invented the airplane and their bitter rival Glenn Curtiss competed for the public's attention in air shows around the nation, when Glenn L. Martin, John K. "Jack" Northrop, Allan Loughead (Lockheed), and Donald Douglas were designing their first airplanes, a quiet Canadian-born journalist and entrepreneur had the vision to make aviation an industry. Armed with his deep knowledge of the intricacies of the financial world through his work as a reporter for the *Wall Street Journal* and as editor of *World's Work*, and later through his vast experience gained as the head of C. M. Keys & Company, his own investment firm, Clement Keys understood the need for serious investment for a new enterprise such as aviation to succeed.

Using the case of the tortuous development of the U.S. railroad network in the 19th century as both a positive and a negative model, Keys firmly believed that the new technology of aviation held a great deal of promise, but only if its growth was well organized and well financed by raising capital while rejecting debt financing that crippled many railroads. Keys was among the first to recognize the critical importance of viewing aviation—especially commercial aviation—as a system with many diverse moving parts working cooperatively. In one of his most telling comments, and one that is still commonly quoted, usually without attribution, is his statement that "aviation is $\frac{9}{10}$ on the ground."

Keys became actively interested in aviation as a business during World War I. He shepherded Curtiss Aeroplane through its financial crisis during the war, saving the company and acquiring control in 1920. Keys continued to fight long and hard to keep the company afloat when the military contracts dried up after the war. At a time when many of his competitors were failing, he kept his business alive through astute decisions and creative persistence. Notably, he used the fame of air racing and the development of the famous Curtiss D-12 engine to promote the company and gradually build up a backlog of military contracts.

Keys was the driving force that saved the entire industry from the ruinous competition created by the military's penny-wise, pound-foolish system of competitive bidding whereby a company that had spent considerable capital in designing a new aircraft could be underbid by a competitor and lose the production contract they needed to recoup their investment. To do so, Keys fought hard to create the laws that would shape the foundation of the industry, which bore fruit in 1926 with the Air Commerce Act, the Naval Aircraft Activation Act, and the Air Corps Act, all of which transformed aviation in the United States, providing a strong legal and regulatory groundwork and a steady source of income through government contracts.

Once the system was established, Keys sought to make aviation a thriving commercial enterprise. He recognized the efficiencies gained by consolidation, combining Curtiss with Wright Aeronautical to build aircraft and engines as Curtiss-Wright and creating North American Aviation as an investment trust for aviation stocks. In 1929, Keys used North American Aviation to gain control of Pitcairn Aviation as well as its air mail routes along the East Coast. Keys renamed the company Eastern Air Transport—the direct predecessor to Eastern Air Lines.

Keys was one of the first to recognize the opportunity offered by the government for creating a system of private contract air mail carriers offered by the Air Mail Act of 1925. He also recognized the need for adequate financing and raised unprecedented sums of capital from numerous investors to ensure that his new National Air Transport (NAT) could safely and profitably carry the mail between New York, Chicago, and points south. In early 1930, Keys lost a proxy fight with Frederick Rentschler and his United Aircraft & Transport Corporation; Rentschler was another visionary who saw aviation as big business. As a result, NAT was joined with Boeing Air Transport, thereby creating a single transcontinental enterprise, soon to be named United Air Lines.

Surprisingly, Keys was not distressed by his loss of NAT, for he was absorbed in building a revolutionary passenger airline, Transcontinental Air Transport (TAT). With the substantial backing of the Pennsylvania Railroad and no federal air mail subsidy, TAT offered 48-hour transcontinental air service. At a time when airplanes only flew during the day, TAT combined rail service at night with day service by airplane, cutting transcontinental travel by half. Carefully planned and brilliantly organized, TAT offered excellent service to its wealthy clientele. But this success came at a cost. TAT lost prodigious amounts of money, a loss that was compounded by the collapse of the stock market in 1929. Desperate for help to stem the hemorrhaging of cash, Keys found relief through Postmaster General Walter F. Brown, who, during his reorganization of the national air mail system, saved TAT by forcing a merger with part of Western Air Express and awarding a lucrative air mail contract. The merger created Transcontinental and Western Air—the original name for TWA.

Curtiss-Wright, TWA, Eastern, United, and North American—these are five of the largest companies in U.S. aviation history, and all of them were closely tied to Clement Keys. So why are his contributions largely unknown? What happened?

Edward (Ted) Young explained it to me. I came to know Ted through my years as the curator of Air Transportation at the Smithsonian Institution's National Air and Space Museum. When I met him, I learned that he was formerly a commercial banker; he was now a financial analyst based in Hong Kong and had developed an avid interest in aviation. He had published several excellent books on World War II aviation in the Far East but had gained a particular interest in the business aspect of aviation history that complemented my own interest. His thorough background in the financial world gave him an unparalleled understanding of the intricacies of Wall Street and the tumultuous 1920s and 1930s. Over the years we have exchanged information in many conversations. Concurrently, I was completing my book about the role of the federal government in creating the airline industry in the United States. In studying the expansion and industry-wide consolidations of the industry during the late 1920s and early 1930s, I was puzzled that I was unable to learn what happened to Clement Keys after 1932. All of my primary sources noted that Keys increasingly took time off to recover from stress, but they said nothing of the origins of that stress. I shared these sources and my question with Ted who, with the expert eye of a banker and his own interest in Keys piqued, examined the sources and discovered that Keys was undone by his careless use of call loans that unexpectedly came due. Following the Crash, Keys owed millions of dollars, most of which he repaid, but not all. At the cost of his career, he struck a deal to write off his remaining debt.

The present excellent volume is a labor of love that was years in the making. During the creation of this work, Dr. Young retired from banking and entered academia, earning his doctorate in history. Armed with an academic's understanding of history and a banker's knowledge of finance, he was well equipped to undertake the task of interpreting the complex life of Clement Keys. The product of his prodigious effort is this highly readable and thoroughly researched account that sheds light on this critically important, yet little known and very human, man of vision who was undone by his hubris.

PREFACE

The 1920s were a formative period for the American aviation industry. To survive and grow, the miniscule aviation industry that emerged from World War I needed to develop a consistent market for its products.[1] While World War I had demonstrated the airplane's military importance and its potential as a means of transportation, the airplane "still had to demonstrate its utility in the postwar world."[2] As historian Roger Bilstein has noted, "particularly in the case of civil aviation, planes had to work for a living and yield a profit in the bargain—a dictum as true for air transport manufacturers as for the airline operators who flew them."[3] It took much of the decade to develop a market for airplanes and to ensure that the sale of airplanes to military and civil users and the airplane's commercial applications would yield something of a profit.

In 1920, the few firms that survived the war faced an uncertain future. The military had only limited demand for new aircraft, and there was no government policy for military aviation. Sales of surplus military aircraft fulfilled the minimal demand from the civilian market, and there was no legal and institutional structure to support commercial air transportation. For most people in America flying seemed a risky proposition best left to thrill seekers. In 1921 the aviation industry built 302 airplanes with a value of $6.6 million; apart from one or two halting efforts, commercial air transport did not exist.[4] By the end of the decade, the foundations of the American aviation manufacturing industry and commercial air transportation were in place, though the following decade would present many challenges. In 1929, the peak year of aircraft production that would not be equaled until the start of World War II, the aviation industry built 6631 planes; they were sold in the United States and abroad and had a value of aircraft, engines, and parts of $70.3 million.[5]

The transfer of the air mail service from the U.S. Post Office to private companies through the 1925 Air Mail Act was the beginning of commercial air transportation; by the end of the decade there were 36,321 miles of established airways across the country, with ninety-seven service providers carrying 173,405 passengers (the number would double over the following year) flying 25,141,499 air miles.[6] The Army Air Corps Act and the Navy Act of 1926 had established government policies for military aviation, while the Air Commerce Act provided a legal framework for commercial aviation. The aviation boom that followed Charles Lindbergh's successful solo flight across the Atlantic to Paris saw tens of millions of dollars invested in aviation securities, until the market collapse in October 1929, providing financing for the aviation industry's expansion in the final years of the decade.

During the decade, a small number of men helped transform a vision for aviation into a business. Among them was Clement Melville Keys, an improbable candidate to become a leading figure in the aviation industry. A Canadian by birth, Keys studied Classics at the University of Toronto, spent several years as a teacher, and then came to New York City in search of opportunity. He became a journalist and an expert on the railroad industry, moving on to become an investment advisor and a financier. He came into aviation by chance and remained in the industry for over a decade, becoming, in 1929, president of the Curtiss-Wright Corporation, one of the two largest groups in the American aviation industry and one of the industry's leading entrepreneurs.[7] He helped form three of America's leading airlines: National Air Transport, which became a component of United Air Lines; Transcontinental Air Transport, the predecessor to Transcontinental & Western Air Lines; and Eastern Air Transport, as well as airlines in China and Latin America.

Through his involvement in the aviation industry, Keys combined the key roles of entrepreneur, bringing new products and services to the market, and financier, arranging the financing for his ventures once investors had accepted the aviation industry as a potentially profitable investment.[8] In addition to his involvement in aircraft manufacturing and air transportation, Keys set up several investment trusts and raised substantial funds for investment in the aviation industry. He was active in promoting the industry and argued for government policies and regulations that would promote a sustainable manufacturing industry and commercial aviation. At his peak, he was chairman, president, or director of twenty-seven aviation-related firms. Although he was widely known at the time, his sudden departure from the industry in 1932 meant that his contributions would not be as well recognized as those of some of his contemporaries.

I first encountered Clement Keys while reading Professor William M. Leary Jr.'s book *The Dragon's Wings: The China National Aviation Corporation and the Development of Commercial Aviation in China.* Keys was instrumental in organizing the China National Aviation Corporation. In his book Professor Leary included a brief biography of Keys, describing his improbable rise from a Classics teacher to his leading role in the aviation industry. At the time I read the book in the early 1980s, I had begun a career as a commercial banker in New York City after deciding not to complete a PhD in Political Science. I was intrigued with the parallel between Keys's shift in career and my own move from academics to the world of business.

In an endnote to his book, William Leary noted that "little has been written about the career of this remarkable pioneer in aeronautical financing."[9] While working as a banker, I still retained my lifelong interest in aviation history and a strong desire to write on aspects of aviation that had yet to be explored. Leary's endnote sparked my interest in Clement Keys and planted the idea that perhaps one day I could write a biography of this important participant in forming the American aviation industry.

In 1984 I made my first foray into the Clement Melville Keys Papers at the National Air and Space Museum (NASM). I began reading and gathering material about his life and times during the early decades of the 20th century. A few years later, I had the good fortune to meet William Leary during one of my research trips to NASM. I talked with him about my idea of working on a biography of Keys. He urged me to pursue this project, suggesting that my experience as a banker and as a financial analyst would give me insights into Keys's contributions to the American aviation industry.

In 1987 I moved to the London office of my then employer, Moody's Investors Service, the bond-rating agency. In the fall of 1989, Bill contacted me and asked me to write an entry on Keys for a volume on the airline industry, as part of the *Encyclopedia of American Business History and Biography*, a work that Bill was then editing.[10] I wrote back listing all the reasons why I couldn't possibly write something on Keys from London. His response was, "Great! I'll expect your draft by next June." This was my first attempt to describe Keys's life and accomplishments.[11] It was the beginning of a long effort.

Many people have helped me along the way, including several members of Clement Keys's extended family who have provided invaluable support for this project. My greatest debt among them is to Edith Keys Stoney, Clement Keys's daughter, who generously donated her father's papers to the National Air and Space Museum. Keys was a prodigious correspondent whose papers document in great detail his activities during the 1920s. A full biography of Keys could not have been written without these papers. Through her gift,

Mrs. Stoney has enriched the history of American aviation. Edith Stoney's daughter Juliet F. Stoney Johnson, Clement Keys's granddaughter, and her grandson, Richard D. Johnson, Clement Keys's great-grandson, welcomed my efforts to write a biography of Keys and graciously shared their memories and family photographs. C. Roy Keys Jr. and his father, C. Roy Keys Sr. (Clement Keys's nephew), provided background on the Keys family and graciously gave me several family photographs. They kindly accompanied me on a trip to Keys's birthplace in Chatsworth, Ontario, and introduced me to Karen Foster at the Gray Roots Museum who gave me useful information on the history of the region. Donald Furin, a great-nephew of Keys's second wife, Indianola, gave me several photographs of Keys.

Professor William Leary was a constant source of encouragement until his untimely passing in 2004. We corresponded regularly on my progress, with Bill prodding me to start writing. I deeply regret that he did not live to see this book. I believe he would have been pleased with the result.

The NASM staff encouraged me to work on Keys's biography and gave their enthusiastic support to my work. Phil Edwards, then Chief of Research Services, was my guide to the Keys papers and other resources in the NASM library. The staff at the NASM archives at the Silver Hill facility arranged for me to view the Keys papers during several research trips. Over many years Dr. Tom Crouch, Dr. Von Hardesty, Dr. Dominic Pisano, and Dr. Robert van der Linden gave me invaluable assistance. They arranged for me to receive several Short-Term Visitor Grants to NASM, which enabled me to work through the Keys papers. My conversations with Dom Pisano and Bob van der Linden on the development of the American aviation industry in the 1920s, and particularly Bob's own work on the growth of the airlines during this period, helped refine my understanding of Keys's contributions to the industry. I am deeply grateful to Bob for agreeing to write a foreword for this biography.

Several libraries and archives were important to my research. I would like to express my deep appreciation to Ms. Reeve Lindbergh for giving me permission to view her father's papers at the Sterling Memorial Library at Yale University and to Judith Ann Schiff, then Chief Research Archivist, for facilitating my visit to the Library. The Missouri Historical Society and the Hagley Museum and Library contained valuable information on several of Keys's aviation activities. I also benefited from visits to the Glenn Curtiss Museum in Hammondsport and the Museum of Flight in Seattle. The staff of the Textual Records branch at the National Archives and Records Administration helped me locate key correspondence between Keys and Major General Mason

Patrick and Rear Admiral William Moffett. Barbara Hanson, president of the United Airlines Legacy Foundation in Chicago, gave me access to the minutes of the Executive Committee of National Air Transport.

In addition to photographs from the Keys family, Nicole Davis at the Museum of Flight gave me permission to use several photographs from the Museum's extensive collection. Michael Poerio found a photograph of the Aerial Experimental Association in the Glenn Curtiss Museum archives. Debbie Seracini at the San Diego Air & Space Museum provided a rare photograph of the China National Aviation Corporation. Jessica LaBozetta of the American Heritage Center at the University of Wyoming kindly provided a photograph of Keys as a member of the 1919 American Aviation Mission to Europe.

I would also like to thank my good friend Philip Wall, who kindly read several draft chapters. My fellow volunteers at the Museum of Flight, David Brown, Steve Ellis, John Little, and Dennis Parks, have all been supportive and have encouraged me to keep working on my biography of Keys. Steve Ellis's questions and comments helped me refine my arguments.

I am grateful to Thom Lemmons, Editor in Chief at Texas A&M University Press, and Madelyn Weston, assistant project editor, for accepting this biography and for all their assistance in arranging publication. The comments from the University Press's anonymous readers were invaluable.

My sisters, Katherine and Upasana Young, and my children, Evan and Sarah, have followed my progress on this project, and each in their own way has given me their constant support. Their faith that I would complete the book meant a great deal to me.

I have dedicated this book to my wife of fifty years, Candis Elaine Litsey. She has lived with this project almost as long as we have been married. She has listened over many years to descriptions of Keys's life, given me perceptive and thoughtful comments on his career, closely edited several early drafts, and patiently supported me during my long years working on this biography. I could not have completed this book without her loving support.

Aviation as a Business

INTRODUCTION

DURING THE 1920S, AVIATION IN AMERICA BECAME A BUSINESS. AT THE beginning of the decade, however, the industry seemed to hold little promise. The miniscule American aviation industry was fighting for its very survival, with barely enough demand from the military to keep the few surviving manufacturing firms alive. The productive capacity that had been built up with such effort, cost, and controversy during the war had been reduced to a fraction of its peak production. Inundated with war-surplus airplanes, the industry had no market for its products. Commercial air transport was far from a reality, a mere dream in the minds of aviation enthusiasts. There were no laws governing aviation, no regulations, no comprehensive government policies, and no capital willing to take the risk of investing in a field that seemed so precarious. Flight and the airplane had captured the imagination of many, but little more than that. Aviation in America in these early years was "unorganized, unregulated, unsafe, and unprofitable."[1] The future of the industry appeared precarious.

By the end of the decade, however, the aviation industry had carved out a place in the American scene. The industry's survival was no longer in question; it had developed "the necessary infrastructure to sustain continuing growth and global success."[2] While the automotive and railroad industries still dwarfed aviation, the transformation in the aviation industry's performance during the decade was remarkable. From a low of $6.6 million in 1921, the value of aircraft produced had grown to $71.1 million by 1929, while production had increased from 302 aircraft built in 1921 to 6034 military and civil aircraft in 1929.[3] The newly established commercial airlines had carried 5782 passengers in 1926; in 1929 they carried 159,751 passengers.[4] Total investment in all aspects of aviation in the second half of the decade had amounted to more than $800 million, compared to around $5 million in 1920.[5]

The infrastructure supporting the aviation industry was physical, legal, organizational, and financial.[6] In 1929 there were 132 firms, of differing sizes, manufacturing airplanes for the military air services, commercial airlines, and private flyers located in states across the country, 34 domestic airlines operating a network of 24,874 route miles, and 948 commercial and municipal airports in operation.[7] A legal framework for the aviation industry emerged in the middle of the decade, with passage of the Kelly Air Mail Act in 1925, which transferred the air mail from the U.S. Post Office to private operators.[8] The Air Commerce Act of 1926 set up regulatory standards for aviation, while the Air Corps and Navy Acts of 1926 authorized five-year programs for building military aviation.[9] New firms emerged to manufacture airplanes, engines, and accessories, to establish commercial airlines for mail and passengers, and to provide a range of services affecting all aspects of aviation.[10] Toward the end of the decade, a merger movement got underway to consolidate firms within the industry, creating large aviation combinations in the form of holding companies and investment trusts that linked manufacturing firms with airlines and other types of firms in the industry.[11] As the industry expanded, other organizations emerged through the "institutionalization of aviation," including industry and research organizations, universities, media, and financial institutions.[12]

The ability of the firms in the aviation industry to access the financial markets was crucial to their development. Charles Lindbergh's flight to Paris in 1927 triggered a surge of public interest in aviation. The burgeoning demand for aircraft and aircraft engines from the military, the new commercial airlines, and the rapidly growing number of private flyers improved the conditions and outlook for many firms in the industry. "At last," as one report on the industry put it, "aviation had become a *business*. Always alive to new developments, the business men and financiers of America have for the past year or two been fully aware of what is going on in this new industry. Aviation stocks are making their appearance in increasing numbers, aviation investments are approaching a substantial basis and, while still speculative, become capable of analysis in the same manner of other securities."[13] The "Lindbergh Boom," as it was called, saw the aviation industry raise hundreds of millions of dollars through the sale of shares in the financial markets, fueling the industry's expansion.[14]

The historian Thomas P. Hughes has described these combinations of technology and social organizations as large technological systems where physical artifacts—in this case airplanes, engines, and airports—combine with different types of organizations and influences, such as manufacturing firms, airlines, investment banks, research facilities, the media, legislative actions, and government regulation.[15] An invention is the precursor to a large technological system. Different types of individuals play key roles, at different times, in large technological systems.[16] It is often an inventor–entrepreneur, a man like Thomas Edison, who begins the process of developing the invention to have wider applications and seeks to develop a market for the product and a profitmaking business.[17] In later stages of development, as the technological system expands through growth, competition, and consolidation, it is often the manager–entrepreneur and financier–entrepreneur who drives the system forward.[18] Hughes labeled the individuals who carried out this integration of technological artifacts and social organizations on a large scale "systems builders."[19] Hughes states that the key characteristics of the system builders are "the ability to construct or to force unity from diversity, centralism in the face of pluralism, and coherence from chaos."[20]

The Austrian economist Joseph Schumpeter wrote extensively on the critical role of the entrepreneur. In his book *The Theory of Economic Development*, he describes the important role of the entrepreneur in the formation of "new combinations," a process that was a "fundamental phenomenon" of economic development, which Schumpeter defined as including the introduction of a new good or service into the economy, a new method of production, the opening of a new market or a new source of raw materials, or the formation of a new form of organization within an industry.[21] It was the role of the entrepreneur to carry out these new combinations, bringing together the productive factors needed and, critically, arranging for their financing. This was not by any means an easy process. It was not simply a question of identifying the possibility or conceiving of a new combination in whatever form but overcoming the inertia of everyday life and even social opposition to the new endeavor. The task of the entrepreneur was to break up old traditions and create new ones.[22] As Schumpeter stated the problem:

> In the breast of one who wishes to do something new, the forces of habit rise up and bear witness against the embryonic project. A new

and another kind of effort of will is therefore necessary in order to wrest, amidst the work and care of the daily round, scope and time for conceiving and working out the new combination and to bring oneself to look upon it as a real possibility and not merely as a day–dream. This mental freedom presupposes a great surplus force over the everyday demand and is something peculiar and by nature rare.[23]

Entrepreneurs were individuals who built businesses. Their role was "to supply the coordination and effort that brought a new process or product to the point of adoption and commercialization."[24] Entrepreneurs, "acting on their own, or inside firms, did more than make decisions in the face of change. They responded creatively to change, seeking to shape and use it for their own purposes. Though Schumpeterian entrepreneurs did not have to risk their own capital, entrepreneurship was inherently a risk-taking activity in other ways, one so painful and fraught with uncertainty that it justified the financial returns it generated when it succeeded, even when they were very large."[25]

During the 1920s, the American aviation industry had its entrepreneurial systems builders. These manager–entrepreneurs and financier–entrepreneurs gradually replaced the earlier pioneer aviators as the heads of manufacturing firms and nascent airlines, bringing in more professional management, larger organizational structures, and developing close business relationships with the financial community.[26] They were at the forefront of setting up the large aviation groups toward the end of the decade and were comparable to the great railway men—Alexander Cassatt, James J. Hill, and Edward R. Harriman—from earlier in the century. Contemporary observers singled out three men as the leading entrepreneurs in the American aviation industry: William E. Boeing, Richard F. Hoyt, and Clement M. Keys.[27] Boeing, together with Frederick Rentschler of the Pratt & Whitney Company, set up United Aircraft & Transport Corporation, the first of the large aviation groups. Richard Hoyt, an investment banker and chairman of the Wright Aeronautical Corporation, built a group of manufacturing companies around Wright. Clement Keys was president of the Curtiss Aeroplane & Motor Company and was active across a broad range of aviation and airline companies as well as in aviation finance.[28] In 1929 Keys and Hoyt merged their two groups to form the Curtiss-Wright Corporation, second to United Aircraft & Transport Corporation in size and breadth of activity. Despite their importance to the development of American aviation during the 1920s, historians of aviation have not served these system

Clement Keys in the late 1920s. His improbable rise from
a teacher of Latin at a Canadian boys' school to one of
the leading figures in the American aviation industry is the
subject of this biography. (Courtesy C. Roy Keys Jr.)

builders well. Boeing is familiar through the company that retains his name, but Hoyt and Keys are known to very few.

Of these three men, Clement Keys is the most remarkable. From modest beginnings, Keys rose to became one of the leading figures during the formative years of the American aviation industry. Keys, like Samuel Insull, was a systems conceptualizer. His vision for aviation was broader than that of many of his contemporaries. He saw the value of the airplane for national defense, as a means of transporting passengers and high-value commodities, overcoming geographic barriers and saving time, and, for some, a vehicle for their own enjoyment of flying. Keys wanted to create a group of related companies that would integrate all these different aspects of aviation both horizontally and vertically. He wanted manufacturing companies that could build airplanes and engines for the military, for air transport, and for private flying; companies that would provide training for prospective pilots; companies that would manage airports for commercial airlines and private pilots; companies that would provide sales and services; and companies that would arrange financing for all these ventures, both at home and abroad.

Above all, Keys understood that aviation was a business. "Too many glowing words are written and uttered," he said, "about aviation, its growth, its future. This romantic element had obscured somewhat the fact that *building* aviation is a difficult business task, to be done by business men, using best business principles."[29] He argued that "the most important element in aviation is nothing in the world but the usual business common sense that underlies all industry."[30] His fundamental operating principle was that nine-tenths of aviation was on the ground, not in the air, that organization, infrastructure, and executive management were more important than actual flying.[31]

Throughout the 1920s, Keys was actively involved in the aviation industry as a tireless promoter of aviation, as an entrepreneur organizing new combinations of manufacturing, airline, and aviation service companies, and as a financier of aviation. Born and raised in a small town in Canada, trained in the classics, a journalist and then investment advisor and banker by profession, he came into aviation by chance during World War I and stayed with it by choice. In a remarkable ten-year period, Keys rescued the Curtiss Aeroplane & Motor Corporation from bankruptcy and from it built one of the largest aviation groupings in America, with interests spanning the manufacture of military and commercial airplanes and engines, aviation exports, air services, air transportation in America and abroad, and finance. He helped organize what became three of the largest airlines in America: National Air Transport, Transcontinental Air Transport, and Eastern Air Transport, predecessors of United Airlines, TWA, and Eastern Airlines.

But like Icarus, Keys rose to great heights, only to fall through his own hubris. An error in judgment forced him to leave the aviation industry after the onset of the Great Depression. Today his achievements are barely known. What follows is an account of how Clement Keys came to realize this vision and turn aviation into a business.

1

INTO AVIATION

From Teaching to Journalism

WHEN CLEMENT MELVILLE KEYS ENTERED THE AVIATION INDUSTRY AT THE end of 1916 as a director of the Curtiss Aeroplane & Motor Corporation, he was forty years old. He brought to his new position the experience, skills, and expertise he had accumulated over fifteen years as a journalist, investment advisor, and investment banker. He was born on April 7, 1876, in the small rural town of Chatsworth in Western Ontario where his father, Reverend George Keys, served as an Anglican minister. George and Jessie Keys had a large family, which was not uncommon at the time. Clement had two older sisters. Four more sisters and two brothers would follow.[1] Keys's youngest brother, Roy Keys, would later work with him at Curtiss. Clement Keys spent his childhood in Chatsworth and nearby Clarksburg, developing a love of reading, sports, and a passion for fishing which, like reading, he would pursue throughout his life.[2]

Clement Keys completed his high school education in June 1893 at the Lindsay Collegiate Institute in Lindsay, Ontario. He won a scholarship to the University of Toronto and a place in the Honor Classics program at University College just after he turned seventeen.[3] His achievement is an early indication of his high intelligence, drive, and capacity for hard work. At the time, the Honor courses were stepping stones to higher advancement in the professions and in teaching.[4] The Honor Classics focused on the language, literature,

Clement M. Keys as a young man, possibly
in his early years as a journalist with *The Wall
Street Journal*. (Courtesy C. Roy Keys Jr.)

and civilization of Greece and Rome; over four years the Honor student was
given "a graduated and progressive course of reading in literature, history, and
philosophy" covering the foundations of Western Civilization.[5] Study of the
Classics, together with that of the Greek and Roman languages, demands a
certain type of intellect: a mind with strong analytical ability, a capacity to
absorb and retain large amounts of information, attention to minute detail,
and above all, critical thinking.[6]

Keys attained his highest marks in his Honor Classics course, but he took
other classes as well. During his four years at the University of Toronto, he
passed examinations in English, Mathematics, History, and Geography.[7] He
graduated from the University of Toronto in June 1897 with a Bachelor of Arts
degree. His family history, education, and inclination steered him toward
one of the professions as a career, a not uncommon choice for University of
Toronto graduates. For reasons that he never explained, Keys decided not to

follow his father into the ministry. Instead, in the spring of 1898, he accepted a job teaching English and Latin at Bishop Ridley College in St. Catherine's, Ontario, on the Niagara peninsula directly opposite Toronto.

Founded in 1888 as an Anglican Church school for boys, Ridley College was modeled on the best British public schools with the aim of providing for the moral, intellectual, and physical development of the young men in its charge, which would help mold them into "men of sound character."[8] Keys quickly established a place in the life of the school through his participation in sports; at Ridley, athletic prowess conferred a special status even on a master.[9] Keys served as a referee during rugby football games, played in rugby games between the masters and the sixth form boys, and coached the fencing team. One of his fencing students was Frederick W. "Casey" Baldwin, who graduated Ridley College in 1900 and went on to the University of Toronto to study mechanical engineering. Eight years later, "Casey" Baldwin would introduce Keys to the new world of aviation.

Although he was successful as a teacher, in the spring of 1901 Keys informed the Ridley headmaster that he had decided to leave at the end of the term. He had won the respect of his students and worked well with the Headmaster and his fellow masters. Had this position satisfied him he might well have remained at Ridley College. Teaching offered a secure future and a respected profession, but for Keys remaining at Ridley offered no challenge. He was thinking of going in a different, less certain direction. Keys decided to become a writer and not a writer in Canada, but in New York.

He had examples to follow. During the last decade of the 19th century, when Canada was in the grip of a deep recession, thousands of young Canadian men emigrated to the United States to seek advancement in a land with more opportunities than Canada could provide. Within his class at the University of Toronto, several of his classmates had gone to New York to "take a shot at being a writer."[10] This was a bold decision: Keys had no contacts in the writing world in New York, no reputation as a writer, and little money.

Keys arrived in New York City sometime during the summer of 1901 hoping to make a living as a freelance writer. Some of the first articles he wrote were a series on birds and bird life that appeared in the Sunday edition of the *New York Times* over the fall and winter of 1901–1902.[11] But soon Keys went in a new direction. As he related to a journalist much later, one day in early fall Keys heard that there was a panic on Wall Street. Word was out that Thomas Lawson, a noted Boston stockbroker and stock speculator closely affiliated with the Amalgamated Copper Company, had suffered spectacular losses in the crash of Amalgamated's stock, dragging other investors down with him. Curious to see what a Wall Street panic looked like and always on the lookout for a good story, Keys went to Wall Street expecting to see crowds of people yelling and shouting outside the Stock Exchange. To his surprise, there was no disturbance; there were no more people on the street than on any other normal business day. But he was intrigued; what had happened with Lawson and Amalgamated Copper, and why? Wall Street and its ways were a complete mystery to him. He decided to learn this new world of finance.[12]

Keys applied successfully for a job as a reporter with the *Wall Street Journal* and began working for the *Journal* sometime in 1902. He never discussed what drew him to finance and business, but he was quite clear on why he became a financial journalist, as he later explained:

> I took that step deliberately, as a matter of training. It seemed to me that when young men enter finance by way of the bond houses, selling securities, they absorb the necessary background of financial knowledge—but with the slight bias of the seller's enthusiasm. On the other hand, the financial reporter must see clearly, without bias; no self-interest colors the opinions which he forms. As a financial writer, I thought I would learn to size up companies and financial issues clearly.[13]

Keys joined the *Journal* knowing nothing about business or finance. He threw himself into this new world, absorbing new information like a sponge. During the day he worked at the *Journal*, learning the skills of a reporter, becoming familiar with the many companies in the news, the men who ran them, and the workings of Wall Street. At night he studied on his own, teaching himself accounting, corporate finance, and the dynamics of the stock and bond markets.[14] Not surprisingly, he proved to be a diligent student and a fast learner.

In 1902 there was no better place for a student of business and finance than New York City and few better places to observe this world than from the *Wall Street Journal*. With offices at 44 Broad Street, just below the New York Stock Exchange, the *Journal* sat at the very center of business and finance. When Keys joined the *Journal*, he began working for Thomas F. Woodlock, soon to become editor-in-chief. He could not have asked for a better teacher. Woodlock had joined the *Journal* as a reporter in 1892. Woodlock developed an interest in the railroad industry, and in 1895 he published *The Anatomy of a Railroad Report*, the first book to describe how to analyze the financial condition of a railroad. By 1900 he had become a recognized expert on the railroads and their financial condition in America. He was a superb writer and editor, one of the leading financial journalists of his day, and with nearly ten years of experience had a firm grasp of financial journalism as a discipline and a business.[15] Woodlock assigned Keys to covering the railroad industry.

The great railroads of the day—the Great Northern, the Union Pacific, the Pennsylvania Railroad, and the New York Central, among others—were, using Thomas Hughes's definition, large technological systems. They incorporated a multitude of components—the physical infrastructure of the railroad line, operating equipment for passenger and freight service, maintenance and repair facilities, administrative structures, financing and advertising functions—all connected into a network to provide a reliable service under centralized control that sought to ensure optimum performance and attain operational and financial objectives.[16] By the end of the 19th century, many of the speculators, investors, and charismatic men involved in the railroads had given way to what Hughes calls "system builders," men like James J. Hill, Edward H. Harriman, and Alexander Cassatt. These men were professional railroad managers or financiers who built the administrative structures that managed their railroads and the connections with leading financial institutions that provided the capital necessary for continued investment and expansion.[17] During his years as a journalist Keys followed these railroad systems and their systems builders, studying their methods, successes, and disappointments. He would build a comprehensive knowledge of the railroad industry.[18]

As the largest single industry in America, the fate and financial condition of the railroads were of vital interest not just to participants in the stock and bond markets, but also to financial institutions, farmers, merchants, manufacturers, regulators, and politicians across the country. In 1903, railroad bonds comprised 88.5 percent of the bonds listed on the New York Stock Exchange and 50.3 percent of corporate stocks.[19] *The Wall Street Journal* provided its readers with news and commentary on the industry, financial data, and financial analysis. News of the railroads featured prominently in the *Journal*. A regular column titled "Railroad Affairs" covered news and events concerning the industry and railroad companies across the country.[20]

The *Journal* reported on the ebb and flow of railroad business, financial results, new bond and stock issues, system improvements, the activities of the Interstate Commerce Commission, as well as general business developments, such as the annual grain harvests, that could have a bearing on railroad earnings. Railroad earnings were a regular feature in the paper, reported on a monthly, quarterly, and yearly basis.[21] When a railroad issued an annual report, the paper published a synopsis of the financial statements. Stock market movements were of keen interest to many. In its "Review and Outlook" commentary, the *Journal* provided insights into recent trading and prices of the most active railroad stocks. Railroad news came from many sources: company press releases, reports from news bureaus across the country, filings with the Interstate Commerce Commission (ICC) and state regulatory commissions, stockbrokers on the Street, and interviews with railroad officials.

Investors needed more than just news and financial statistics. In the stock and bond markets at the turn of the century, investors had little access to analysis of securities or companies, and even fewer sources of reasonably objective opinion on the relative financial soundness of railroads or other corporations. The public had to take a great deal on faith, which left the market open to constant manipulation.[22] For most people, determining the value of a particular stock or bond or the financial strength or weakness of a railroad was a difficult task. In his book on railroad reports, Thomas Woodlock stated that for most people, a railroad's financial report ". . . is a meaningless mass of unintelligible figures which are undecipherable and generally confusing."[23] The *Journal* provided its readers with formal analysis of the financial condition of selected companies and railroad systems, detailed descriptions of bond offerings, and responses to readers' inquiries concerning their investments or investment opportunities. In these features, the *Journal* drew explicit conclusions about a company's relative strength and the value of its stocks and bonds, frequently giving its readers direct investment advice.

Keys likely began his career at the *Journal* working on the "Railroad Affairs" column, compiling news from railroads around the country, which provided a good introduction to the railroad industry. In 1903, after Thomas Woodlock had become editor of the *Journal* at the end 1902, Keys was promoted to Railroad Editor. Over the next three years Keys gained recognition as an expert on railroad finance and management. During this formative period, he made an intensive study of the railroads as transportation systems and as organizations, gaining insight into railroad administration, management, and financing, as well as an appreciation for the vital importance of transportation to commerce and society. The breadth of the *Journal*'s coverage of the railroad industry gave Keys a unique opportunity to observe and compare the strong railroads and the weak ones, and to learn what contributed to the success of a railroad, and what led to failure. This knowledge would provide a foundation for his later work on the development of air transportation.

A transportation system, be it by road, water, rail, or air, is deceptively simple. It entails the movement of people and goods from one point to another, over a certain distance, with a relative degree of certainty, comfort, reliability, and safety. The "system" used to provide the service consists of the mechanism that provides the service—a bus, train, boat, or plane—and the infrastructure needed to support the service and sustain the mechanism used to provide it. As a transportation system expands geographically in its frequency of service, in the volume of passengers and freight carried, and in the technological complexity of the mechanism used to provide the service, the need for coordination and control expands concurrently.[24] Management of this complex, interlinked system with thousands of moving parts—what one contemporary observer described as "a mechanism so vast, so complicated, and so delicate"—was no easy task.[25] The success or failure of a railroad depended on a host of often interconnected variables, many of which were outside the railroad's control, where cause and effect had to be carefully calculated down to the last dollar. Keys would bring this understanding of the dynamics of transportation systems to his later efforts to build successful commercial airlines.

In the words of the contemporary observer, "the extreme intricacy with which we are dealing, cannot be emphasized too much. In the practical management of a railroad, so wide is the area and so intricate the relation of all the causes and all the effects, that if we depart from existing types or methods in any subordinate part of the operations, it may be at risk of grave consequences that cannot be foreseen. . . . Disturb any existing service by some change, however slight, and it may make the difference between a good share of a competitive business and none at all."[26] Managing a railroad required, above all else, relentless attention to detail. This was a lesson not lost on Keys. It would be the foundation of his approach to commercial air transport systems.

Keys learned from Thomas Woodlock how to analyze railroad companies, to identify their strengths and weaknesses. He would have learned that many railroad failures in the 1893–1897 depression were due to excessive debt. The lessons of those failures were clear: the railroads that survived were those that had a high level of operating efficiency and thus lower operating costs, a strong business franchise, conservative financing with low levels of debt and interest payments, and strong, stable earnings. Those companies that lacked these elements often failed. Excessive debt and fixed charges could cripple a company, leading to bankruptcy and a painful reorganization. The cost to the bond investors, who often lost all of their principal and interest in these railroad reorganizations, left a lasting impression on many who witnessed the process.[27]

Keys studied the railroad men who built and ran these large technological systems. The character, capacity, and management style of these leaders of the railroad industry would be a recurring theme during his years as a journalist. As a reporter and railroad editor, Keys came to know the leading railroad men of his day, men like Alexander Cassatt of the Pennsylvania, William K. Vanderbilt of the New York Central, George Gould who controlled the Gould Lines, B. J. Winchell who ran the Rock Island Line for William H. Moore, Stuyvesant Fish of the Illinois Central, and others. The two men who featured most prominently in the pages of the *Journal*, however, were James J. Hill and Edward H. Harriman, the leading railroad men of the era, whom the *Journal* described as "men of gigantic strength, indomitable will, and marvelous accomplishments."[28]

Hill was known as "the Empire Builder" for his remarkable achievements developing railroads in the Northwest. After two decades of arduous labor, Hill and his able managers had built the Great Northern Railway into one of the greatest railroad systems of its day, pioneering routes from the Midwest to the Pacific Northwest, while at the same time driving for ever greater efficiency. Hill became renowned for his attention to the details of all phases of railroad operation.[29]

Harriman was a phenomenon in the railroad industry.[30] Unlike many of the industry's leaders, Harriman had not spent years as a railroad executive. He rose from office boy to head of his own investment banking firm, E. H. Harriman & Co., becoming one of the leading figures on Wall Street by the 1880s. He burst onto the national railroad scene when he became chairman of the Union Pacific Railroad in 1898 to undertake its reorganization. Harriman's genius was in improving the capacity and efficiency of the railroads under his supervision. Like Hill, Harriman paid remarkable attention to detail. In the space of just a few years he transformed the Union Pacific into one of the most efficient railroads in America, perhaps the most remarkable achievement in railroading in that era. In reporting on Hill and Harriman, Keys had an opportunity to compare and contrast two men of exceptional ability and achievement. They would prove an inspiration to him.

From Journalist to Investment Advisor

By 1905, Keys had established a reputation as one of the outstanding reporters on the *Journal*'s staff.[31] That year he began to contribute to the magazine *The World's Work*, one of the leading magazines of the day. In early 1905, Walter

Hines Page, founder and editor of *The World's Work*, commissioned Keys to write an article for *The World's Work*, titled "A 'Corner' in Pacific Railroads," examining how Edward H. Harriman, president of the Southern Pacific and Union Pacific railroads, acquired control of most of the railroad lines crossing the Rocky Mountains to the Pacific.[32] Keys followed this article with one in the June issue of *The World's Work* on the men who had built the Chicago, Rock Island & Pacific Railway into a major American railroad system in the space of just a few years.[33] For the August issue, Keys returned to the railroad competition between the transcontinental railroads, describing the struggle between James J. Hill and his Northern Pacific Railroad, Edward Harriman and the Union Pacific, and the Canadian Pacific Railway Company.[34] His final article for Page that year, "As Many Railroad Methods as Railroad Kings," was a study of the different management methods employed at the great railroads of the day, comparing and contrasting Cassatt of the Pennsylvania Railroad, Vanderbilt of the New York Central, Hill, Gould of the Missouri Pacific, and Harriman of the Union Pacific.[35]

In the fall of 1906, Walter Page offered Keys a position as Financial Editor at *The World's Work* to cover the transportation industry and the world of finance.[36] The offer was a great opportunity for Keys. While the *Journal* had an excellent reputation as a financial paper, it was not well known outside of financial and business circles. In contrast, *The World's Work* was a national magazine, with a circulation of over 70,000 and growing. Keys would have his own byline, something he did not have at the *Journal*. In October 1906, Keys left *The Wall Street Journal* and moved uptown to the Doubleday, Page & Co. offices at 133 East 16th Street where Page produced *The World's Work*.

Moving to *The World's Work*, Keys gained the steadier pace of producing monthly articles. He shifted to writing in greater length and depth about the subjects he covered as financial editor. Walter Page's focus was on the growth of American commerce, particularly transportation and communication.[37] During his years writing for *World's Work*, Keys wrote extensively on the American railroad industry and the leading railroad men of the day, continuing his examination of railroad management methods. *World's Work* gave Keys an opportunity to explore new areas in American business and commerce, particularly the American financial system.

As financial editor of *The World's Work*, Keys took on responsibility for the magazine's monthly investment advice column. When Page started *The World's Work* in 1900, no national magazine gave practical advice on investments. The investment column soon became a regular feature in the magazine, providing the "disinterested, expert, free advice" that Page wanted to give his readers.[38] Keys found that far too many men and women approached investing with too little experience, too little knowledge, and too great a desire for high returns on their money, and accordingly paid the price. The stock and bond markets of the day were not for the innocent or the faint-hearted. As one contemporary writer described them, "without training or study or intelligent consideration, these outsiders, or 'lambs' as they are often called, stake their money on the mere chance of the rise and fall of prices. . . . No wonder they are generally all sheared in the Street."[39]

For *The World's Work* columns, Keys gave sound, conservative, commonsense advice, focusing on the problems of the small investor with only a small amount of capital to invest. He centered on risks to the investor. Where safety of principal was the investor's primary consideration he advocated a conservative investment strategy. He invariably recommended high-class bonds, which paid a reasonable return with very little risk. To the investors who wrote to him asking where they could find higher returns on their money, he repeated over and over again that higher returns came only with higher risk. Higher risk investments were for the businessman who could afford to absorb a loss, not for the average investor.

Keys was critical of the lack of knowledge of finance displayed by the legions of investors then entering the stock market. The financial education of the American public was, in Keys's view, in its infancy. The larger banks on Wall Street had little interest in the smaller investor, and even when they did, the advice they gave was not always in the investor's best interest. What investors needed was a banker who was "a man of conscience and of judgment; and his desire to sell bonds or stocks, and so make much money, is tempered by his desire to really serve the interests of the men who entrust to him their hardearned money."[40] There was an opportunity here, and within a few years, Keys would reach for it.

After four years as Financial Editor of *The World's Work*, Keys had solidified his career as a journalist; he was now one of Walter Hines Page's key staff members on the magazine. But despite this success, Keys had begun to think about a new challenge. Early 1911 became a turning point for him. As Keys later related, Page approached him with an offer to become editor of the magazine.[41] After some consideration, Keys declined it, intent on pursuing a different direction. In February 1911, he wrote Page a letter explaining that he had decided to stop working for a salary and that he was going into business for himself. As Keys told Page, "I have made no plans, talked business with nobody in a definite way, and cannot say what I am going to do."[42] For the second time in his life, Keys turned his back on a promising career to take on a new challenge. He had left teaching to become a journalist, and he was now leaving journalism for a new career in the world of finance.

In the decade that Keys had worked as a journalist, as a product of America's rapid economic growth, a small revolution had occurred in the investment business. With more money at their disposal, the American middle class's interest in investment grew apace; the number of individual investors in securities increased from 4.4 million in 1900 to 7.4 million in 1910.[43] What these individual investors needed above all was advice. In his years as a journalist, Keys had written his investment columns in a language that the investing public could readily understand. He had responded to hundreds of inquiries, in the process making numerous contacts in the business world and among individual investors. This experience would be the basis for his business. In the fall of 1911, Keys established C. M. Keys & Co. as an investment advisory firm, with offices in a building on 35 Nassau Street in Manhattan, two blocks north of Wall Street and the New York Stock Exchange, with one clerk and a secretary. He had only his own savings as capital, his expertise, and his willingness to work hard.

At the end of 1911, Keys began placing small advertisements in the financial sections of *The World's Work*, *Literary Digest*, and later the *New York Times*, listing C. M. Keys & Co. as "Investment Counsel." The advertisements stated that his business was "to advise and act as broker for individual investors, banks, institutions, businesses, and those handling trust funds."[44] His advertisements offered short, pithy investment advice, speaking to safety, diversity, and the benefits of unbiased investment counsel, with nothing to sell but service. He repeated the lessons he had given in his investment advice column: "every chance for large profit is balanced by an equal chance for large loss."[45] The business of giving investment advice quite naturally led Keys to buying and selling railroad, public utility, and industrial bonds on behalf of his clients. As a bond broker, Keys worked with the large Wall Street banking houses, participating in their bond underwritings as a sub-underwriter. This gave him more practical experience in corporate finance. As the business of C. M. Keys & Co. expanded, Keys took on several large corporate clients, including the Willys-Overland Corporation, the automobile manufacturer. Keys came to know John Willys, an automotive pioneer who had organized the company.[46] They would later work closely together.

As tensions escalated in Europe during the summer of 1914, the financial markets in Europe and America shared a growing uneasiness. While appalled at events in Europe, most American businessmen saw a European war as an opportunity for American industry and agriculture. As the only major industrial nation not at war, the United States could provide what Europe needed.[47] Keys and others knew that European nations would have to liquidate their large holdings of American stocks and bonds in order to purchase the materials they would need to support their war effort, giving American investors who had ready cash available the chance to pick up these securities at bargain prices. This was, Keys believed, the "silver lining" to the clouds of war.[48]

The war in Europe generated an increasing amount of business for C. M. Keys & Co. As predicted, England, France, Russia, and other European nations turned to America for food, raw materials, and manufactured goods. To finance this production and the expanded transportation needs it generated, American companies and the railroads issued over $1 billion in new bonds during 1915.[49] There were plenty of opportunities to select among quality bond issues, and for those who wanted to support the British and the French in their fight against Germany, to acquire their government bonds. Providing investment advice to his clients and buying and selling bonds on their behalf kept Keys busy. Of particular note, in early 1916, a prospective client contacted Keys. The man had recently sold his company for a substantial amount of money and wanted to know if Keys could help him with investment advice. His name was Glenn Curtiss.

Into Aviation

Keys had his first exposure to aviation while working at *The World's Work*. On July 4, 1908, at Hammondsport, New York, Glenn H. Curtiss made the first official flight greater than one mile in the *June Bug*, winning the *Scientific American* trophy. The August issue of *The World's Work* featured dramatic photographs of Curtiss in the *June Bug* and the airplane in flight. In an editorial in the issue, Page stated that "the problem of aerial navigation is solved," giving credit to Curtiss and Dr. Alexander Graham Bell's Aerial Experiment Association.[50]

Bell had organized the Aerial Experimental Association in 1907 and had recruited Glenn Curtiss for the Association because of Curtiss's experience with lightweight gasoline engines. Three other young men became members at Bell's invitation: Lt. Thomas Selfridge, of the United States Army and two young Canadians, recent mechanical engineering graduates of the University of Toronto, J. A. D. McCurdy and Frederick "Casey" Baldwin, Keys's former student at Ridley College. When Keys learned that Casey Baldwin was a

A photograph of the members of the Aerial Experimental Association.
Left to right: Glenn H. Curtiss, Frederick W. (Casey) Baldwin, Dr. Alexander
Graham Bell, Lieutenant Thomas E. Selfridge, and J.A.D. McCurdy.
Keys had taught Casey Baldwin at Ridley College in Canada.
Baldwin introduced Keys to J. A. D. McCurdy, who later introduced
Keys to Glenn Curtiss. Thus, Keys's acquaintance with Baldwin led to
his entry into aviation. (Glenn Curtiss Museum, Hammondsport, NY)

member of the Aerial Experiment Association, he arranged for a *World's Work*
reporter to visit Hammondsport with a letter of introduction to Baldwin.
Later that fall, Baldwin and McCurdy went to New York for a visit, where
Baldwin introduced McCurdy to his old teacher. Keys and McCurdy struck
up an enduring friendship.[51]

At the end of 1910, Glenn Curtiss set up his own enterprise, the Curtiss
Aeroplane Company, and a sister company, the Curtiss Motor Company, in
Hammondsport, New York. By the summer of 1914, Curtiss had established
a reputation in America and Europe for building airplanes, flying boats, and
airplane engines. With the outbreak of war, Curtiss received orders from
Great Britain, Italy, and Russia for flying boats and a new training aircraft.[52]
At the end of 1915, the Curtiss Aeroplane Company, together with the Curtiss
Motor Company, had contracts for airplanes and engines amounting to

between $13 million and $15 million. On December 16, 1915, the *Wall Street Journal* reported that the Curtiss Aeroplane Company had announced the signing of a new contract with the British government for an additional $15 million.[53]

In less than a year, the Curtiss Aeroplane Company had grown from a single small factory in Hammondsport to an organization with two factories in Buffalo, New York, a factory in Canada, and an engine factory in Hammondsport. Glenn Curtiss was an inventor, far happier tinkering in the informality of his workshop than in running a factory.[54] He brought in production managers from the auto industry, which helped address the need for greater standardization and efficiency, but his company also had to raise capital to fund continued expansion. Curtiss had limited experience with finance. He needed men with broader business experience, familiar with the demands and disciplines, both organizational and financial, of large companies.

Fresh capital was available. Investors were keen to take advantage of a rapid increase in profits from war-related production. As the largest manufacturer of airplanes in America, and with confirmed contracts in the millions, the Curtiss Aeroplane Company was an attractive investment. In the fall of 1915, the New York investment banking firm of William Morris Imbrie & Co. approached Glenn Curtiss with an offer to purchase the Curtiss Aeroplane Company on behalf of a syndicate of banks and other investors. Curtiss accepted, and on December 31, 1915, he announced that he had sold control of his company to the William Morris Imbrie syndicate.[55] On January 14, 1916, the syndicate incorporated a new company, the Curtiss Aeroplane & Motor Corporation. The syndicate formed a Board of Directors with businessmen and bankers from Buffalo and New York City.[56] Curtiss was appointed president, but he now had a board of experienced businessmen to advise him.

Glenn Curtiss was now a rich man. For selling his interest in his company, he received $5 million in cash (roughly equivalent to $92 million today), $3.5 million of the preferred shares, and 75,000 shares of common stock.[57] With no idea about how to deal with his new wealth, Curtiss consulted with J. A. D. McCurdy, who recommended that Curtiss go to New York City and contact his friend Clement Keys for investment advice, which Curtiss readily did.[58] This began a business and personal relationship between Curtiss and Keys that would last until Curtiss's death in 1930, a relationship that would lead directly, though not immediately, to Keys's involvement in aviation. Keys's initial meeting with Glenn Curtiss did not result in much more than providing advice on investing his money, though Keys appears to have impressed Curtiss with his financial acumen sufficiently for Curtiss to call on him later in the year.

During 1916, Keys became heavily involved in corporate finance, working with a number of large corporate clients, in addition to dealing with an increasing volume of brokerage business in stocks and bonds. In early December, Glenn Curtiss contacted Keys and asked him if he could help the Curtiss Aeroplane & Motor Corporation avoid a damaging default. When the Imbrie syndicate incorporated the new company, it had issued $3 million in 6% notes, maturing on January 1, April 1, and July 1, 1917.[59] During 1916 the Curtiss Corporation had to cancel and restructure several contracts for the British Royal Naval Air Service. Since the airplanes ordered under the new contracts would not be completed until the latter part of 1917, the Curtiss Corporation would not have sufficient cash to pay off the $3 million notes maturing between January and July. Keys was pessimistic about the chances for success and told Curtiss so, but he agreed to try. Keys arranged for Curtiss to come down to New York to meet with a group of bankers. The negotiations were difficult, and the discussion was heated.

Keys and the banking group worked out an agreement to refinance the Corporation's debt on more favorable terms, extending the maturity of the new debt to better match the Corporation's ability to repay. Under the terms of the agreement, Curtiss Aeroplane & Motor Corporation sold $2 million in 6% Serial Notes to a banking syndicate that William Morris Imbrie & Co. had organized; these notes were to be repaid in installments of $400,000 from 1918 to 1923. An additional $2 million 6% notes, maturing in 1926, were sold to private investors. With the funds in hand, the $3 million in serial notes maturing in 1917 were called for redemption, leaving the Corporation with an additional $1 million for working capital.[60]

Glenn Curtiss apparently wanted to continue to benefit from Keys's financial experience and expertise. At the end of December, Keys was elected a director of the Curtiss Aeroplane & Motor Corporation, made a member of the executive committee, and joined Glenn Curtiss and James Imbrie on the Curtiss Aeroplane & Motor Corporation's Voting Trust, which controlled the company's stock.[61] He may not have realized it at the time, but this would be a turning point in his life. His appointment as a director was to be his entry into aviation. For the next fifteen years, Keys would be directly involved with the affairs of the Curtiss company and the formation of the American aviation industry. What Keys brought to the Curtiss Corporation was his experience in corporate finance and his contacts on Wall Street. He knew how and where a company could obtain financing.

America Enters the War

From the time of America's entry into the war in April 1917 until the Armistice in 1918, Clement Keys was responsible for arranging financing for the Curtiss Aeroplane & Motor Corporation's exceptionally rapid wartime expansion. It would prove to be a task that, though successful, involved much frustration as Keys and Curtiss management struggled with often conflicting demands from the American government. For the first time in his life, Keys would be directly involved not just in financing Curtiss but also in the active management of a company through a period of unprecedented challenges. And, for the first time, he would negotiate directly with government agencies.

Ironically, soon after his appointment to the Curtiss Board, Keys participated in negotiations for a possible merger of the Curtiss Aeroplane & Motor Corporation with the Wright-Martin Aircraft Corporation, a company formed in 1916 through a merger of the Wright Company, which Orville Wright had sold to a syndicate in 1915, and the Glenn L. Martin Company of Los Angeles.[62] The Wright-Martin Aircraft Corporation was then gearing up to produce the French Hispano-Suiza airplane engine at a new plant in New Brunswick, New Jersey, and building several Martin aircraft for the United States and the government of the Netherlands East Indies. The Corporation experienced delays in getting the Hispano-Suiza into production and needed funding, having nearly run through its own working capital raised the year before.

A merger of the two corporations, with new financing, would have created a well-capitalized firm with extensive capacity for airplane and engine production. In an interview with the aviation magazine *Aerial Age* in early March, Keys stated that negotiations between Curtiss and Wright-Martin had been going on for some time, but that no definitive agreement had been reached. Keys explained the rationale for the discussions, saying that "the sole basis of the negotiations has been a strong mutual desire to put the aeroplane manufacturing interests in a position to meet a national need and emergency. We all need added strength at this moment, and we are doing our best to get it."[63] However, no merger took place. *The Wall Street Journal* reported that the talks had been broken off as the two companies had not been able to come to an agreement on the value of their respective assets and terms of a merger.[64] The merger of the two great names in American aviation, in which Keys would play an even greater role, would not take place for another twelve years.

That same month Keys made one of his most significant contributions to the Curtiss Corporation's effort in World War I. Dr. Charles Walcott, director of the Smithsonian Institution and chairman of the Executive Committee of the National Advisory Committee for Aeronautics (NACA), organized a meeting in Washington between officers of the Army and Navy and representatives of the aviation industry to discuss ways to accelerate the preparedness of the industry for war.[65] Some fifteen companies, involved in various aspects of the aviation industry, attended the meeting. Brigadier General George O. Squier, chief of the Signal Corps who was responsible for the Army's Aviation Section, and Rear Admiral David W. Taylor, chief of the Bureau of Construction and Repair which was responsible for aircraft procurement for the Navy, represented the military. Howard E. Coffin, a prominent automotive engineer, a forceful advocate for American aerial preparedness, and a member of the Council of National Defense, who would play a pivotal role in America's aviation program in World War I, attended as well. Keys was invited to the meeting along with several other managers of the Curtiss Aeroplane & Motor Corporation.[66]

During the day-long meeting, the participants discussed the state of the aviation industry and the prospective needs of the Army and Navy over the next several years. In his opening address, Walcott submitted an estimate, based on discussions with the military, that called for an annual production of 2000 combat planes, 2000 training planes, and 2400 aviators a year, with the assumption that this tentative program could be accomplished by 1919.[67] Walcott acknowledged that the aviation industry, such as it was, was unable to meet the needs of the military services. The industry was woefully unprepared for the task ahead. Only some twelve companies had the manufacturing facilities and capacity to build aircraft that could possibly meet government requirements, and of these twelve, only five had ever built more than ten aircraft. By the end of 1916, the Army's Aviation Section had received a total of 149 airplanes and had approximately 302 airplanes on order; the U.S. Navy had even less.[68]

It was estimated that, based on current orders, the industry would deliver approximately 600 airplanes during 1917; expanding production into the thousands would be a monumental task. On March 29, the NACA sent a letter to the secretaries of War and the Navy to convey the results of the meeting with the industry, stressing the importance of implementing a multiyear procurement program so that the manufacturers could invest the capital required to expand their facilities with confidence.[69] Describing the inadequacy of the aviation industry in America, the letter stated that "At present, the industry may be divided into two parts, the Curtiss Aeroplane & Motor Corporation, and others."[70] The implications for the Curtiss company were clear: if America went to war, the Curtiss Aeroplane & Motor Corporation would be flooded with orders.

What the government was proposing was production of aircraft on an unprecedented scale. Keys realized that the Curtiss Corporation lacked the experience of quantity production on the scale contemplated. The Corporation needed to bring in experienced production managers, and quickly. The automobile industry was one of the few comparable industries that did have experience with quantity fabrication and production, though events were to prove that there was a vast difference between building automobiles and building airplanes. Keys turned to John N. Willys, president of the Willys-Overland Automobile Company, whom he knew through his banking business. Willys had personal experience with the challenges of the rapid expansion of industrial production. In 1908 he took over and reorganized the Overland Automotive Company, renaming it the Willys-Overland Company. He built his company's automobile production from 4000 cars a year in 1910 to 150,000 cars a year in 1915, with factories in Michigan, New York, and Ohio. Keys, apparently with Glenn Curtiss's approval, persuaded Willys to join Curtiss Aeroplane as a director; he was elected to the Board on April 10, 1917.[71]

Keys was right to be concerned about the demands that would soon be placed on the Curtiss Corporation and the need for John Willys's experience in production. Shortly after the United States' declaration of war against Germany, the Wilson administration struggled to develop an airplane program as part of a massive mobilization of American industry. On April 12, the NACA submitted a more ambitious plan for aircraft production than its first proposal in March; this new plan called for the production of 3700 planes in 1918, 6000 in 1919, and 10,000 in 1920.[72] The government established the Aircraft Production Board on May 16, with Howard Coffin as chairman, to oversee production. Within a month the government abandoned the NACA plan for one that called for even more airplanes. In response to a request from the premier of France for more American aircraft, the Army Signal Corps' Aviation Section devised a plan to produce an astounding 22,000 airplanes by the summer of 1918, consisting of some 10,000 primary and advanced training planes and 12,000 combat planes, with additional reserves.[73] The sense of urgency that pervaded Washington at the time and the abiding faith in American productive ability swept away all caution and concerns with the limited capacity of the American aviation industry and the immensity of the burden that was being placed on it. With the approval of Secretary of War Newton Baker, the Army requested an appropriation from Congress for $600 million, shortly increased to $640 million, to begin the airplane program. Congress accepted the program, which passed the House with unanimous consent and was approved in the Senate on July 24, just two months after the French premier's cable was received.

By the summer of 1917, the volume of business at the Curtiss Buffalo and Hammondsport factories had increased so rapidly that the company was feeling the strain. The Corporation was behind in its deliveries of flying boats to Britain as the U.S. Army was about to place substantial orders for Curtiss JN-4 training planes.[74] In the midst of this frenetic activity, the Aircraft Production Board told the Curtiss Corporation that it needed to strengthen the Corporation by bringing into senior management men with stronger manufacturing experience and building up the Corporation's working capital.[75] Keys was soon deeply involved in negotiations with John Willys for a combination of the Curtiss Aeroplane & Motor Corporation with the Willys-Overland Company in order to improve Curtiss's manufacturing capacity.[76]

In an interview with the *Wall Street Journal*, Keys explained the rationale for the combination of the two companies. The Curtiss Aeroplane & Motor Corporation was the leading American company in the experimentation and development of airplanes. When it became clear that the country would require the large-scale production of standardized airplanes, it became equally apparent that the Curtiss Corporation would have to greatly expand its manufacturing capacity or be relegated to a subordinate position. The directors of the Curtiss Corporation realized establishing a relationship with the Willys-Overland Company, which had broad experience in producing both automobiles and engines, seemed the quickest way to bridge the gap between development and production. By joining with Willys-Overland, the Curtiss Corporation would gain access to manufacturing expertise and new sources of supply.[77]

Under an agreement worked out with the Curtiss Corporation, Willys-Overland purchased a controlling interest in Curtiss for a period of three and a half years. John Willys immediately put in place new senior management. Willys became president of the Curtiss Aeroplane & Motor Corporation, while Glenn Curtiss became chairman of the Board, relinquishing day-to-day control over the company. James E. Kepperley, a vice-president of the Willys-Overland Company, joined the Curtiss board as a director. Keys remained a Curtiss director, and Willys appointed him vice president for finance. With equal speed the Curtiss board authorized the issue of 63,000 shares of common stock at $35 a share, raising $2 million in new working capital.[78] On July 14, Willys wrote to Howard Coffin to describe the changes he had made at Curtiss to strengthen management and improve the company's financial position. He told Coffin that "what I want to impress on you is that the Curtiss Company, beginning next Monday, is a different institution. In three to four months, no other concern will be able to compete with them in the cost of manufacture and ability to get a large volume."[79]

New orders soon flooded into Curtiss. The government had decided that instead of designing new aircraft, the American aviation industry would build suitable European aircraft types already in production. The Bolling Mission, sent to Europe to select the aircraft to be built in America, recommended the British De Havilland DH-4 day bomber and Bristol F2B two-seat fighter, the French SPAD fighter, and the Italian Caproni tri-plane bomber.[80] As expected, the Curtiss Corporation received the bulk of the orders for combat planes. In a contract dated September 19, 1917, the Army Signal Corps gave Curtiss verbal confirmation that it would place an order for 3000 SPAD pursuits and 500 Caproni heavy bombers, a contract worth $30 million. The formal written contract for the SPADs and Capronis came on October 8, 1917.[81] At last, five months after the declaration of war against Germany, John Willys and his colleagues had the official go-ahead to begin work on the great undertaking of building airplanes for an American air army in France. Or so they must have thought at the time. What followed over the next ten months was a pattern of conflicting orders, cancellations, and confusion that brought only intense frustration and next to no combat airplanes for the Army.

Working day and night, Curtiss engineers at the new North Elmwood factory in Buffalo had completed 90 percent of the design drawings, and the company had begun purchasing materials to begin production when, on November 7, the Aircraft Production Board told the Curtiss Corporation to stop all work on the SPAD. The Board had received advice from American officers in Europe that two-seater fighters would do all the air fighting in the future. Curtiss was left with a set of design drawings for the SPAD representing six weeks' intensive work, a large quantity of raw materials, newly hired employees, and a huge, new factory that was now idle.[82] Shortly thereafter, the Aircraft Production Board canceled the contract for Caproni bombers.[83]

In response to the cancellation of the SPAD contract, Keys drafted a letter to Howard Coffin, which John Willys and James Kepperley also signed, outlining all the problems and frustrations the Curtiss Aeroplane & Motor Corporation had encountered working with the government. In the letter, Keys stated that while neither the Aircraft Production Board nor the Curtiss Corporation had control over the causes of many of the delays encountered, the end result was that "we have not, at this time, an order to build a single machine of combat type. Far from being able to crowd our great new plant to the doors with fighting machines for the service of American and Allied forces in the field, the Corporation is unable at this time to write a single definite bill of material, set up a single jig for construction, or even to send working drawings to the dozens of vendors and contractors who wait to cooperate with us in this task."[84]

In a tone of restrained anger, Keys outlined all the efforts the Curtiss Corporation had made to live up to its responsibilities, not the least of which was building and financing the new North Elmwood plant for government production. The Curtiss Corporation, he pointed out to Coffin, had kept faith "under our broad unwritten agreement of cooperation in every possible way with your Board and the United States in its aviation program," strengthening the management of the Corporation, supporting the training of other contractors and government inspectors, providing detailed plans to subcontractors, and rapidly building training planes. The delays and confusion would mean, he argued, that the possible maximum production the Curtiss Corporation could achieve by July 1918 had been cut by 25 to 40 percent. Keys and his colleagues wanted it made perfectly clear to the Aircraft Production Board, the Signal Corps, and the government that the Curtiss Aeroplane & Motor Corporation was in no way responsible for the delays.[85]

With American observers in Europe urging production of the two-seat fighter, the Aircraft Production Board and the Army Signal Corps shifted Curtiss to manufacturing the Bristol F2B. Ironically, an example of the Bristol fighter arrived at the Curtiss Buffalo factory on the day the Aircraft Production Board canceled the SPAD contract.[86] The Signal Corps' Aviation Section, however, insisted on redesigning the Bristol fighter, replacing the airplane's 200 hp Rolls-Royce Eagle engine with the 400 hp Liberty-12 engine to maximize use of the Liberty, regardless of the fact that the Liberty was too big and too powerful for the Bristol.[87] On January 10, 1918, Curtiss received a contract for 2000 Bristol fighters with 1200 sets of spare parts worth $14 million.[88] Although reluctant to build the redesigned Bristol with the Liberty engine,

Curtiss-Bristol Fighters in various stages of construction at the Curtiss Aeroplane factory in Buffalo. The aircraft proved to be overweight and the Liberty engine too powerful. On July 20, 1918, the government canceled its contract with Curtiss for 2000 Bristol fighters. For Keys, this was a valuable lesson. (Library of Congress LOT 6028)

the Curtiss Corporation had no choice but to comply and it managed to complete its first example at the end of January.

The first production examples of the Curtiss-built Bristol began test flights at Wright Field, in Dayton, Ohio, in early May. The plane was a failure; it was overloaded, overpowered, and unsafe to fly.[89] In two successive accidents on June 10 and July 15, 1918, both the pilot and the observer were killed. On July 18, the Army Air Service, which had replaced the Signal Corps Aviation Section, stopped production of the Bristol. Two days later, the Air Service canceled the Curtiss contract after twenty-seven airplanes had been built.[90] By then, the Curtiss Corporation was already at work on a new contract to build 1000 British SE-5 pursuits. In a reversal of previous decisions, the Aircraft Production Board decided that after all there was a need to build single-seat fighters in America, and so, on the recommendation of the British War Mission in Washington, the SE-5 was chosen to be used as an advanced trainer. By the middle of July, plans and four examples of the SE-5 were at the Curtiss plants in preparation for production. In the space of a year, the Curtiss Aeroplane & Motor Corporation had received instructions and contracts to manufacture four different types of European combat aircraft, and it had managed to complete a mere twenty-seven Bristol fighters, which had to be scrapped.[91] The failure to meet the goals of the aviation program for the Army created a public outcry and a furor in Washington in the spring of 1918, leading to sensational accusations of mismanagement, vast waste, graft, and conspiracy.[92] This led to a Senate investigation of aviation production and a separate Department of Justice investigation under former Supreme Court Justice Charles Evans Hughes.

The controversy over the failure of the aircraft program for the Army overshadowed Curtiss's record of successfully building flying boats for the U.S. Navy. The Curtiss Corporation's experience with the Navy was the reverse of the frustrations the company experienced with the Aircraft Production Board and the Army Signal Corps. In October 1917 the Navy had ordered substantial numbers of Curtiss HS-1/2 single-engine flying boats and larger H-16 twin-engine flying boats from Curtiss, with production of these aircraft allocated to other companies as well. Curtiss also built training aircraft for the Navy. In June 1918, the first Curtiss HS-1s were assembled in France and began flying patrols. Production reached a peak in September when the Navy received thirteen twin-engine and thirty-eight single-engine flying boats each week. By the summer of 1918, when the Curtiss Corporation had yet to complete a successful combat aircraft for the army, Curtiss flying boats were on active service with the Navy in Europe.[93]

Stress and Strain: Addressing Financial Problems at Curtiss

By the fall of 1917, Keys was having to cope with the Curtiss Corporation's increasingly precarious financial position. The rapid pace of building up the production of training and combat aircraft, combined with the delays and contract cancellations, made it difficult for the corporation to finance its activities on a prudent basis. The frenetic pace completely overstretched the Corporation's limited accounting and control systems that were in place when America went to war. Between November 1916 and December 1917, the total assets of the Corporation doubled from $13.1 million to $26.4 million. The total number of employees grew from 2081 in January 1917 to 12,337 by December, an increase of 493 percent. Over the same period, monthly shipments for the Army and the Navy grew from a little over $300,000 a month to over $2.8 million a month.[94]

Financing this expansion was a constant challenge. As vice president for finance, Keys was responsible for raising funds for the Curtiss Corporation, but market conditions were difficult. In October 1917 Keys had helped arrange the $4 million mortgage on the new North Elmwood plant with several local banks in Buffalo, and he had even taken $500,000 of the amount for the account of C. M. Keys & Co., but apart from this borrowing and the issue of common stock the previous summer, the Curtiss Corporation had not obtained any additional funding from the financial markets.[95] When production began to accelerate after the Army placed its first contracts at the end of June 1917, Curtiss found itself in a financial bind. The Corporation had to make payments to vendors for purchases of raw materials and pay its employees before it could get reimbursed from the Army or the Navy.

Many of the Curtiss vendors were small suppliers in the Buffalo area who could not afford to extend any credit; even some of the larger suppliers demanded immediate payment for purchases. As the inventory of steel, wood, fabric, and the hundred and one other items needed to build airplanes grew by the day, and as the number of employees grew equally as rapidly, the Curtiss Corporation had an ever greater amount of money tied up in inventory and receivables from government contracts that it could not readily convert to cash, while at the same time growing amounts of accounts and notes payable were coming due. Because the Corporation's cash reserves were insufficient to pay its bills promptly, many of its accounts payable to its vendors were in arrears. This lack of liquid working capital would persist for much of the wartime period, causing no end of problems. The Curtiss Corporation's reputation with its vendors was already problematic when America went to war, and it worsened as the months went on, to the point where some vendors simply refused to sell to Curtiss. This caused a shortage of critical parts, which led to further disruptions in production, to the annoyance of the Army and the Navy.[96]

All these problems came to a head after the SPAD contract was canceled in early November. The Curtiss Corporation now had its huge new North Elmwood plant standing idle, large amounts of raw materials in stock or on order that could not be used, and thousands of employees on the payroll with little or no work to do but who had to be kept on in anticipation of production of the Bristol fighter, no contract to cover these costs, and little likelihood of obtaining financing. After reviewing the Curtiss Corporation's financial position and the points Keys brought out in his November 22 letter, the Aircraft Production Board finally recognized that Curtiss was expanding beyond its capacity to finance on its own. Having Curtiss factories operate at peak efficiency was vital to the success of the aircraft program. To achieve this efficiency, the credit standing of the Curtiss Aeroplane & Motor Corporation with its vendors had to be maintained.[97] The quickest solution to the Corporation's financial problems was simply for the government to provide the Corporation's financial requirements for the duration of the war.

Keys assumed the task of negotiating an agreement with the government, working with the Aircraft Production Board, the Army, the Navy, and the War Credits Board in Washington. In the final agreement that Keys negotiated, the War and Navy Departments agreed to advance $11.9 million to the Curtiss Aeroplane & Motor Corporation. This advance repaid $3.5 million in earlier government advances, retired the $4 million mortgage on the North Elmwood plant, and redeemed $3.4 million in outstanding notes from 1917. In return, the Curtiss Corporation agreed to give demand notes to the War and Navy Departments for $11.9 million and a first mortgage on all its real estate, plant, and equipment. Curtiss would repay the advance through deductions on its government contracts, with interest at 5 percent. As Keys had feared, the requirement to repay existing debts would leave the Curtiss Corporation with only $1 million in new working capital, which he knew was insufficient, but there was little choice. He signed the contract with the War and Navy Departments on January 18, 1918.[98]

At the same time that he was negotiating with the Army and the Navy for the financing, Keys was working to meet the Curtiss Corporation's immediate funding needs. In December Keys had managed to negotiate an advance of $2 million from the Army against receivables on Army contracts, and then he hurriedly arranged a 30-day loan with the Irving Bank in New York for several hundred thousand dollars, based on his personal pledge that the loan would be promptly paid at maturity. Keys had spent the first week of January 1918 in Washington negotiating with the Army and the Navy. On his return to New York, he learned from J. F. Prince, the Curtiss Corporation's Treasurer, that as of January 3 the Corporation had $4.6 million in cash and receivables and $7.2 million in advances, accounts payable, and notes payable coming due. More critically, there was a cash balance of $470,000 against notes due in the next ten days of $453,000 and a dividend payment of $210,000. Despite the pressure he was under, Keys noted laconically that this situation " . . . brought to my mind that within the next week something must happen."[99]

The Corporation managed to make these payments, but the situation remained challenging. On January 25, Prince wrote to Keys informing him that the money from the Army and the Navy had been received, but stating that "our payables, of course, to a great extent, are considerably past due, and I am wondering if you can give me any idea of about when we may count on securing new funds."[100] In early February, Keys managed to obtain a larger line of credit with the Irving Bank of $400,000, and he told Prince that he was trying to arrange additional lines of credit with other banks in New York. On February 16, 1918, Keys published notices in the financial papers calling in all the existing Curtiss Aeroplane & Motor Corporation notes, and he instructed Prince to "ease up on payments as much as needed" because he could not approach the banks until he had confirmation of the Corporation's financial position as of January.[101]

In conjunction with the call for repayment of the Curtiss Corporation notes, and with a likely view to reaffirming Curtiss's reputation in the financial markets, Keys issued a statement to the press clarifying the Corporation's financial position. Keys said the call was part of a general plan worked out with the government to address the Corporation's financing requirements while it was working on large government war contracts. Keys noted how the Corporation's rapid expansion had exceeded the working capital available to it and that the Army and the Navy had decided on this plan to relieve the financial strain on the Curtiss Corporation as well as to improve the Corporation's working capital.[102] In an article on the Curtiss Aeroplane & Motor Corporation's progress in *The Wall Street Journal* a few days later, the paper noted that "from a financial standpoint the company is now on a sound foundation," as evidenced by its calling its notes and by the government's cooperation in meeting the Corporation's financing requirements.[103]

In mid-April 1918, Keys submitted a proposal to the War Credits Board for additional advances to Curtiss, recommending a refinancing of all existing notes through an advance of $12 million, to be secured with a trust deed instead of a mortgage, in which the Curtiss Corporation would agree not to pledge any of its assets up to the amount of the advance. The switch from a $15 million mortgage to a $12 million trust deed would free up an additional $3 million in working credit that could be made up through bank financing. Keys further recommended that $3 million of the advance be made immediately, so that Curtiss could address its working capital problem.[104] At the same time, Keys continued to work with the banks in New York to arrange additional lines of credit.

Over the summer, the Corporation pursued a number of alternatives to solving its financial problems. During June, at the request of the Board of Directors, Keys explored the possibility of listing the Curtiss Aeroplane &

Motor Corporation stock on the New York Stock Exchange. A listing had many advantages, not least of which was the benefit to the Corporation's perceived credit standing. Keys recommended that the Corporation seek a listing as soon as possible, but the Board apparently decided not to pursue the idea.[105] Although Keys's negotiations with the War Credits Board resulted in a tentative agreement for the Board to advance an additional $4 million to Curtiss, the Corporation chose not to make an application for the advance.[106]

For all the work Keys did that summer, the real solution was to obtain an increase in production. The Curtiss North Elmwood plant was still operating well below its capacity, a situation that cancellation of the contract for the Bristol fighters did nothing to improve. Resolution of the problem was slow in coming. With the contract for 1000 SE-5 fighters underway, Curtiss received a new contract from the Navy for 60 F-5L flying boats, an evolution of the earlier Curtiss flying boats.[107] During September discussions took place with the Bureau of Aircraft Production, the successor to the Aircraft Production Board, for additional contracts for the Army that would fully utilize the capacity of the North Elmwood plant. In early October, Keys accompanied John Willys and James Kepperley to Washington to negotiate contracts with the Bureau. Keys advised Willys that one of the more important factors for the Curtiss Corporation would be the terms covering the cancellation of orders at the end of the war, and urged that they attempt to negotiate "a good clean cancellation clause that would protect both us and the Government."[108] The contract the three men negotiated with the Bureau of Aircraft Production covered orders for 4000 USD-9A airplanes, a development of the De Havilland DH-9A, and 2,000 USB-1 aircraft, a revised version of the ill-fated Bristol with a Wright-Hispano 300 hp engine; when these new contracts were added to the contract for the SE-5s, the combined total amounted to $60 million, the largest airplane order awarded during the war.[109] The final contract was signed on November 8, 1918, three days before the Armistice and a year after completion of the North Elmwood plant. During that year, and prior to the final contracts for the USD-9A and USB-1, the Army had placed orders with the Curtiss Corporation for 6500 combat airplanes and had received a total of twenty-seven Bristol fighters and a single SE-5 due to the nearly constant delays, interruptions, and cancellations.

For Keys, Willys, Kepperley, and the other senior managers at Curtiss, the year and a half they had spent working with the Aircraft Production Board, and its successor, the Board of Aircraft Production, had been one of intense frustration. In April, Keys had written to Commander G. C. Westervelt, a superintendent for aircraft construction with the U.S. Navy's Bureau of Construction and Repair, saying that "I do not know any leader in the aeroplane industry who would not have been glad to have been out of it from September 1917 to the present moment. I personally would at any time during this period have been delighted to have had an opportunity to deliver or cancel contracts and throw the whole industry into the hands of the people who, knowing nothing at all about it, have exercised full authority over it. I can see in it neither profit, honor, nor the feeling that comes from jobs well done."[110]

The Curtiss Engineering Corporation

One problem the Curtiss Corporation confronted as it tried to expand its production of airplanes in the first half of 1917 was the fact that the experimental design department in Buffalo was also acting as the production engineering department. This was quite a different role for which the experimental design department had little experience. Keys and some of the other senior managers realized that this weakness had to be addressed before the company could take on the large orders that were planned following the declaration of war.[111] For some time they had planned to separate these two functions as soon as it was practicable, taking Glenn Curtiss and some of the engineers and putting them in a separate company that could concentrate on developing airplanes and airplane engines. When the Willys-Overland Company assumed a controlling interest in the Curtiss Corporation, this created an opportunity to set up the new venture. During August, while he was wrestling with the Aircraft Production Board about financing the North Elmwood plant, Keys was heavily involved in getting the new company up and running. The Curtiss Engineering Corporation, as it was named, was incorporated on August 13, 1917, with Glenn Curtiss, John Willys, James Kepperley, K. B. McDonald, and Keys as directors; Keys also became treasurer of the Corporation.[112] Keys reported to Howard Coffin at the Aircraft Production Board that Glenn Curtiss had bought 20 acres of land at Garden City, Long Island, where the company's factories and experimental shops would be built. Keys also noted that construction was underway and that the Curtiss Aeroplane & Motor Corporation was prepared to appropriate $1 million for the new company.[113]

During 1918, Keys became heavily involved with the work of the Curtiss Engineering Corporation at Garden City, Long Island. While afflicted with some of the same pressures, problems, and frustrations that he found at the Curtiss Aeroplane & Motor Corporation, his experience at Garden

City working with the U.S. Navy provided him with a different perspective on the relationship between government agencies and private companies. At the Curtiss Engineering Corporation, Keys witnessed what could be achieved when the military, instead of dictating to a manufacture what it should build as the Army aircraft program had done, allowed the manufacturer's own engineers to meet the military's requirements using their own expertise and designs.

The Curtiss Engineering Corporation had been designed for experimental work, but because of the wartime demands for production and the need to finance its operations, the Corporation now also had two large production contracts for the Navy. One contract was for fifty H-16 large flying boats, and another contract was for eighty MF flying boat trainers. The Garden City factory faced the same problems that had developed in Buffalo, what Keys called "the old difficulty of a bad mixture of engineering and manufacturing," which he attributed to a lack of clearly defined authority in the manufacturing area.[114] Keys appears to have stepped into the vacuum. Although he was only the treasurer of the Corporation, Keys was soon making production decisions, which he had never done before. James Kepperley wanted Keys to continue to be heavily involved in Garden City. "I wish," Kepperley told him, "you would get into this thing and go to the bottom of it as the reports at this time coming to the organization here do not help matters as it looks as though somebody was trying to pass the buck."[115]

Keys wrote to Kepperley that he wanted to strengthen the engineering department to make it both more efficient and productive, while at the same time revising production and strengthening management at the factory. With the approval of the Navy, Keys set up an Engineering Department to separate development work from manufacturing.[116] Keys arranged for a capable manager from one of the Curtiss Engineering Corporation's subcontractors to assume responsibility for the Garden City factory, though he asked Kepperley if he could spare an additional factory manager from Buffalo for Garden City.[117] By the end of April, the factory had attained a production rate of two MF flying boats per week and had managed to complete nine MF flying boats that month, though deliveries of the H-16 were behind schedule, with only one airplane delivered during April against a schedule calling for ten.[118] During that same month, Keys drew up a draft agreement between Curtiss Aeroplane & Motor Corporation and Curtiss Engineering establishing the basis of cooperation and profit sharing on experimental and production orders.[119]

The Curtiss 18T prototype with the 400 hp Kirkham K-12 engine. Curtiss engineers developed the 18T independently. In August 1918, the 18T achieved a speed of 163 mph, making it the fastest airplane in the world at the time. (2008-03 -31_image_658_01, Peter M. Bowers Collection, Museum of Flight, Seattle, WA)

While working on the Navy contracts, the Curtiss Engineering Corporation also took on an experimental aircraft engine project. In 1916 Charles Kirkham, chief engineer for aircraft engines at Curtiss, had started work on a large, powerful engine to compete with the Hispano-Suiza engine that the Wright Aeronautical Corporation was planning to build under license. By the end of 1917, Kirkham was testing a revolutionary engine designated the K-12, a liquid-cooled inline 12-cylinder engine. The K-12 engine was the forerunner of a family of 12-cylinder engines that would bring the Curtiss Corporation fame and fortune during the 1920s. Kirkham then designed an airplane to take advantage of his new engine's power and small frontal area. His design was a two-seat, tri-plane fighter, with twin machine guns for the pilot and two flexible guns for the observer following the pattern of the Bristol fighter.

Kirkham's fighter, designated the 18T, made its first flight in early July, demonstrating a top speed of 150 mph and an impressive rate of climb. This performance attracted the interest of the Army, which borrowed the first prototype from the Navy for tests. On its first official flight, on August 19, 1918, the 18T attained a speed of 163 mph, making it the fastest airplane in the world at that time.[120] Despite its remarkable performance, neither the Army nor the Navy placed production contracts for the 18T. The K-12 was still an experimental engine and was not ready for production, while the tri-plane configuration had gone out of favor.[121]

The success of the K-12 engine and the 18T made a deep and lasting impression on Keys. Through its own limited efforts and with the help of the Navy, the Curtiss Corporation had come up with an engine and an airplane that was far superior to any of the European airplanes the Army had ordered into production, a full 40 mph faster than the Bristol Fighter. Keys came away convinced that if only the Army had supported Curtiss and its own designs, the results would have been far more successful than the Army's own aviation program. A year later, testifying before Congress, Keys said, "if we had been put on that [the K-12 engine] in July 1917, instead of being put on the Bristol and all such phony machines, we would have been in production in January 1918, and producing machines in quantity production in the spring of 1918."[122]

The End of the War

When the end came, it came quickly. A week after the Armistice was signed on November 11, 1918, the government ordered the Curtiss Aeroplane & Motor Corporation to stop work on most of its contracts, canceling orders for the SE-5, the USB-1, the USD-9A, and 350 twin-engine flying boats for the Navy, leaving Curtiss with contracts for 144 training planes and 60 F-5L flying boats.[123] John Willys protested to the Bureau of Aircraft Production that the Curtiss Corporation was not being treated fairly, but to no avail; demobilization was the order of the day, and with the war at an end, the government had no more need for airplanes.[124] The Corporation quickly began winding down production, laying off 4000 employees at the Buffalo factories in the week following notice of the contract cancellations. From April 1917 to March 1919, Curtiss delivered 5221 airplanes and over 5000 airplane engines to the Army and Navy.[125]

There was much to be proud of in this record. The Curtiss JN-4 and its Navy equivalent, the N-9, formed the backbone of the Army and Navy's pilot training, while the Curtiss H-12, H-16, and HS-1 and 2 flying boats operating from bases in the United Kingdom and France gave sterling service in the war against German submarines. The Curtiss flying boats were American aviation's one real success during the war. But the failure to produce a combat plane for the Army still rankled. Keys remained convinced that had it not been for the Army's cancellation of the contract, the Curtiss Corporation could have attained quantity production of the SPAD fighter by the spring of 1918, and if the Army had only supported Curtiss's own aeronautical engineers, the Curtiss Corporation could have designed and built its own airplanes for service in France.[126]

Keys had made his own significant contributions to the Curtiss Aeroplane & Motor Corporation's wartime effort. He had quickly understood that Curtiss Aeroplane lacked managers with the production experience necessary to fulfill the greatly expanded orders that came from the Army and Navy once America entered the war. He was instrumental in getting John Willys, with his automotive production experience, to join Curtiss. Keys had primary responsibility for financing the Curtiss Corporation's rapid growth in assets during the war, using his banking contacts and direct negotiations with the government to ensure that Curtiss had the necessary funding to expand facilities, purchase raw materials, and pay workers. Despite a lack of direct experience in managing a manufacturing company, in the final months of the war Keys found himself drawn into making key decisions at the Curtiss Engineering Corporation.

Through his work at Curtiss, with all the challenges and frustrations he and the other managers at Curtiss had to face, Keys had concluded that the fundamental problem of the American aircraft program was a lack of knowledge and a lack of experienced engineers. Writing fifteen years after the war, Keys stated his belief that "our weakness in this period was not a weakness that could be remedied by money borrowed in large quantities; it was a weakness of essential personnel, essential knowledge of material, and essential knowledge of a highly technical design and manufacturing job. Money was available in whatever quantities we needed, but these much more difficult elements were not available at all."[127] His experience with the wartime production program, which entailed a nearly four-fold expansion in output at Curtiss "left one thought indelibly settled in my mind, namely, that every nation must have in being at all times the nucleus of engineering and designing, settled production and equipment there for, and experienced departments in fundamental factory details for the handling of such a task"—in other words, "an engineering and even a production industry for both motors and planes capable of rapid, efficient, and well led expansion."[128] This belief would become a guiding principle for Keys as he and the Curtiss Aeroplane & Motor Corporation, and the entire American aviation industry, struggled to build a firm foundation for a business in the years after World War I.

2

THE STRUGGLE FOR SURVIVAL

What Future for Aviation?

ON A COLD EVENING IN JANUARY 1919, CLEMENT KEYS MADE HIS WAY TO THE Waldorf Astoria Hotel in Manhattan to attend a celebratory banquet held by the Manufacturers' Aircraft Association, the American aviation industry's trade association formed in 1917. A thousand members of the Association and their guests filled a grand ballroom decorated with American flags and aircraft propellers. The assembly was a veritable "who's who" of American aviation; almost all the leading figures in the aviation industry and in Army and Navy aviation were in attendance. Glenn Curtiss was there, with John Willys, James Kepperley, and Charles Kirkham joining Keys as representatives from the Curtiss Aeroplane & Motor Corporation. Frank Russell, a Curtiss vice president and president of the Association, served as toastmaster for the banquet.[1]

The banquet was initially intended to be a victory dinner, a celebration of the aviation industry's achievements in the war, but in view of the rapid cancellation of contracts and demobilization of the industry following the Armistice, the organizers saw the dinner as an opportunity to confer on aviation's future.[2] The most enthusiastic proponents of aviation hoped for the rapid development of commercial aviation, from air transport to private flying. Articles in aviation journals and the popular press in the months following the war spoke confidently of air transport networks that would, within a few years, link America with the rest of the world. Air travel would become as commonplace as travel by rail, private flying would flourish, and the airplane would find myriad uses.[3]

Of course, there were problems to overcome, but for many in the aviation industry, the spurs to development of commercial aviation seemed to be straightforward. Aeronautical engineers had to design and develop airplanes that were suited to the needs of commercial air transport and the private aviator, with an emphasis on safety, reliability, and comfort. Landing fields had to be built across the nation, particularly near major cities. And importantly, a supportive government policy toward the aviation industry was vital. The industry needed a set of national laws governing air transport which would cover air routes, the registration of aircraft, and the licensing of pilots, as well as a comprehensive program covering both commercial air transport and military aviation for national defense. The industry needed leaders who were enthusiastic about the possibilities of aviation, had a vision of aviation's future, and were willing to commit their energy and resources to making their vision a reality. Recalling the great railroad builders of the generation before, on January 1, 1919, *Aviation and Aeronautical Engineering* wrote in an editorial that:

> The commercial side of aeronautics is now dependent on the coming of the far sighted industrial leaders, who, like the late James J. Hill, will lay out an empire and resolve to be its masters regardless of the difficulties. Mr. Hill and the other early railroad pioneers were thought to be headed for financial ruin when they embarked on projects of transportation through unsettled western lands, but their judgment and the ultimate success of their extensions never could be shaken. To the men who resolve to make the empire of the air theirs, and who will insist on breaking down all bars to success, will come the everlasting credit due great commercial leaders of the century.[4]

At the banquet John D. Ryan, who had succeeded Howard Coffin as director of the Aircraft Production Board, gave a more sobering assessment of the aviation industry's prospects. While optimistic that those assembled before him would one day develop aviation along broad lines, Keys noted Ryan's more cautious outlook:

> I think that the great manufacturing organizations should not be destroyed, or allowed to fall into disuse. I don't advocate the building of great numbers of aircraft for military purposes. I don't think it

is necessary. But I think that sufficient encouragement and employment can be given to the well-developed aircraft factories of this country, to keep them in business, to induce them to make every discovery, to do everything they can to promote the science of aircraft. The organization should be kept intact, the men who are able and who have done this thing, should be kept together as far as possible. It would be a small expense, and God knows, it might be a great measure of economy someday.[5]

A year and a half later, Keys would be confronted with this very challenge.

From War to Peace

With the rapid cancellation of almost all government contracts after the Armistice, the Curtiss Aeroplane & Motor Corporation had less need for new financing. Although Keys retained his positions with the Curtiss Aeroplane & Motor Corporation and the Curtiss Engineering Corporation, over the next several months he was less involved in the day-to-day management of both companies. His principal tasks were negotiating the liquidation of government contracts, arranging for the government's purchase of surplus material from Curtiss, and overseeing the repayment of the government mortgage and advances with the proceeds. Soon after the Armistice, Keys began negotiating for the government's purchase of the North Elmwood plant once the final production of flying boats for the Navy had been completed.[6]

John Willys set about preparing Curtiss Aeroplane for the future. He was optimistic about the future of aviation, firmly believing that commercial aviation would soon develop. Speaking at the Aeronautical Exhibition in New York City in March 1919, Willys commented that one "does not have to be a seer to realize that the days of commercial aviation are here. I firmly believe the present year will witness the airplane and flying boats taking their place with the steamship, automobile, and railroad as a means of transportation. It will not be long before aerial navigation will be generally established and accepted by the public."[7] A week after the Armistice was signed, Willys had called a meeting of the Board of Directors to plan the rapid demobilization of the Curtiss workforce and the business policies for the postwar era. Willys was convinced that the company should focus on developing small, inexpensive airplanes for the individual pilot to use for sport flying or instruction and what he called an aerial limousine capable of carrying several passengers in relative comfort.

The three-engine Curtiss Eagle airliner intended, with the single-engine Curtiss Oriole, for what John Willys hoped would be a postwar boom in aviation. The boom did not materialize, and Curtiss sold few Eagle airplanes. (4352, Record Group 18WP, National Archives and Records Administration (NARA))

In April 1919, Curtiss Aeroplane's first postwar airplane designed exclusively for the commercial airplane market, the single-engine, three-seat Curtiss Oriole, made its trial flight. That same month Curtiss Aeroplane began marketing the Curtiss Seagull, a civil version of the U.S. Navy's MF flying boat trainer, advertised as "swift as a seabird," for pleasure or business.[8] The more ambitious project was the Curtiss Eagle, a small three-motored airliner carrying six passengers intended for the airlines that many believed would soon be forming. In July the Corporation began building 150 new Curtiss Orioles and 75 Curtiss Seagulls for inventory. Describing the decision a few years later, Keys said "the people then in control of the Curtiss Company decided that the way to sell airplanes was to build them, and have them, so to speak, on the shelf to deliver when wanted, and we built them."[9] This decision was not one he agreed with.

Keys had a reason to be skeptical. For the majority of Americans, the airplane remained "a startlingly spectacular novelty."[10] Despite the rapid advancement of aviation during the war, the airplane was still too new, too uncertain, too linked with its image as a weapon of war to attract much interest outside the aviation fraternity. While willing to take a flight for the thrill of it, the ordinary person "feels after his flight that he has just escaped a great danger. . . . The return to earth is hailed with relief."[11] This attitude toward flying affected both the demand for private airplanes and the potential for commercial air transport. In a 1919 report to the National Advisory Committee on Aeronautics, Dr. W. F. Durand stated that with the equipment then available air transport "will be more expensive, will be less safe and less reliable, [and] will provide less comfort while in route."[12] He was not optimistic that the traveling public would trade safety, reliability, and comfort for a savings in time.

There were other developments that did not bode well for the future. Both the Army Air Service and the Navy demobilized rapidly after the Armistice. There were more than enough airplanes in storage to equip this markedly reduced force, and there was no need for replacements. Having won the war to end all wars, and with no conceivable threat to America, the Congress was determined that the Army and the Navy be returned to their small, regular peacetime establishments. To the dismay of many in the aviation industry, Congress slashed appropriations for both Army and Navy aviation, granting the Air Service $25 million and the Navy $15 million for the Fiscal Year 1920. This was a response to a demand for economy in government that would last until 1926.[13] With no need for the airplanes held in storage, the War Department began selling the surplus.[14]

Facing the threat of surplus airplanes flooding the market, Keys and two other members of the Executive Committee decided to lay off those workers who could be terminated immediately and to make plans to shut down the plants, at least temporarily.[15] Keys even proposed that Curtiss Aeroplane be liquidated, as it was obvious to him that the Corporation could not build new airplanes for the private market and successfully sell them in competition with government surplus airplanes.[16] His proposal was not accepted. Keys then suggested to John Willys and Glenn Curtiss that if Curtiss Aeroplane could purchase surplus Curtiss aircraft and engines from the government at a competitive price with the surplus Curtiss material coming from other sources, then the Corporation ought to stay in business.

Keys was asked to estimate how much it would cost to remain in business for one year. After a thorough but hurried study, Keys reported to Willys, Curtiss, and the other directors that it would likely cost the Corporation $1 million to continue pursuing the commercial aviation market for another year. It was a gamble to find out if a commercial market would develop. Keys later commented that "it took a lot of nerve to spend $1,000,000 in a company that had never paid a dividend on its common stock. . . . It took a lot of foolishness, I thought, in a sense to appropriate any money to spend so as to keep the art alive, but that is what we determined to do."[17] The directors instructed Keys to go to Washington to meet with the War Department and bid for the surplus Curtiss airplanes in a block, if the government would agree.[18]

Over the next several months, Keys negotiated with the War Department and the Army Air Service to acquire surplus Curtiss JN trainers and engines. Keys agreed to purchase 2727 surplus JN trainers and 4608 Curtiss OX-5 engines for a total of price of $2.7 million.[19] With this purchase, Curtiss Aeroplane hoped to gain an ability to control the flow of surplus Curtiss airplanes into private hands. The company stated clearly that it was purchasing the airplanes to protect its market at a time when the aviation industry was in its infancy and struggling to build a business.[20]

Even more threatening for the future of the aviation industry was a return to the military's prewar methods of contracting for equipment. Traditionally, the Army and Navy purchased equipment through competitive bidding, with the winning bid going to the lowest responsible bidder. With the legacy of the scandals and accusations of fraud and unwarranted profits surrounding the aviation industry, in the immediate postwar years military procurement officers insisted on competitive bidding to avoid any potential congressional criticism.[21] The Army and Navy now treated airplanes as any other commodity, negotiating design contracts for experimental airplanes with a manufacturer, but then putting the production contract out among the industry for competitive bidding. The original designer did not retain rights to the design.[22] The competitive bidding system, while meant for the public good, had the unintended consequence of driving out any incentive to spend money on experimentation and development. It would prove to be highly destructive to the industry.[23]

An element of uncertainty surrounding the industry was the lack of government policy toward either commercial or military aviation. Congress held the key to any comprehensive aviation program, but there was no indication of when the Congress would act, or what position it would take toward aviation. Writing on the outlook for the Curtiss Aeroplane & Motor Corporation, *The Magazine of Wall Street* commented that the common stock of the Curtiss Corporation was a purely speculative investment and " . . . must be bought largely upon faith in the future of the aircraft industry. Whether the investor has this faith in the future or not is perhaps a question of temperament. Certainly, it would be unsafe to make any definite predictions."[24]

With his knowledge of Wall Street, Keys would have been keenly aware that in the postwar climate aviation companies would be unlikely to find the capital they would need for future growth. An article in the May 1919 issue of *Nation's Business*, a leading business magazine, noted that the aircraft manufacturers had less than a million dollars' worth of government business in hand, and that the market for airplanes from the private sector was likely not more than a few million dollars in total.[25]

The American Aviation Mission to Europe

While John Willys was actively preparing Curtiss Aeroplane for what he hoped would be a growing market for commercial airplanes, Keys accepted an opportunity that turned out to be a formative experience. In early May 1919, Newton Baker, the secretary of War, directed his assistant secretary, Benedict Crowell, to go to Europe to study the organization of aviation in England, France, and Italy so that developments in Europe could be included in the War Department's formulation of a policy on military aviation to present to Congress.[26] Crowell asked Navy Captain Henry C. Mustin, Colonel Halsey Dunwoody of the Army Air Service, and Lt. Colonel James Blair from the Army General Staff to join the mission as representatives from the military. From the aviation industry Crowell selected Howard Coffin, S. S. Bradley, General Manager of the Manufacturers Aircraft Association, George Houston, president of the Wright-Martin Airplane Corporation, and Clement Keys from Curtiss Aeroplane & Motor Corporation.[27]

Leaving New York on May 22, the Mission's members spent six weeks in Europe. Keys studied all phases of military, naval, and commercial aviation in France, Italy, and Britain. He visited the factories and met the heads of the leading airplane and engine manufacturers, firms like Breguet, Nieuport, and Hispano-Suiza in France, Caproni, Ansaldo, and F.I.A.T. in Italy, and Handley-Page, Bristol, and Rolls-Royce in England. He toured all the principal government aeronautical laboratories, had long discussions with the heads of the technical departments responsible for aviation in the military, and witnessed flights of several types of aircraft. While his principal focus was on technical development, Keys likely attended some of the meetings that George Houston arranged in order to investigate the development of commercial air transport in Europe. In France and England Houston met with the government officials responsible for civil aviation and the new airline companies that were beginning to provide services between major cities and across the channel.[28]

The American Aviation Mission en route to Europe in May 1919. *Left to right*: Samuel Bradley, representing the Manufacturers Aircraft Association; Lt. Col. James Blair, U.S. Army; Assistant Secretary of War Benedict Crowell (*standing center*); Howard Coffin (*seated center*); Capt. Henry Mustin, U.S. Navy; George Houston, president of the Wright-Martin Aircraft Corporation; and Clement Keys. (Photograph of the 1919 Aviation Mission, Manufacturers Aircraft Association records, Accession 6858, Box #776. American Heritage Center, University of Wyoming)

The members of the Mission prepared a report and presented their findings to Secretary of War Newton Baker in late July. The Mission's report and recommendations addressed the critical issues that would be debated over the next several years within the military, the Congress, and among the proponents of aviation, a debate in which Keys would play an active role. The Mission reached several "inescapable" conclusions that argued for an active role for government in the development of aviation. First, in any future war, victory would go to the power that was the first to achieve aerial supremacy and maintained it. Second, for reasons of economy, no nation would be able to maintain in peacetime an air force adequate to the needs of defense but would have to rely on a reserve of trained personnel and productive capacity created through the development of commercial aviation, including air transportation. Third, it was impossible to suddenly create an air force to face a national emergency; a minimum of two years would be needed to achieve quantity production. Fourth, since the healthy growth of commercial aviation and air transport was vital to the security of the nation, it was equally vital for the United States government to adopt a comprehensive policy promoting the development of all aspects of aviation. The air transport industry required regulations governing the use of air routes, the licensing of pilots, the certification of airplanes, and other measures that would aid the growth of air transport. The report noted that the lack of a definite policy on aviation had been detrimental to the American aviation industry since the Armistice. In a statement that Keys might have written, the report said, "No sensible businessman is justified in keeping money invested in the aircraft industry under the conditions which have maintained in the United States since November 11."[29]

The urgent need was to provide support to the rapidly disappearing aviation industry in order to preserve at least some of the structures that had been built up at such great expense during the war. The industry did not need a large volume of business to survive, the members thought, but it would require a "steady and dependable demand; otherwise private capital and enterprise will not long remain interested."[30]

The recommendations regarding technical development, which Keys most likely authored, sought to establish a role for the technical departments of the government that would not compete directly with private firms in the design of complete aircraft or engines. This practice would be particularly detrimental to firms like Curtiss, who maintained a sizable design staff. If the government designed its own airplanes and engines, it would effectively drive out private design bureaus, thus limiting its sources of supply and stifling creativity. To sustain a healthy productive capacity within the aviation industry, the report recommended a "well-defined and continuing program of production for military purposes, over a period of years." The aviation industry would continue to voice this recommendation until the Congress passed the five-year Army and Navy aircraft procurement programs in 1926.[31]

Participation in the American Aviation Mission gave Keys a perspective and a standard with which to measure the achievements of Curtiss Aeroplane and the American aviation industry that few others had. His experiences in Europe and his participation in drafting the Aviation Mission's report stimulated his own thinking about the development of military and commercial aviation in America, and particularly the role that the government should play in aviation's development. These issues would take up much of Keys's time in the years ahead.

Testifying before Congress

After their return from Europe, Keys and other members of the Aviation Mission testified before several congressional committees. Several members of Congress were anxious to tackle the complex issues surrounding aviation's place in the military and in commerce. In the opinion of many newspapers around the country, the "stodgy inertia" in Congress with regard to aviation had "ceased to be absurd, merely, or merely scandalous, and ... becomes downright perilous."[32] While Great Britain had appropriated $350 million for aviation and France $270 million, the U.S. Congress had allocated a paltry $50 million, prompting the New York *American* to say "America, the land which gave birth to the airship, would stifle its growth just as it had reached the critical stage and leave to foreigners the benefits of its continued improvement."[33]

In his testimony before Congress, Keys stressed that aviation should be a business like any other, where firms competed to provide the best product and where they could earn a sufficient return on their capital. In August Keys went before the Senate Subcommittee of the Committee on Military Affairs that was considering a bill to establish a Department of Aeronautics to administer military and civilian aviation.[34] He returned in November for hearings before a House Select Committee on Expenditures in the War Department (the Frear Committee) and again in December for hearings before the House Subcommittee of the Committee on Military Affairs relating to a United Air Service (the LaGuardia Committee).[35]

Keys argued that it was in the nation's best interests that aviation become a self-sustaining industry through its commercial development and not rely solely on the military if the nation wanted to avoid the production delays experienced in World War I. Demand from the military alone would be insufficient to sustain an industry that could provide sufficient productive capacity in time of war. Keys was against the government taking on activities that would place it in competition with private aircraft manufacturers. The government should not, in his view, design complete aircraft and build them in government factories. Far better, Keys said, to have broad competition within the industry in order to push development.[36]

Keys was convinced that government production would be damaging to the aviation industry. In response to a question from Representative Fiorello LaGuardia during his hearings concerning the Naval Aircraft Factory, Keys replied, "of course, as an industrial man, I think it is absolutely unfair in an infant industry to establish a factory big enough to take care of the entire peacetime demand, if it were operated sensibly, on an industrial basis, and then expect the industry to stay alive and cooperate with the Government."[37] Nor did Keys want subsidies for the industry, as he believed this would drive out capital. The government, in his view, should purchase only those airplanes that it actually needed and would make use of, distribute its orders among a group of manufacturers, and pay the cost of production plus a "reasonable business profit."[38]

Keys was candid in his assessment of the aviation industry's prospects, noting that 80 to 90 percent of the productive capacity built up during World War I had already been liquidated. When asked about the outlook for the industry if it had to depend solely on Army and Navy orders, he replied, "Oh, it is not any kind of an industry. Of course, I am not staying in it myself. It does not look like the kind of industry to put capital into and I would not invite anybody to go into it. . . . I would not want to go into the business commercially, and would not want to go into any aviation industry at this moment. We know that we have no Government military or naval policy adopted, and no commercial business developed. I think the commercial business is the best end of the three, looking at it at this moment."[39]

By December, when Keys was called to testify before the LaGuardia hearings, he was even more pessimistic about the outlook for commercial aviation. Describing Curtiss Aeroplane's decision to find out if there was a commercial aviation industry, Keys said that " . . . it cost us $1,000,000, and I do not know today whether there is such a commercial industry or not. We know there is some demand for airplanes, but to make airplanes in small quantities and sell them at prices any man will pay for an airplane has got to show a margin of profit, or else there is not an industry, and we do not know that yet."[40]

The Worsening Situation at Curtiss Aeroplane

By the end of 1919, Keys had good reason to be pessimistic. Curtiss Aeroplane had shrunk to a fraction of its wartime peak. Employment had declined dramatically from 17,850 to 760, and the number of factories had shrunk from nine to two; only the Church Street factory in Buffalo and the Curtiss Engineering Corporation factory on Long Island remained in operation. Keys stated that while the Corporation hoped to increase production in the spring of 1920, these operations would be curtailed if there was no prospect of additional government business. The Corporation had managed to pay off more than $13 million in government debt and accounts and bills payable, shrinking total assets from $26 million to $16 million, but a substantial amount of funds was still tied up in the remaining government contracts. These had yet to be resolved, and the sale of the North Elmwood plant to the government was still pending. Worse, the Corporation was losing money.[41]

By the early months of 1920, Curtiss Aeroplane had achieved only marginal success in selling its new commercial aircraft. The associated losses had become an alarming problem. For 1919 the Corporation had recorded a loss of $176,891 on total sales of $11,805,141.[42] To add to the growing list of problems, Curtiss now faced an escalating challenge from the Internal Revenue for back taxes. In November 1919, the Office of the Commissioner of Internal Revenue had informed the company that it owed an additional $2,480,239 in excess profits tax on its wartime income. Then, in early February 1920, the Internal Revenue revised its estimate of the amounts the Corporation owed the government, increasing the amount to $3,495,914.[43]

At a meeting on March 26, 1920, the Executive Committee of Curtiss Aeroplane asked Keys to become chairman of a special Finance Committee to deal with the rapidly deteriorating situation at Curtiss.[44] Before Keys could complete a review of the Corporation's financial position, the Corporation received a letter from the Internal Revenue that threatened disaster. The Internal Revenue Service now said the Corporation owed $4,732,862 in taxes on excess profits.[45] To make matters even worse, the Treasury Department requested that the War Department withhold payment of any money due to the Curtiss Corporation pending resolution of the additional tax assessment.[46]

What Keys reported to the directors during the first week of April was unsettling.[47] He had found that the Corporation's claims on the government had now become a vital source of working capital. Keys estimated that Curtiss could expect to receive $1.8 million from sale of the North Elmwood plant to the government, and an additional $1 million or more from the final resolution of war contracts, but these payments were now frozen until resolution of the tax claim. Keys later said that he recognized that unless the Corporation could achieve a settlement with the government and free up payments, the Corporation would in all likelihood default on its obligations and end up in receivership.[48] Keys insisted that the Finance Committee be given complete control over all activities of the corporation, until the financial situation had been resolved. The Executive Committee agreed to this emergency measure. With that, Keys set out to work.[49]

The immediate need was to persuade the Treasury Department to free up payments from the government. Keys told the War Department that in the absence of a settlement with the government, the Curtiss Corporation would likely go bankrupt.[50] Keys then met with Internal Revenue and the Treasury Department to work on the tax issue. To Keys's relief, the War Department told the Internal Revenue Service that "the Army was extremely interested in seeing that nothing prevented the Curtiss Company from going ahead in business."[51] The Treasury Department assured Keys that it would "do everything it could do to prevent any undue strain on the Curtiss Company's finances."[52] Keys submitted a plan whereby Curtiss would provide security to the Treasury Department for the taxes owed to allow the release of government payments due Curtiss. Fortuitously, that same week Congress finally approved the War Department's $1.8 million purchase of the North Elmwood plant from Curtiss. Convinced that obtaining security from Curtiss would be too complicated, the Treasury officials proposed an alternative solution: Treasury would assess a tentative tax on Curtiss of $2 million and hold the proceeds from the North Elmwood plant sale as a preliminary assessment. Curtiss could submit a claim for a rebate and begin resolution of the tax issue. In the interim, the Treasury would agree to release of other payments to Curtiss. Keys readily agreed to the plan.[53]

Keys went from one problem to another, shuttling back and forth between New York and Washington. After completing negotiations with Internal Revenue and the Treasury Department, Keys received a letter confirming that the Internal Revenue had filed a tentative assessment of $2 million to cover back taxes and would accept the $1.8 million from sale of the North Elmwood plant. The sale was completed on April 24, and the proceeds were turned over to the Treasury Department, freeing up the other government payments owed to Curtiss. The tax issue was complicated, involving the calculation of net worth and invested capital during the war years, the valuation of inventories, and the amount of appropriate depreciation on both inventories and patents. Though it would take another two years to finally resolve, Keys had removed one of the swords hanging over the Curtiss Corporation. But there were still others.[54]

Keys had to arrange financing for Curtiss Aeroplane. To meet the Corporation's pressing need for cash, Keys agreed to advance $150,000 of his own funds to Curtiss, taking as collateral an amount of Curtiss stock sufficient to give him a controlling interest in the company. Keys negotiated a $500,000 line of credit with the Irving Bank, with $200,000 due in 60 days on June 27 and $300,000 due in 90 days on July 27. He arranged for an additional line of credit for $250,000 from a bank in Buffalo to be repaid at the end of July. Although successful in getting the bank financing, this arrangement provided only a temporary reprieve.

But even with partial resolution of the problems with the government, the outlook for Curtiss Aeroplane's finances looked ominous. Keys outlined for the directors the sources of capital available to the Corporation up to September against its liabilities and budget for operations. The bank loans, due in July, had been used to pay off current loans and overdue trade accounts, leaving a balance of only $167,000 available for working capital. Keys expected a total of $550,000 to come in from settlements with the government, but nothing more for at least another six months. He was convinced there was no possibility of obtaining financing from the public markets. The results from operations to date were "exceedingly unsatisfactory," with the results for March and April particularly distressing. Revenues amounted to $180,000 against a budget estimate of $580,000 for the two months. The operating budget called for spending $500,000 a month through September. "In my judgment," Keys stated, "such a budget under the conditions prevailing, would be suicide and it is necessary immediately to take whatever steps are necessary to stop it."[55] In June Keys announced that due to the influx of surplus airplanes into the market, the Curtiss Corporation had abandoned plans for manufacturing commercial airplanes.[56]

It may have been some consolation to Keys and the other directors when a few days after announcing they had abandoned their plans to build commercial aircraft, Curtiss Aeroplane won an Army contract to build fifty Orenco D pursuits by underbidding the designer, the Ordnance Engineering Corporation under the War Department's system of competitive bidding.[57] In addition, at the end of June, Curtiss received a small contract from the Navy to build two experimental Curtiss CT-1 torpedo planes.[58] But the prospects for winning additional government contracts appeared slim. While Congress had passed the National Defense Act of 1920, which established the peacetime structure of the Army and provided a legal basis for the Army Air Service, Congress remained in a parsimonious mood. The Army received $33 million in the Fiscal Year 1921 appropriations for the Air Service, which was only a small increase over the prior year, while the Navy was granted an appropriation of $20 million for naval aviation, an amount that Secretary of the Navy Josephus Daniels claimed provided only half of what naval aviation needed.[59]

Through June and into July, Keys worked to negotiate a settlement of Curtiss Aeroplane's claims on the government. Keys was soon conversant with the complex accounting and cost allocation issues surrounding the Corporation's claims, writing detailed letters arguing for favorable treatment.[60] Curtiss Aeroplane claimed reimbursements from the government, but it owed payments to the government as well. Keys chose a particularly aggressive response: it threw into the mix a $4 million claim for damages against the War Department for its failure to provide data and drawings necessary to manufacture the SPAD, Bristol, and Caproni aircraft, and for forcing Curtiss to prematurely prepay its $4 million note issue. Keys proposed the following settlement: Curtiss would waive all its claims for damages if the War Department would agree to convert the payments due to notes from Curtiss to be paid in 1921 and 1926. The War Department accepted Keys's proposal and submitted an outline of a general settlement to the War Claims Board in early July, pending a final hearing and decision.[61]

John Willys Abandons Curtiss Aeroplane & Motor Corporation

The accumulation of ills at Curtiss Aeroplane no doubt weighed heavily on John Willys. By the end of June 1920, Curtiss Aeroplane had accumulated a deficit of $1,445,238, with little prospect of reversing the trend.[62] Commercial aircraft sales were in the doldrums, military sales were small and dependent on the unpredictable system of competitive bidding, the Corporation was running short of financing with a sizable tax bill to be negotiated with the Government, and the matter of the Curtiss claims against the War Department was still pending. At the June meeting of the Curtiss Board, Keys had informed the directors that the Corporation did not have the money to pay the dividend on the preferred stock or make the regular contribution to the preferred stock sinking fund.[63] Willys-Overland's period of control of Curtiss would not be up until the end of 1920, but John Willys had already decided that it was time to withdraw from aviation.[64] There were growing problems at Willys-Overland. Willys had overexpanded in expectation of a booming car market. The postwar depression, which was just beginning in the summer of 1920, caught Willys offguard; within months, Willys-Overland would be in severe financial difficulties.[65]

On Tuesday, July 27, 1920, the $300,000 loan from the Irving Bank came due. It was not paid. That Friday, a representative of the Willys-Overland Corporation met with Keys in his capacity as chairman of the Curtiss Finance Committee. He informed Keys that Willys-Overland did not intend to pay back the Irving loan, nor any other debts coming due. Willys-Overland could no longer carry Curtiss and intended to put the company into the hands of a receiver. As Keys recalled the conversation, he told the Willys-Overland representative, perhaps recalling John Ryan's comments at the January 1919 banquet, that

> it would be a pity to let the one big engineering group in aeronautics in the United States go to pieces, because if it did I did not see how anybody else would build up for many years another group of engineers capable of coping with and keeping abreast of engineering developments in Europe in both motors and planes. I said that if the Curtiss Company went out of business it would be, I thought, a national disaster so far as aviation was concerned.[66]

Perhaps there was something Keys could do for Curtiss. He asked the Willys-Overland representative to give him the weekend to think it over.[67] Keys spent the weekend at his New Jersey home reviewing the Curtiss Corporation's prospects and his own future if he was to continue his association with the company.

Keys was fully aware that the Curtiss Corporation's financial condition was perilous. In addition to the overdue bank loans, the Corporation's trade creditors were about to take action to enforce payment of their obligations. There was little prospect that the Corporation could raise additional capital from the markets. A strict accounting based on the true market value of the Corporation's assets would have shown that the Curtiss Corporation was close to being insolvent. If the Corporation was forced into receivership, not only would the pending negotiations with the government over taxes owed and payments due become hopelessly complicated, but a sale of the Corporation's assets would also be insufficient to cover all the Corporation's debts. The preferred stockholders and the common stockholders would be wiped out.

The outlook for the few remaining aviation companies seemed equally bleak. Keys had only just shut down the Curtiss effort to build a commercial business, and with a large supply of surplus aircraft available, future sales appeared doubtful. Commercial airlines were a project for the future at some undetermined time. The government was the aviation industry's only remaining customer, but the prospects for additional military business were slim due to congressional constraints on military spending. While Curtiss might have a good chance of winning a decent share of business from the military, the system of competitive bidding meant there was no certainty that even a leading manufacturer like Curtiss could win against other potential lower bidders. A year and a half after the end of the war, the United States still lacked a government policy regarding either commercial or military aviation. There appeared to be no ready market for Curtiss airplanes and airplane engines.[68]

There were also good reasons for Keys not to deepen his personal involvement with the Curtiss Corporation. When he thought about the situation, he was not at all sure that he could do something about it.[69] The myriad problems at Curtiss would demand an inordinate amount of his time and energy, with no certainty of success. While he had demonstrated his ability to cope with complicated problems of corporate finance, Keys had little experience managing a company, even one as small as the now-shrunken Curtiss Corporation. Saving Curtiss Aeroplane, if he could, would be the biggest challenge he had ever faced. The situation would call on all his skills as a financier, and more. There was no little risk involved. Failure would damage his reputation in the market and quite possibly harm his own business. The safer, perhaps more prudent choice would have been to turn away from Curtiss, and from aviation entirely, and remain with his own banking business.

Keys's experience with the American Aviation Mission renewed his conviction that Curtiss airplanes and engines were equal to anything Europe could produce. He later said that "I knew very well that we had in the Curtiss Company itself, apart from the rest of the industry, plenty of ability to hold our own with Europe, in fact, to beat them in the air."[70] Keys also believed that key to a strong aviation industry was development of commercial aviation and air transportation. Having spent years studying the railroads, Keys understood the dynamics of transportation as a system and as a business. He clearly saw the airplane's potential to revolutionize transportation through speed and a capacity to overcome the physical obstacles that limited road and rail. It might take years to bring to fruition, but it would happen. Keys wanted to participate in aviation's development.[71] He could have easily walked away from Curtiss, but he didn't.

By the end of the weekend, Keys had decided to do what he later called "a foolish thing."[72] He would take control of Curtiss and try to save it. Motivating him may have been his sense of the opportunity that aviation presented. As a journalist, the men Keys had admired most were the system builders, the men who had started out in new fields of endeavor, who had attempted what others had deemed impossible, and who through intense effort had built a new edifice of transportation or industry where none had existed. Prime among this group of men were Alexander J. Cassatt, James J. Hill, and Edward H. Harriman, the great system builders of the railroads. For Keys, aviation was a new field waiting for a Hill or a Harriman to build an aviation empire; *Aviation* had sounded this call in its first editorial of 1919.[73] Sitting in his New Jersey home that August weekend, Keys may well have asked himself: could he achieve in aviation what Hill and Harriman had done with the railroads? Was this his chance?

On Monday evening, Keys met with the Board of Directors of Curtiss Aeroplane and representatives from Willys-Overland and offered to buy Willys-Overland's holdings of common stock for cash. Keys agreed to meet the debts coming due in return for a controlling interest in Curtiss. Willys-Overland and the other Curtiss directors agreed to his proposal. Keys would take over the company and run it.[74]

Getting Out of Danger

When Clement Keys walked out of the meeting that Monday evening, he had more immediate challenges than most businessmen face in a lifetime. He had to pay off the Irving Bank loan to forestall receivership as well as find a way to pay the other debts coming due; he was in the midst of difficult negotiations with the War Claims Board and the Internal Revenue which might or might not save the company he had just agreed to take over; and, not least, he had, somehow, to keep the Curtiss Corporation alive as a going concern despite all the company's problems and the aviation industry's uncertain outlook.

On Tuesday, August 3, Keys borrowed $250,000 on his own credit, then extended an advance to Curtiss Aeroplane for this amount to help pay off the $300,000 loan to Irving Bank; the deposits the Corporation held at the Bank made up the remaining $50,000 owed.[75] He then went to Washington to meet with the War Claims Board. Before meeting the Board, Keys met with Secretary of War Newton Baker. Keys advised Baker that he had just bought control of Curtiss. Explaining the critical situation at Curtiss, Keys pleaded with Baker to get the Army and the Treasury Department to release the funds that were due to the Corporation. He promised Baker that he would personally oversee operations at the company and that he would devote every dollar toward launching an engineering program to develop new types of airplanes and engines that would put the United States on par with European aviation, bringing back world records to America. Keys took this promise as a personal obligation.[76]

At the War Claims Board meeting, Keys explained that the Curtiss Corporation had exhausted its working capital, forcing the company to curtail its manufacturing program, default on its War Department contract, and cancel dividend payments on its preferred stock. It was highly unlikely that in its current condition Curtiss could obtain new financing from banks or the public market. The Corporation's only source of new working capital was the payments due from the War Department in settlement of the claims at issue. He told the Board that the amount of working capital that would come through the settlement would likely be sufficient for the Corporation to continue in operation, albeit on a much smaller scale. The Board learned that Secretary of War Baker and Major General Charles Menoher, then chief of the Air Service, held that continuation of the Curtiss Aeroplane & Motor Corporation as a going concern was of "vital necessity in the interest of the Government and of aviation as an industry."[77] The secretary of War wanted the Air Service to provide all the financial assistance it could legally provide to Curtiss to help the company avoid bankruptcy.[78] The Board deliberated over the next few days before finally recommending the settlement, including the payment to Curtiss of $611,345.[79] Keys later told Newton Baker that payment of these funds saved the Curtiss Company.[80]

After his meetings in Washington, Keys returned to New York to work out what assets Curtiss Aeroplane could pledge as security for his advance to the Corporation. He then paid off the most pressing trade accounts with his own funds.[81] It had been a busy week. By Friday, the Curtiss Corporation was out of immediate danger, though Keys would have to wait another two weeks for the badly needed payment from the War Department.

With the Corporation's financial situation temporarily stabilized, Keys negotiated transfer of a controlling interest in Curtiss. He agreed to purchase 100,000 shares of Curtiss Aeroplane & Motor Corporation common stock from Willys-Overland, out of a total of 218,000 shares outstanding, giving him effective control over Curtiss Aeroplane and the Curtiss Engineering Corporation. He bought the common stock for $4 a share, which was a good price given that the preferred shares were selling below $15 a share. Keys apparently raised the required $400,000 purchase price on his own.

Once the stock purchase from Willys-Overland was completed, Clement Keys became president of the Curtiss Aeroplane & Motor Corporation. He announced this news to S. S. Bradley, his colleague on the American Aviation Mission to Europe and now secretary of the Manufacturers' Aircraft Association. Keys informed Bradley that he had recently purchased the Willys-Overland interests in Curtiss Aeroplane. He told Bradley in no uncertain terms that he was now in control of the Curtiss Corporation " . . . from now until the time when somebody takes it away from me."[82]

Plans and Possibilities

Keys had no illusions about what he was undertaking. Writing in November 1920 to Frank Allen, president of the State Savings Bank & Trust Company of Moline, Illinois, who represented a group of midwestern Curtiss Corporation stockholders, Keys acknowledged the enormity of the responsibility he had taken: "I recognize that I have a difficult job on my hands, and I should like very much to have the cooperation of the stockholders at large; but I know perfectly well that I cannot expect such cooperation unless I prove in time to deserve it."[83] Keys tried to sound positive. "Personally," he said, "I feel quite confident that if we have a reasonable adjustment with the Treasury, we shall be able to make a good future for this company, and build a solid little industry out of it."[84] A month later, Allen wrote Keys to tell him that while many of the Curtiss stockholders he represented had written Curtiss off, a number of them were beginning to have renewed confidence in the company's financial future. "I had the privilege of talking with Mr. Earl, the vice president of the Willys-Overland Co.," Allen wrote, "and he had many complementary things to say about you and your business ability and expressed very keen interest in your progress and felt sure of your success in handling the business of the aeroplane enterprise."[85]

Keys had four tasks to accomplish: restructure the Curtiss Corporation's balance sheet, resolve the tax issues with Internal Revenue, get whatever business was available, and concentrate all effort on carefully selected research and development. In the current environment, the military services were the only customers who could place meaningful orders with Curtiss. Keys decided that Curtiss would take whatever contracts the corporation could get from the Army and Navy where there was a chance to break even.[86] But without secure sources of financing, the Curtiss Corporation would need to operate on a smaller scale. Keys would have to shrink the Corporation's assets to the level of military business Curtiss could reasonably expect to win. To continue in business he had to reduce the company's inventory of unsold airplanes and raw materials, thereby reducing the need for working capital.[87] At the same time, Keys needed to rebuild the company's credit standing in the market, particularly with trade creditors. Resolving the tax issue with the Internal Revenue was critical. The Corporation had a claim with Internal Revenue for a refund of $1.8 million. If Keys could get Internal Revenue to agree to the refund, he would obtain a badly needed source of working capital.

Keys needed to build Curtiss's reputation for engineering excellence, but in its present financial condition the Corporation could not maintain a large engineering staff devoted solely to research and development. This was a dilemma that all airplane manufactures faced at the time. The design and development of aircraft required that highly skilled engineers and workmen be kept on the payroll if the manufacturer was to maintain any capacity for production. But few companies then had enough business to support the engineers and skilled workers they needed.[88] The Curtiss engineering program needed to focus on what would bring financial success and to achieve the world aviation records that Keys had promised Secretary of War Baker he would pursue. Fortuitously, one of the many projects the Curtiss engineering staff had been working on since the end of the war would be the answer to Keys's requirement: Charles Kirkham's experimental 400 hp K-12 airplane engine.

While in Europe, Keys found that none of the European engine manufacturers he visited had an engine that could equal the K-12 in terms of power, weight, and small frontal area. Keys realized that the 400 hp K-12 had considerable potential as an engine for high-speed airplanes. Upon his return to the United States, Keys had urged Willys to expand the K-12's development, but Willys was reluctant to spend money on extended development, particularly with no support from either the Army or the Navy.[89] Development of the engine did continue slowly and, now designated the C-12, the engine demonstrated its capabilities at the 1920 Gordon Bennett races in France.[90] Enthusiastic about the potential for the C-12 engine, Keys decided to focus on developing the C-12 for high-speed racing planes.[91] Speed had always captured the imagination, and speed records were both popular and much sought after. These were the records Keys aimed to bring back to the United States. More importantly for the future of Curtiss, Keys understood that high-speed racing planes were effectively prototypes for pursuit planes for the Army and the Navy.[92]

By a fortunate coincidence, Ralph Pulitzer, the newspaper publisher, established the Pulitzer Trophy Race to promote aviation. Keys asked the Curtiss engineering staff if the company could build an engine and a racing plane that could win the Pulitzer race in 1921. When the engineers said they could, Keys gave the go-ahead to proceed. Keys then went to Washington to meet with Navy officials, urging them to get into the racing business in competition with the Army. Despite their limited funds, both the Army and the Navy were interested in seeing further development of the C-12 engine. The two services decided to share the cost of developing a new model of the engine. In October, the Navy agreed to purchase a new model of the C-12 at a cost of \$31,000; the Army would cover the costs of testing the engine at McCook Field.[93] It was not much, but it was a start.

With an engineering program underway, Keys did what he could to clean up the Curtiss Aeroplane & Motor Corporation's balance sheet. Through the sale of the North Elmwood plant to the government and the write-off of other assets Keys managed to reduce the balance sheet from \$16 million to \$10 million at the end of 1920.[94] Keys used a War Department payment of \$611,000 to repay his \$250,000 advance to Curtiss and some of the more pressing trade debts. There was nothing he could do about the accumulated losses, which were horrendous. For the full year, the Curtiss Aeroplane & Motor Corporation recorded a loss of \$1,756,583.[95] Still, the company had survived its brush with receivership, and Keys could at least hope that 1921 would bring better fortune.

Holding On

The only good news for Keys in early 1921 was a successful settlement with the Internal Revenue. Instead of owing some \$4.3 million in taxes, Curtiss was due a refund of \$1,346,124.[96] Losses, however, continued through April. On the commercial side, Keys pointed to "an absolute lack of demand throughout the winter season," while government orders were in abeyance until the new Harding administration worked out its priorities.[97] Keys's initial optimism about the future of Curtiss Aeroplane was fraying, though he refused to give it up completely. Writing to Frank Allen, Keys noted that "it may be that I am unduly disappointed with the results of the fight I have been making for this company. Most of the people in the company seem to think we have done pretty well to keep out of trouble of a financial sort. The main factor of the disappointment, however, is the lack of business, and it is not easy to see just how this is going to be remedied. Possibly it will be, but it is not in sight at the present time."[98]

The lack of business caused Keys to take a hard look at the Curtiss Corporation's engineering effort at the Garden City factory. Despite his commitment to engineering, Keys was worried about the cost of carrying a large engineering department when the only business coming in was from small engineering and development contracts for the Army and Navy, what Keys called "engineering jobs," as opposed to the larger production contracts. Without production orders, the company would likely experience heavy losses. Keys told Frank Russell, in charge of the Garden City factory, that he thought the work on the Navy torpedo plane, the Curtiss CT, "for which we will be paid about $85,000, and on which our costs are somewhat over $120,000, has fairly well demonstrated for about the 20th time, the impossibility of maintaining a strong and well-balanced engineering plant on engineering jobs alone."[99] He instructed Russell to drop all engineering expenses not directly related to work on Army and Navy contracts, or on programs the Executive Committee had approved. Russell was to bid on as many competitive contracts as possible, but only as many as he could accommodate with his existing engineering staff. Maintaining an organization simply to participate in competitive bids made no business sense.

In early June, Keys issued a statement containing one of the first public criticisms of the government's system of competitive bidding and the damage it was causing the aviation industry. The statement appeared in several of the leading aviation journals of the day.[100] Keys said that the Curtiss Corporation after the war adhered to Glenn Curtiss's pioneering policy of trying "to meet the nation's requirements in military aviation; to conduct scientific research, and to construct new commercial types according to the demand."[101] Conditions in the industry since the war, however, had made it impossible to continue development of all aspects of aviation on a broad front. The U.S. government had "adopted the idea that aeroplanes, and even aeronautical motors, should be bought only under competitive bids, with a partial exception in the case of experimental models."[102]

Keys remarked, disingenuously, that he had no desire to criticize these steps or "to suggest that they were intended to oppose a weak industry to the very powerful competition of the Government."[103] But he wanted the public to know the damage the government's policies were causing. "It is, however, necessary for a clear understanding of the position of the aeroplane industry," Keys said, "to state that they necessarily brought about a very great curtailment of all development expenditure by the industry."[104] No company could afford to make that investment if the government was to take over the product and submit production to competitive bidding. "I think it is fair to say," concluded Keys, "that because of these conditions, over which the industry has no control, not only the Curtiss Company but all other forward looking institutions of this art have curtailed their efforts, economized their resources, and foregone their ambitions for the art in order to adapt themselves to the policies of their Government."[105]

That June the Army Air Service and the Navy gave out several contracts through competitive bidding. The Air Service finally selected manufacturers

The Curtiss NBS-1 Night Bomber. Under the Army Air Service's system of competitive bidding, Curtiss underbid the Glenn L. Martin Company for a contract to build fifty NBS bombers that Martin had designed. Though the system did give work to other firms, Keys believed it was ultimately destructive to the industry. (2008-03-31_image_640_01, Peter M. Bowers Collection, Museum of Flight)

for a new bomber the Glenn L. Martin Company had designed in 1919 as the Martin MB-1. That same year the Air Service initiated bidding for the production contract under the system of competitive bidding, but it delayed selection until 1920. Now designated the NBS-1, the Martin Company lost out to Curtiss on a contract to build fifty NBS-1 bombers. Curtiss was the low bidder on the contract, offering to build the NBS-1 at $17,490 each, compared to Martin's bid of $23,925.[106] At the end of June Curtiss won a second contract through competitive bidding to build thirty-four examples of the Navy-designed TS-1, the Navy's first carrier-borne fighter plane.[107]

Even with these contracts in hand, Keys ordered further cuts in the engineering staff at Garden City. Keys knew that Curtiss could not make money on the NBS-1 bomber contract, but he went ahead in order to get at least some business for the company. He instructed Frank Russell to eliminate all costs that enabled other companies to submit bids lower than Curtiss. That meant cutting back the engineering research effort to the bare minimum, just as Keys had predicted would happen in his earlier statement outlining the negative impact of competitive bidding.[108]

Despite cuts in other areas, Keys had the Garden City factory continue work on the high-speed engine and racing planes. With an improved C-12 engine now designated the CD-12 in a Curtiss-designed racing plane, Curtiss test pilot Bert Acosta won the 1921 Pulitzer race on November 3, 1921, with a speed of 176.7 mph.[109] Three weeks later, Acosta flew the Curtiss Racer to an official American speed record of 197.8 mph. The Curtiss Racer's performance demonstrated the CD-12 engine's potential, but more work needed to be done to turn the engine into the real winner Keys hoped it would become.[110]

Thanks to the strict economies that Keys had imposed and the Army and Navy contracts that had come in, Curtiss Aeroplane reported a net profit of $101,207 for 1921 as opposed to the $1.5 million loss a year earlier.[111] During the year Keys kept the company free of bank debt and reduced the outstanding trade accounts. The company had $1.7 million in government contracts on its books. More importantly for what Keys wanted to do for the future, the amount of cash available at Curtiss had grown from $43,424 at the end of 1920 to $994,880 at the end of 1921, a remarkable achievement.[112]

Keys was relatively new to the role of chief executive, but having observed, analyzed, and advised scores of companies, he had an idea of what made an effective executive. "As I see it," he wrote to Frank Allen later that year, "the executive head of a corporation should not handle much detail, but should have a broad vision in which the various functions of the corporation are all embraced, each in its proper perspective. He should know the relative importance of selling, manufacturing, finance, etc., and if he does not have a proper conception of the relative importance of all of these different functions, the corporation will almost certainly be less efficient than a corporation the heads of which have such a perspective."[113] To this list Keys would add the importance of delegation. The critical task was first to have a vision for the company and then to ensure that competent and trusted subordinates worked harmoniously to achieve that end. This was far easier to say than to do.

With more cash now available, Keys proposed to the Board of Directors spending whatever money was necessary to build the engineering effort up to the highest standards possible.[114] The Board agreed and accordingly instructed the Executive Committee to work out an engineering program

The Curtiss D-12 engine, developed from Charles Kirkham's twelve-cylinder K-12 engine, was the finest liquid-cooled engine of its day. The D-12 powered several Curtiss racing planes and later Curtiss pursuit, attack, and observation aircraft for the U.S. Army Air Corps and U.S. Navy. (3B-13953, Record Group [RG] 342FH, National Archives and Records Service [NARA])

for 1922 and to budget for additional personnel and equipment that would improve the Engineering Department's efficiency. Keys wanted to ensure that the Engineering Department could meet government demands with new Curtiss designs for aircraft and engines. The Corporation was still heavily involved in development work on the CD-12 engine; if this proved successful, as Keys hoped, then orders for new aircraft designed around this engine were sure to follow.

The result of this effort was the Curtiss D-12 engine, a "mechanical masterpiece" that set the pattern for all liquid-cooled airplane engines for the next twenty years.[115] On May 27, 1922, the Army Air Service placed an order with Curtiss for two racing airplanes for the 1922 Pulitzer Race. Designated R-6, these aircraft had the new D-12 engine. On the first flight of the R-6 at Garden City, Army Air Service pilot Lt. Lester Maitland reached 223 mph in a full power pass over Curtiss Field, faster than the official world's speed record of 205 mph.[116] The Army Air Service had announced that it was participating in the high-speed races with the intention of developing a potential pursuit plane superior to those in service.[117] This is exactly what Keys had gambled on. To hammer the point home, Curtiss took out a full page ad in the journal *Aviation* to coincide with the Pulitzer races, advertising the D-12 as "The Premier All-American Pursuit Engine."[118]

Keys went out to Detroit for the Pulitzer races. The results were a triumph for Curtiss, vindicating Key's commitment to the company's engineering effort. On October 14, 1922, the Curtiss racers with the D-12 engine swept the field, capturing first, second, third, and fourth places in the Pulitzer race. Lt. Russell Maughan won first place in an Army R-6 with the winning speed of 206 mph, 20 mph faster than the Navy's Curtiss CR racers and 30 mph faster than the nearest competitor. As the journal *Aviation* said, "regarding the Curtiss engines which figured so prominently in the Pulitzer race, little could be said that would speak more eloquently than their magnificent performance."[119] Four days after the Pulitzer race, Brigadier General William Mitchell set a new world's speed record in the R-6, averaging 224 mph over a 1-kilometer course.[120] The aviation records that Keys had promised Newton Baker he would bring back to the United States were starting to come in.

The Road to Recovery

During 1922, Curtiss Aeroplane barely managed to eke out a profit. Keys wrote Frank Allen in December that operations for 1922 would "show badly."[121] "I do not know just what the figures will show," he said, "but they will not be good, owing to the fact that practically all the business for 1922 was competitive business, and not business on Curtiss products."[122] There continued to be problems building the NBS-1 bombers at Garden City, where the bombers were under construction. By August, deliveries of NBS-1 bombers were behind schedule. Worse, completing the bombers required additional labor above and beyond the initial estimate to complete the contract. The Curtiss Executive Committee's effort to exercise control over production at Garden City was inadequate, due in part to a lack of uniform information on the rate and progress of production. Keys blamed himself for the Executive Committee's failure, having allowed himself to get distracted with other business. Keys was learning, painfully, that delegation was all well and good, but it would not prevent problems from emerging. Without adequate information and diligent follow-up, problems could not be identified and addressed.[123]

Nevertheless, Keys was optimistic for the coming year. He told Frank Allen that the company had received a contract for $2.5 million to build D-12 engines at the Buffalo plant. Curtiss had also won a competitive bid from the Navy to build six examples of a Navy-designed scout-bomber-torpedo plane designated the Curtiss Scout (CS-1). Work on the CS contract for the Navy looked like it would be profitable. But there was more to his hope for better prospects for Curtiss. "We are also putting into the air next week," he wrote Allen, "a new type of pursuit ship, which, I believe, is the most advanced aeroplane product that has been built in America."[124] Keys refrained from promising great results. "I do not know what the possibilities of business are in this ship," he said, "but we propose to make the best of them."[125] The new pursuit ship represented a change in policy for the company. As Keys explained to Allen, "it has taken a long time to get started on the policy of building Curtiss products in our factories rather than entering open competition for other people's products. What the ultimate outcome of the success of this policy will be, I cannot say, as it is not profitable to prophesy in a trade as uncertain as this."[126]

In May 1922, Keys and his engineers had discussed with the Army Air Service the possibility of building a pursuit plane around the D-12 engine.[127] Air Service officers explained that while the Service could not enter into a contract for development of a pursuit, there was a clear need for a new pursuit with excellent performance. If the Curtiss Corporation wanted to undertake the development of such an airplane, the Air Service would give serious consideration to purchasing several examples for testing. Keys and the other directors, knowing the promise of the D-12, apparently decided to gamble on getting a production contract. The Curtiss engineering staff completed work on the new pursuit in December. Initial flight tests at the factory began in January 1923. The new ship soon demonstrated a higher speed, a better rate of climb, and a higher ceiling than the Boeing-built Thomas-Morse MB-3A pursuits then in service.[128]

In early March 1923, Frank Russell wrote to Major General Mason Patrick, who had replaced Major General Menoher as chief of the Air Service, to advise him that Curtiss Aeroplane was anxious to demonstrate the new pursuit's capabilities to the Air Service.[129] At the end of the month, Keys added his own plea. "My thought about the Pursuit ship," he wrote Major General Patrick, "can be very simply stated. It is that if it is not possible, for any reason, for the Service to buy from a contractor, who has, at great risk to his capital, created a new type of ship that is an outstanding type, and obviously superior to anything in

The Curtiss XPW-8, developed from Curtiss racing planes, first flew
in January 1923. The Army Air Service purchased this airplane in
April and two more prototypes. This was the first in a long series
of Curtiss pursuit planes. (3B-32449, RG 432FH, NARA)

its class in existence, then the creation of new and advanced types must definitely cease in the industry. . . . If this process of creation is to end, for any reason, I think it will be a disaster, the effect of which will be felt for a long time to
come."[130] Patrick agreed, and arranged to send an officer from McCook Field
in early April to test the new airplane. After reviewing the McCook Field officer's favorable report, on April 27 the Air Service agreed to buy the prototype,
designated the PW-8, and shortly thereafter ordered two more examples for
further testing.[131] Though it was not a production contract, it was a start.

In March, the Curtiss Corporation released its annual report for 1922.
Profits were meager at $16,169, but there was a profit. With the losses of
$300,000 on the NBS-1 contract, the results could have been worse. But the
outlook for the coming year was promising. Keys reported to the stockholders
that orders on the books of the company at the end of 1922 stood at $3,752,009,
an increase of nearly $2 million over the prior year.[132] The success of the D-12
engineering program enabled the company to resume engine production on
a larger scale, allowing the reopening of part of the Churchill factory for the

first time since the end of the war. With knowledge of discussions with the Air Service concerning the new Curtiss pursuit, Keys outlined the company's policy for the coming year. Without going into detail, he told the stockholders that management intended " . . . to carry on a substantial engineering program for the development and invention of improved types of aeroplanes, motors, and appliances, as only by such a policy can the permanent place of your corporation in the aeronautical art be assured."[133] After reporting on business prospects, Keys presented a plan to the stockholders for a complete reorganization of the company to deal with the problem of capitalization.[134]

Since his takeover of the company, Keys had known that the Curtiss Aeroplane & Motor Corporation would have to be reorganized. The Corporation's $5.3 million in 7% preferred stock was simply too much for the company to carry with its present and prospective level of business. While Keys was hopeful that the profits would improve over time, he doubted whether they would be sufficient to provide a reasonable dividend on that amount of preferred stock. The Corporation had not paid a dividend on the preferred since January 1920; at its present level, the preferred stock would remain a nondividend paying stock for many years to come. To provide a return to the preferred stockholders, Keys needed to reduce the Corporation's preferred stock outstanding to an amount where the Corporation's earnings could cover the dividend requirement. The question facing him was how to achieve this reduction.

Keys found a creative solution to the problem. He proposed splitting the Corporation's assets between two corporations, organized for different functions. The current Curtiss Aeroplane & Motor Corporation would take the manufacturing facilities and inventories of raw materials belonging to Curtiss Aeroplane and Curtiss Engineering and would reduce its capital to $2.5 million in cumulative preferred stock, plus the existing common stock, and would continue as a manufacturing company. A new company would be incorporated to take the inventory of unsold airplanes and the older Curtiss patents, with a capital of $2.8 million noncumulative preferred stock. Keys was confident that the manufacturing company could earn enough to provide a dividend on the smaller amount of preferred stock. The second company would pay off its preferred stock by gradually liquidating its inventory of airplanes and engines and from royalty payments on the patents it owned. Under his scheme, holders of the existing Curtiss preferred stock would be offered shares in a new preferred stock issue for the manufacturing company and certificates of beneficial interest in the liquidation company. He believed that his plan would help reestablish the credit standing of the Curtiss Aeroplane & Motor Corporation and enable it, finally, to meet its needs for working capital.[135]

Keys announced the reorganization plan on March 12, 1923, in a special letter to all stockholders. The Curtiss Assets Corporation would purchase from the Curtiss Aeroplane & Motor Corporation its surplus aircraft and engines and the rights to all royalties on Curtiss patents. A new manufacturing company named the Curtiss Aeroplane & Motor Company would purchase all the manufacturing assets of the Curtiss Aeroplane & Motor Corporation in return for $2,731,500 in new 7% preferred stock and 218,060 shares of common stock, the same number of shares of outstanding common stock of the Corporation. As a form of inducement to existing preferred stockholders, Keys effectively promised that a dividend would be paid on the Curtiss Aeroplane & Motor Company's new preferred issue during 1923.[136]

To Keys's relief, the reorganization plan was well received. A solid majority of the preferred and common stockholders approved the plan and signed proxies in its favor. Even Wall Street seemed pleased, pushing the older preferred stock issue above $30 a share for the first time in months.[137] The *Wall Street Journal*'s "Answers to Inquiries" column, which Keys had once worked on, noted approvingly that with the liquidation of older airplanes and engines segregated in another entity, the plan would enable the new manufacturing company to devote itself solely to business development.[138] At a special meeting of the stockholders on April 20, 1923, the plan received unanimous approval.[139] The new Curtiss Aeroplane & Motor Company, Inc. and the Curtiss Assets Corporation were incorporated under the laws of New York State on May 12, 1923 and took over the manufacturing business of the Curtiss Aeroplane & Motor Corporation as well as the inventory of unsold airplanes and older patents, respectively, on July 1.[140]

What the newly reorganized Curtiss Company needed most was more business. The Buffalo factory was hard at work producing D-12 engines for the Army and the Navy, but there were frustrating delays, partly due to weaknesses in the Curtiss engineering organization, which Keys could only address if there was the certainty of a higher volume of business.[141] The Garden City factory was working on experimental contracts from the Army and Navy, but the NBS 1 bomber and TS-1 pursuit contracts were nearing completion, which would leave the Garden City factory short of work after July. Though Keys was hopeful of winning a contract from the Army for pursuits based on the PW-8, he knew the Army and the Navy would not confirm any orders until after Congress had approved the appropriations for the next fiscal year.

When Keys heard from Frank Russell in July that the Army Air Service was not likely to give Curtiss a production contract for pursuits for some months to come, he wrote a strongly worded plea to Major General Mason Patrick. "I have felt," Keys wrote,

> for the past year and a half, that we in the Curtiss Company have been making a good fight for aviation in the United States. We have carried through to completion, the Martin Bomber contract, delivering I believe, good ships and raising our own standard of quality and construction.... During the same period, we have carried forward our engineering in both motors and planes to a higher point than we have been able to reach at any time in the past, and we hope and believe that we have, in this respect, created for the Army and the Navy, at a very trifling expense to them, types of motor and plane which represent substantial advances over all competitors.[142]

The new Curtiss pursuit, he argued, was in advance of anything else in the world. "If, then, this ship ... must await subsequent development by other designers, while our plants lie idle and our organization is whittled to pieces, of what possible use is it to me or to the Curtiss Company to plan far ahead to create major advances in military aviation, to risk the shocks and losses of Government contracting in a highly technical trade and to devote my own life and the money of my friends to carry on so thankless a task as this?"[143]

Patrick responded positively to Keys's plea. He replied that he knew Keys was aware of his "desire to obtain the best possible equipment for the Service and at the same time to be fair and just to those who must manufacture it."[144] To Keys's undoubted relief, Patrick informed him that he had instructed the chief of the Air Service Supply Division to place a modest order for pursuit planes with the Curtiss Company. In an apparent shift in policy that September, rather than putting the contract out to competitive bidding, the Air Service ordered 25 PW-8 pursuits directly from the Curtiss Company.[145] It was not a big contract, only $400,000, but it would keep the Garden City factory going until July of the next year when the first deliveries were scheduled, while the Buffalo factory was fully occupied building the D-12 engine.

The fall air races brought more triumphs for Keys and Curtiss. At the end of September, the Schneider Trophy Cup races took place at Cowes, England. On September 28, U.S. Navy Lt. David Rittenhouse won the Schneider

The Curtiss R2C Racer with Navy Lieutenant Alford Williams.
On October 6, 1923, Williams won the Pulitzer Race with a speed of
243 mph. Ten days later, Williams set a new world's record in this
airplane with a speed of 266 mph. (20343, RG 18, NARA)

Cup in the Curtiss CR-3 seaplane with a speed of 177 mph, an astonishing 34 mph faster than the winner of the previous year's Cup.[146] Eight days later at the Pulitzer Trophy Races in St. Louis, Navy Lt. Al Williams took first place in the Curtiss R2C-1 with a speed of 243 mph. A little over a month later Williams used the R2C-1 to set a new world's speed record of 266 mph.[147] The Curtiss racers had set world records two years in a row, fulfilling Keys's promise to Newton Baker. The victories brought the Curtiss racers and their D-12 engines wide acclaim. The editors of the English aviation magazine *Flight* expressed their admiration "for the determination, backed by designing and engineering skill, which has enabled America in a few months, figuratively, not only to catch up but to overtake the rest of the world in several spheres of aviation activity, as the number of records now standing to the credit of American aviators testify."[148]

What Keys achieved in three and a half years was remarkable. He was starting to build an aviation business. He took the Curtiss Company from the brink of collapse to a position as one of the world's leading aeronautical firms, with record performances in the Pulitzer races and the Schneider Cup, a world-class airplane engine in the Curtiss D-12, and a brand new pursuit plane for the Army Air Service. Through the corporate reorganization, he restored Curtiss to reasonable financial health. Despite its manufacturing problems, the Curtiss Company was modestly profitable and finally able to provide a return on investment to its patient stockholders. But could this narrow success be sustained?

Victory in the Schneider Cup was immensely gratifying, but Keys remained concerned with the quality of the business the Curtiss Company was getting from the Army and the Navy and the company's performance on these contracts. While the newly reorganized company was on track to have a profitable year, the Garden City factory was losing money on its Army and Navy contracts. Other companies—the Dayton-Wright Company, LWF Engineering, the Ordnance Engineering Company, and Aeromarine Plane and Motor Company—had failed or withdrawn from aviation. Now that the Curtiss Company had superior products to sell, would there be any demand from the Army and Navy, the company's sole customers, and would the system of competitive bidding allow Curtiss to regain the costs of developing the D-12 and the Curtiss pursuit? Or would the Curtiss Company have to shrink even further and abandon its engineering effort altogether, the one effort that held out some hope for a future for the company?

3

A FRAMEWORK
FOR MILITARY AVIATION

The State of the American Aviation Industry

WHEN THE FRENCH UNDER-SECRETARY FOR AIR LEARNED OF A NEW AMERICAN speed record, he commented, "Records, yes, but they are a façade; back of them there is nothing!"[1] His assessment, to the frustration of many in the American aviation industry, was all too correct, for the American aviation industry was still struggling to survive, with most of the surviving firms barely managing to scrape by on the few Army and Navy contracts available. As the yearbook of the Aeronautical Chamber of Commerce of America, which had replaced the earlier Manufacturers Aircraft Association, put it, "it is disturbing to contemplate the truth."[2] Five years after the end of the war, and four and a half years after Keys and the members of the American Aviation Mission had returned from Europe full of ideas and recommendations for building American aviation, America still had no national policy for either commercial or military aviation and no laws governing air commerce.

The American Aviation Mission recognized the importance of aircraft in any future war. The Mission's report concluded that "victory cannot but incline to that belligerent able to first achieve and later maintain its supremacy in the air."[3] The Mission also concluded that for economic reasons no nation could maintain in times of peace the air force it would need in time of war. The answer, the Mission believed, was to create "a great reserve in personnel, material and producing industry through encouragement of civil aeronautics."[4] But the development of commercial aviation would take time, a period of years that no one could predict with certainty. In the interim, the aviation industry would have to rely on demand from the military. The Mission's report made the following recommendation:

> The remaining aircraft production industry should be conserved and kept in a healthy condition by a well defined and continuing program of production for military purposes, over a period of years. This policy should be continued until commercial demand is adequate to support an industry of sufficient proportions to form an effective nucleus upon which can be built a war time production in case of need.[5]

In 1921 Herbert Hoover, newly appointed secretary of Commerce, set up the Committee on Aerial Navigation to work on a national aviation policy looking at issues of national defense and commercial aviation.[6] The Committee reached a conclusion similar to that of the American Aviation Mission: that "commercial flying would not mature sufficiently to play an important role in American aeronautics for some years, highlighting the need for the military to shoulder the development burden in the near term."[7] In its 1921 Annual Report, the National Advisory Committee for Aeronautics (NACA) recommended that, in addition to support for legislation to establish federal regulation of "aerial navigation," the aviation industry should "be kept in such a condition as to be able to expand promptly and properly to meet increased demand in case of emergency."[8] The government, NACA recommended, "as the principal consumer" of aircraft, "should formulate a policy which would be effective to sustain and stabilize the aeronautical industry and encourage the development of new and improved types of aircraft."[9]

Keys's experience with Curtiss Aeroplane & Motor Company since 1919 had reinforced his conviction that with commercial aviation in its infancy, continuity in military aircraft orders was critical to the survival of a sound aviation industry.[10] As the NACA 1921 Annual Report implied, an aviation industry that was not sound and stable could not develop the "new and improved types of aircraft" that commercial aviation would need in the future. Unfortunately, in his view, since 1919 the government had done nothing to promote a sound aviation industry. Instead, the Army and Navy's policy of denying design rights to the aircraft manufacturer and awarding production contracts through competitive bidding as well as the lack of continuity in orders for military airplanes was destructive to the aviation industry.[11] Competitive bidding, he believed, was not only driving capital out of the industry but would ultimately destroy what he termed "the creative function," the ability of companies to develop new and improved airplanes.[12] Keys was by no means the only representative of the aviation industry to hold this view.

As a result of the lack of either government or commercial aircraft orders, in the first few months of 1924 ten of the fifteen main manufacturing and engineering firms had plants that were either idle or doing the bare minimum of work.[13] The lack of sustained demand, together with the fierce competition for the few orders that did emerge from the Army and the Navy, was damaging the entire industry, including the more active firms, such as Boeing, Vought, Consolidated, Martin, and Curtiss. By no means an unbiased observer, the Aeronautical Chamber of Commerce stated in its introduction to the *Aircraft Year Book* for 1924 that the American aviation industry had "passed perceptibly nearer that point where it must inevitably—unless a definite national policy is adopted and adhered to—cease to exist as a practical factor in our national defense."[14]

Among those who believed that a healthy aviation industry was vital to national defense there was a growing sense of alarm over the deteriorating condition of the aviation industry and deficiencies in the military air services. In his final report issued in March 1923, Assistant Secretary of War J. Mayhew Wainwright stated that the situation in the Army Air Service was "most critical."[15] The supply of war surplus aircraft was rapidly diminishing through deterioration and attrition. Congressional appropriations were insufficient to maintain the supply of aircraft the Air Service needed even to maintain its peacetime functions. The aviation industry in America was "practically facing extinction," and unless the government took action to place sufficient orders, "the industry as such will disappear."[16] That fall the Lassiter Board, formed under Major General William Lassiter to investigate the present and future status of the Air Service, confirmed Assistant Secretary Wainwright's conclusions. If appropriations for the Air Service continued at their present level, by 1926 the Service would have less than 300 aircraft available against a requirement of 1,655.[17] The aviation industry, the Board said, "at present is entirely inadequate to meet peace and war requirements, it is rapidly diminishing and under present conditions will soon practically disappear."[18] Both Wainwright and the Lassiter Board recommended a multiyear procurement program, which would enable the Air Service to build up to its authorized strength with sufficient reserves and replacements.[19]

Naval aviation faced a similar problem. In his annual report to the secretary of the Navy, Admiral William Moffett, chief of the Bureau of Aeronautics, estimated that by the end of Fiscal 1924 the Navy's stock of surplus aircraft and engines would be exhausted.[20] Dwight Davis, who replaced Wainwright as assistant secretary of War, put the blame squarely on Congress. The Congress, Davis said, "has reduced our air force below the safety point of national defense. . . . What we are doing is literally starving to death the most vital branch of our national defense. If Congress does not make adequate appropriations in the near future, our air service will be rendered absolutely impotent."[21]

The legacy of the failure to produce airplanes in quantity for the Army during World War I lingered long in the halls of Congress where a "suspicion that only conspiracy, graft, and fraud could explain a failure of such scale" remained for much of the interwar period.[22] In the view of many members of Congress, the industry's poor record reflected a failure to ensure that free enterprise and laissez-faire capitalism were not subverted.[23] To preserve free enterprise and gain the perceived benefits of widespread competition to obtain the lowest bids on aircraft contracts, the Congress insisted that the military forces return to their traditional procurement practices, including denying a manufacturer proprietary rights to a design.[24] In adhering strictly to the laws on procurement, as was Congress's intent, the military services "led aircraft manufacturers into a dog-eat-dog era of destructive competition that penalized the very firms doing most to advance the art."[25]

The Debate over Competitive Bidding

Keys's immediate concern was the survival of the Curtiss Aeroplane & Motor Company. Larger appropriations, multiyear procurement programs, and greater government support for commercial aviation were all badly needed, but for Keys, survival of the aviation industry depended on ending the system of competitive bidding with its assignment of a manufacturer's proprietary design rights to the government. As president of Curtiss, he had a responsibility to press for government policies that would benefit his company. And he had a duty to his shareholders to earn for them a return on the capital they had invested in Curtiss. In this case, he believed that what was good for the Curtiss Aeroplane & Motor Company would be good for the industry. In his view, the government's policies were driving capital out of the aviation industry and destroying any incentive for investment in the research and development that was so necessary to aviation's progress. The end result, he feared, would be a weak, inefficient industry lagging technologically behind other countries and incapable of meeting the needs of national defense. Keys had a strong sense of fairness, a belief that constructive risks taken were entitled to a fair reward, especially if those risks led to an improvement in a technology, a service, or to general progress, however defined. A system of competitive bidding was both unfair and destructive, for it robbed a manufacturer of the fruits of his own labor. After taking control of Curtiss, Keys devoted considerable energy to lobbying for an end to competitive bidding on military aircraft contracts.

In his position as president of Curtiss Aeroplane, the leading firm in the aviation industry, Keys had a unique platform from which to argue for the policies he believed in. Although he sometimes thought his views represented those of the larger firms in the industry, he did not seek to become the public spokesman for the entire industry, nor did he adopt the outspoken public advocacy of Brigadier General William Mitchell, who aggressively campaigned for air power. It was not in his character to seek public recognition. Keys did not play an active role in the Aeronautical Chamber of Commerce, the aviation industry's newly formed trade association organized to advance the cause of American aviation. Instead, Frank Russell, vice president of Curtiss Aeroplane, represented the company.

Keys did express his views publicly, in both speeches and writings, when called on or when he deemed it necessary, but with his natural reserve he preferred to work quietly, behind the scenes, trying to influence those key figures in government who had the authority to make the decisions that directly affected the Curtiss Company. In this effort he would be relentless, and not always tactful, in putting forth his arguments. He became an active correspondent, writing regularly to Major General Mason Patrick, chief of the Air Service, and also to Admiral William Moffett, chief of the Bureau of Aeronautics, the secretary of the Navy, and other senior government officials. In this area he was more active than his peers in the industry, particularly with Patrick, corresponding with him more frequently, and more forcefully, than most.

Keys began writing to Major General Patrick in the fall of 1921, soon after Patrick took over as chief of the Air Service from Major General Charles Menoher in October. Patrick sought the views of Keys and other leading manufacturers concerning the policies and practices of the Air Service. He asked Keys for his frank opinion, which Keys gave him. In a detailed letter, Keys outlined his principal concerns. Most of all, Keys wanted to point out to Patrick the damaging effect of the competitive bidding system. He told

Major General Mason Patrick, who became head of the Army
Air Service in 1921, learning to fly at the age of fifty-nine. Keys
carried on a lively correspondence with Patrick over the system of
competitive bidding. (LC-DIG-ggbain-37460, Library of Congress)

Patrick that "in my judgment, a very considerable contributing cause to the
fact that the art of design and creation is being rapidly wiped out in the aeroplane trade in the United States lies in the fact that anything created by the
independent designer, no matter how much it costs him and how much labor
was spent on it, is apt to become almost immediately an object of competitive
bidding for production."[26] While Keys understood that the Air Service had
no control over the matter, as it related to bidding policies set by Congress, he
doubted that "under this condition, anyone would be so foolish as to spend
large sums of money to push forward the art of aeronautics in this country

until such time, if it ever comes, as there is created a commercial demand for aeroplane invention."[27]

Patrick sympathized with Keys's arguments. Patrick understood, as Keys did, that a vibrant aviation industry with a strong manufacturing base served as the military's reserve in times of war. No nation could afford to maintain in peacetime the large air force it would need in wartime. Patrick believed that the government was ignoring the health of the aviation industry. He wanted to do what he could, with the limited funds and within the constraints Congress imposed on the Air Service, to support the industry and the development of commercial aviation.[28] On the critical question of competitive bidding, Patrick told Keys that "I would get away from this if I could. Unfortunately, Congress for years has been obsessed by the hoary shibboleth, 'Competition is the life of trade.'"[29] He thought that an attempt to create an exemption for the aircraft industry would likely meet with strong opposition from certain segments in Congress. "Nevertheless," he wrote Keys, "I do propose, if I can secure the backing of the War Department, to attempt the impossible and to see whether I can convince some of these omniscient beings that there is a fallacy in this ancient maxim."[30]

Over the next several months Keys wrote to Patrick regularly. The two men soon established a rapport and a mutual respect; though they did not always agree, they came to express their opinions freely and with frankness. For Keys, Patrick was not only one of his most vital customers, the other being the Navy, but also an ally in the struggle to get Congress to change the system of competitive bidding. Maintaining Patrick's goodwill and keeping him advised of the fortunes of the Curtiss Corporation, as well as the harmful impact of the government's policies on the aviation industry, were critical to Keys. In such a competitive environment, any advantage he could gain for Curtiss, however slight, was worth the effort. Patrick, in turn, came to appreciate Keys's expertise and his views on the aviation industry. Keys gave Patrick insights into the condition of the aviation industry as well as arguments he could use in his own efforts to win support for a change in policy toward military and commercial aviation.

Toward the end of 1922, Keys decided to undertake a broader campaign against the policy of competitive bidding. He began with a letter to Major General John J. Pershing, the Army's chief of staff, in response to a report that Pershing had issued on the state of aviation in the United States. Keys told Pershing that between 1918 and 1922 much of the aviation industry had been practically wiped out. Keys warned that "if the tendencies of this period should continue, there will not be in aviation, two years from now, a single unit that is now operating on a production scale."[31] Under the system of competitive bidding, manufacturers could not afford to maintain any reserve production capacity, which the military would need in time of war, nor could they sustain the engineering departments so necessary to the development of the art. "For that reason," Keys wrote, "the fact . . . is that every progressive company in the country has lost money steadily, and must withdraw from the industry as early as possible."[32] All that would likely remain would be small factories without an engineering and development function, capable of taking only small production orders. Keys doubted that the industry's problems could be remedied under the present laws. In fact, he argued, "the tendency of the day, instead of being toward any remedy of this condition, is toward an aggravation of it."[33] Like Major General Patrick, Pershing was sympathetic to Keys's concerns. Pershing, too, understood the need to have an aviation industry capable of quantity production in a national emergency, and that this required having an existing, healthy industry. Pershing replied to Keys saying, "I am quite safe in saying that the War Department realizes the necessity for keeping this aircraft industry alive in time of peace and to this end is willing and ready to all in its power."[34]

Keys also wrote to Edwin Denby, secretary of the Navy, though his first effort was not well received. Secretary Denby had written an article on the aviation industry and national defense in which he argued that commercial aviation was the answer to preventing the industry's collapse.[35] Keys did not think commercial aircraft could be rapidly converted for military uses. As he explained:

> The military aeroplane is a thing of extremes. It aims for maximum speed, maximum climb, maximum maneuverability. It cuts factors of safety in every direction, in order to accomplish these abnormal results. It seems to me that the whole trend of aeronautical engineering will be in exactly the opposite direction, viz., toward a maximum of safety, low landing speed, low upkeep. If my conclusions about this matter are correct, commercial aeronautics, while it will be exceedingly valuable to the country in its maintenance of the trained fabricating and flying personnel, can not by itself keep this country in the first place in military aeronautics.[36]

Keys offered Denby the benefit of his own experience as well as that of the Curtiss Corporation, if that would be helpful to finding solutions for the problems he had outlined. "I believe," Keys wrote, "that I have perhaps had a wider experience in this matter than anybody else in the country, and if that experience can be of value to you, or to other officers of the Government, it is entirely at your disposal."[37]

By coincidence, Keys wrote to Denby when an editorial on the status of the aircraft industry appeared in the January 1, 1923, issue of *Aviation*, whose president, Lester D. Gardener, was a strong supporter of the industry. *Aviation* stated that "the most important problem for 1923 . . . concerns the maintenance of an aircraft industry in the United States. Judging from the 1922 record, unless a definite policy is adopted, there will be a situation created that will be a serious menace to the air defense of the nation."[38] According to *Aviation*, the American aircraft industry was confronting a crisis, a crisis that procrastination and damaging government policies had brought about. The nation needed a private aircraft industry that was financially sound, maintained first-rate production facilities, and was committed to engineering development. Under the system of competitive bidding, the editorial argued, "a properly financed, well-equipped and thoroughly responsible aircraft company . . . is compelled to compete in price with individual bidders who are lacking in financial reserve, responsibility, or productive capacities and who therefore cannot be regarded as a substantial element."[39] The current situation was driving companies out of the industry. "It can only result," the editorial concluded, "in forcing capital now invested in the exceedingly precarious aircraft industry to seek other channels, more profitable and better appreciated."[40]

This was the point that Keys was trying to convey to the military and government officials responsible for aviation. It has been argued that noneconomic rewards motivated those involved in the aviation industry to remain involved despite the dim prospects for the future.[41] The reality is that almost all companies, in almost all industries, are established to make a profit, or in the hope of making a profit, from their activity.[42] At a minimum, to survive companies must at least cover their costs of production. Establishing a company and continuing in operation, however, requires capital. The providers of capital, in the form of loans or capital stock, will in the normal course require a return in the form of interest or dividends that is commensurate with the risk they are taking in investing in the firm.[43] If the return does not meet expectations, capital will be withdrawn and providers will seek other, alternative investments, as the editorial in *Aviation* noted.

Keys had recently received confirmation of this reality from John J. Raskob, chairman of the General Motors Corporation, who owned the controlling interest in the Dayton-Wright Airplane Company. In response to receiving the letter that Keys had written to General Pershing, Raskob informed him that General Motors had decided to liquidate the company and withdraw entirely from aviation. In three years of operations, the Dayton-Wright Company had incurred around $400,000 in losses, which General Motors could no longer sustain. Raskob told Keys that the reality facing the industry was that the Army and the Navy comprised the entire market for airplanes and that "so much difficulty had been found in dealing with the Government that it is considered useless to carry on the work any further."[44]

Keys was correct in his analysis, but his letter to Secretary Denby raised hackles within the Bureau of Aeronautics, and for the following reason: it came shortly after the editorial on the aircraft industry in *Aviation*, which the Bureau linked with Keys, and a speech Frank Russell had given that appeared to sharply criticize the work of the Naval Aircraft Factory. Commander E. S. Land, head of the Bureau's Material Division, was charged with drafting a reply for the secretary. Keys, said Land, "was not only a remarkably clever businessman but is also undoubtedly an expert on commercial aviation. The serious objection . . . to following the advice of Mr. Keys is the fact that he is President of the Curtiss Company and is therefore undoubtedly a prejudiced party."[45] Land believed that Keys was behind the editorial in *Aviation*. "The fine Italian hand of Mr. C. M. Keys," he wrote, "is seen in this editorial."[46] Land viewed Keys's criticism of competitive bidding as an argument for the Bureau of Aeronautics to have an approved list of bidders, an argument that one could also read into Keys's letter to Secretary Denby.

In allocating its limited funds, the Bureau was expected to reward those firms that were, as the *Aviation* editorial defined them, "responsible in financial reserve, competent in production facilities, and progressive in engineering," qualities that defined the Curtiss Aeroplane & Motor Corporation.[47] While Land admitted that there were some advantages to such a system, it would open the Navy to "very violent criticism in that those bidders not on the approved list maintain that such procedure is un-American in its form, contrary to the principles of government under which we exist, unfair to concerns struggling to get into the business."[48] Secretary Denby wrote a rather abrupt reply to Keys's letter. Keys tried to explain that his intention had never been to criticize Navy policy toward aviation, but the damage had been done. From then on, Keys's relationship with Admiral Moffett and the Bureau of Aeronautics was neither as open nor as supportive as his relationship with Major General Patrick and the Air Service.

At the end of December 1923, following the Navy's victories with the Curtiss Racers at the Schneider Cup and the Pulitzer Trophy races, Keys wrote a detailed, six-page letter to Edwin Denby, building on the arguments he had used earlier in the year and supporting his conclusions with more facts. He told Denby that due to the extensive losses the Curtiss Company had sustained on Navy contracts, the Company could no longer provide the Navy with the services it had rendered on the same basis.[49] According to Keys, the Curtiss Company had lost $438,125 on its principal aircraft contracts with the Navy since 1920. Keys admitted that the contract for the six CS Scout-Torpedo-Bombing planes, the first of which was delivered to the Navy in November, had been a disaster for the company, generating a loss of $292,429. Building the CS had been a technological challenge, involving a substantial shift from wood to metal construction. Both Curtiss and the Navy had underestimated the costs of this evolution, but as Keys argued, it was the Curtiss Company and not the Navy that would bear the loss.

Under the current system, any company that committed itself to engineering development, and engineering excellence, was in effect subsidizing the rest of the industry. "The Curtiss Company," he maintained, "simply because of the high standard of its engineering ability and the versatility of its methods, would suffer a very heavy penalty; whereas any and all other contractors who may . . . build ships of this type merely as production companies, would reap what we have sown, and gather to themselves the entire benefit . . . of this high grade engineering development."[50] What Keys wanted from the Navy was support "for that part of an industry which creates new standards of performance, rather than for large production facilities."[51] Keys and the small group of men around him kept going because they had a vision of the future of aviation, when it would, one day, become a large and successful industry. "So far as we are personally concerned," he said, "it is essential to this vision and to our morale that the company not only should cease to lose money on its aeroplane business, but that it should receive a moderate return on its small invested capital, so that when the time comes for us to take our places as pioneers and leaders in a great industry, we need not appeal for capital with a long record of financial failure, but rather with one of reasonable success."[52]

By now, Major General Patrick and Admiral Moffett not only recognized the damage the system of competitive bidding was doing to the aviation industry, but they also wanted to change it. The arguments that Keys and others had been stating repeatedly were largely accepted. Patrick and Moffett began changing the roles of McCook Field and the Naval Aircraft Factory, moving these facilities away from production that competed directly with the industry toward a focus on research and development.[53] Patrick chose to "bend" the competitive bidding system by giving Curtiss Aeroplane, and shortly thereafter the Boeing Company, contracts for pursuits without putting them out to competitive bidding; Moffett did have similar flexibility in the Navy regulations, but he chose not to use it.[54] But other goals that would relieve the crisis in the aviation industry—larger appropriations to replace rapidly obsolescing surplus World War I aircraft, multiyear procurement programs to maintain authorized aircraft strength, a change in the competitive bidding system, and support for commercial aviation—could only come about through congressional legislation.[55] But Congress had to act. In an editorial on December 21, 1923, the *New York Times* commented that "civil aviation languishes in the United States, and, splendid as are the feats of our military fliers, the army and navy services are absurdly inadequate. We have a few up-to-date machines in which the records are made, but a great many obsolete ones that are little better than junk . . . Will the United States fall so far behind in civil and military aeronautics that she will not possess adequate resources for trade and defense? It is for Congress to say."[56]

There was a growing frustration with the continued congressional focus on investigation over legislation.[57] Newspapers around the country were starting to direct criticism toward Congress for its failure to act. Reacting to Major General Patrick's warning that the Air Service was being starved to death, the Dayton *Journal* said that "it is the duty of Congress to provide money and encouragement. Economy purchased at the cost of safety is pernicious. Yet that is what is being done today," while the Scranton *Republican* said the following: "though the inability of certain members of Congress to reason out their clear duty in the matter the air service of the United States is being crippled."[58]

Keys chose this moment to state publicly the concerns for the future of the aviation industry that he had expressed privately to Major General Patrick, Admiral Moffett, and Secretary of the Navy Denby, as well as to openly criticize the government's policies toward the industry. As he later explained to John J. Raskob, "we faced the problem of either seeking to retire from the business . . . or starting a battle with the Government for the right to live and for the establishment of fair methods of contracting in Washington. We determined to make a fight of it and the Curtiss annual report for 1923 . . . began that battle."[59] The annual report was issued on March 8, 1924. Keys told Curtiss stockholders that "the outlook for the entire aeroplane industry is vague and uncertain', and outlined the reasons why.[60] Keys cited three factors: first, the industry's increasing competition from government-operated plants; second, the policy of the government to buy airplanes at what Keys believed was below the cost of production; and third, a strong tendency toward a government monopoly of development work, which he claimed was eliminating private incentive. "In view of these conditions," Keys stated, "a large proportion of the capital invested in the aeroplane engineering industry at the end of the war has been either destroyed or driven out of the industry, and so far as we know, practically no new capital has embarked in the industry or is now available for it."[61] Keys informed the stockholders that the Curtiss Aeroplane & Motor Company would continue to reduce its engineering work force. Keys said, "it is obviously necessary . . . to cease creative effort and abandon definitely the policy which has been the Curtiss policy from the beginning of the art, namely, to regard every experimental contract as a cooperative effort with the government and to accomplish the result sought after regardless of the cost to your company."[62]

The principal problem, Keys said, was that:

> outstanding accomplishment in the industry for the benefit of the Army and Navy does not lead to the placing of substantial contracts with those whose capital has been spent in the accomplishment of great adventures. On the contrary, such accomplishments usually lead to financial losses. As soon as new standards are set, at great expense in effort and money by private companies, their accomplished successes are immediately copied by others who have no great engineering expense to absorb. Consequently, units in the industry organized solely for small production work can and do sell products below the cost to the original creator of the type of motor, plane, or accessory involved. Any important development . . . is immediately broadcast to the trade by the government so that thereafter the value of the invention ceases to its inventor and inures if at all, to the lowest bidder on construction contracts.[63]

As an example, Keys mentioned the company's experience building the CS airplane for the Navy. Keys told the stockholders that "it now appears likely that a quantity order for these ships will be offered to public competition so that the fruit of our expenditure may be gathered by those who do not risk capital to accomplish advances in the art on behalf of your government."[64]

A number of newspapers carried Keys's statement in the Curtiss annual report. The *New York Times* featured extensive quotes from the report under the headline "Deplores Outlook in the Plane Industry," while *The Wall Street Journal* printed the annual report in full.[65] In an editorial a few days after Keys released the annual report titled "Preparedness in the Air," the *Times* contrasted the aircraft program of Great Britain's Labor Government with what was happening in the United States. "Nothing is being done here to stop the decline of the airplane industry," the *Times* said, "It is going from bad to worse." After quoting comments from the Lassiter Board report, the *Times* said, referring to Keys's comments in the Curtiss annual report, "a prominent American manufacturer said the other day he could not remain in business unless he received larger orders from the Government. If we decline to follow the subsidizing example of the French and the British, liberal appropriations must be made for the purchase of modern planes for the military services."[66]

While Keys was "one of the loudest and most articulate critics of Congress' policies," he was only one of several voices deploring the condition of the aviation industry in America.[67] William Boeing, president of the Boeing Aircraft Company, sent a copy of the *Wall Street Journal's* article carrying Keys comments to Major General Mason Patrick. While Boeing said that he had "no comment to make on it excepting that the Curtiss Company is having the same disturbing reaction that we are," he raised the same concerns that Keys had raised with Patrick, namely, that under the system of competitive bidding the Boeing Company's engineering and development efforts benefited other companies that had not made the same investment.[68] Like Keys, Boeing said that the only policy his company could follow would be that of the Curtiss Company: to sharply cut back their development effort and become simply a production facility.

On March 24, 1924, the House passed a resolution creating a special committee to investigate the operations of the Air Service, the Bureau of Aeronautics, and the air mail service, consisting of nine members with Representative Florian Lampert, a Republican from Wisconsin, as chairman.[69] Concern with the state of the military air services, particularly after statements from Major General Patrick that the Air Service was "now entirely incapable of meeting its war requirements," as well as charges that a "nefarious 'aircraft' trust" was causing waste and inefficiency, prompted Congress to set up the special committee.[70] Keys was cautiously optimistic about the future, despite the fact that Representative Lampert's committee, the ninth congressional investigation relating to the state of aviation in America since

1917, appeared to be one more in a seemingly endless series of investigations that led to nothing.[71] In response to a worried letter from Frank Allen asking Keys for a clarification of his remarks in the Curtiss annual report on behalf of alarmed stockholders, Keys replied that he felt it was necessary to inform the stockholders of the conditions facing the company, though it was his personal belief that "the tendencies which have created those conditions have nearly run their course. . . . I believe that constructive forces are at work, for the first time since 1918, and that before long the entire policy of [the] Army and Navy will be dictated by Congress to be that the industry must do the work, and that those factors in the industry which have been constructive from the beginning will get the jobs to do."[72]

Ironically, the day after writing to Frank Allen Keys received an irate letter from Admiral Moffett, who took strong exception to the comments in Keys's letter to his shareholders. Admiral Moffett could not agree with Keys's argument that the government wanted a monopoly of engineering development and the elimination of private incentive. Nor did Moffett agree with Keys's complaint that the government had reserved the proprietary design rights to the "advanced type of plane for the use of the United States Navy," as Keys had referred to the Curtiss CS-1 airplane in his letter to stockholders.[73] Moffett pointed out to Keys that the terms of the CS-1 contract, which Curtiss had accepted, incorporated the sale of the proprietary rights to the Navy. He seemed to imply that Keys had no right to complain. In response to Keys's statement that it appeared likely that a contract for the CS-1 airplane would be put out to competitive bidding, Moffett stated that the Navy had offered Curtiss a price for quantity production of the CS-1, which Curtiss had declined as unsatisfactory. "The Bureau does not want to enter into a controversy as regards this matter," Moffett said, "or any kindred matters. Furthermore, the Bureau is extremely desirous of doing its part to assist the aircraft industry . . . all the Bureau of Aeronautics desires is fair play and a square deal. Constructive criticism is welcomed, but it is entirely impracticable to submerge the identity of the Bureau for the benefit of a single constructor, no matter how valuable this constructor may be."[74]

Much to his frustration, the Bureau of Aeronautics had indeed opened the production contract for the CS-1 to competitive bidding. The Curtiss Company's initial offer was to build twenty-five aircraft at $46,120 per airplane, compared to bids of $38,675 from the Boeing Airplane Company, $29,396 from the Douglas Company, and $40,192 from the Glenn L. Martin Company.[75] The final contract, for thirty-five aircraft designated SC-1 under the Navy's new

The Curtiss CS-1 Scout. To Keys's frustration, the Glenn L. Martin Company underbid Curtiss on the production contract for thirty-five SC-1 aircraft for the Navy, winning a second contract for forty aircraft as the SC-2. (2008-03-31_image_638_01, Peter M. Bowers Collection, Museum of Flight)

system of designations, went to the Glenn L. Martin Company in May whom Curtiss had underbid on the earlier Martin bomber contract for the Army Air Service. Eight months later the Martin Company received a follow-on contract for forty SC-2 aircraft.[76] As Keys had feared, and as he argued would inevitably happen under the policy of competitive bidding, the Curtiss Company, having lost nearly $200,000 on the development of the CS-1, had no chance to recoup its investment. As Keys saw it, the Curtiss Company was supporting the rest of the aviation industry, putting its own capital at risk to fund research and development, the benefit of which went to firms that had not taken the same risk, for no return. His stockholders would have had every right to ask why he had invested the firm's capital in development, which is why he had decided to cut back on the company's creative efforts.

By the end of the summer, Keys was writing to several of his correspondents that he had never seen the Curtiss Company with fewer orders on its books, though he was reasonably confident that there would be more business coming the company's way. A talk with the Air Service's chief of Procurement in October gave Keys greater certainty that Curtiss could expect more consistent orders from the Air Service for both airplanes and engines in the coming year.[77] But until then, much to his frustration, he had no choice but to reduce the Curtiss Company's payroll and overhead once again. He decided to consolidate all of the company's manufacturing activity in the Buffalo plant, leaving only the Engineering Division in Garden City until the plant could be sold.[78]

Keys could not pass up the opportunity to remind Major General Patrick and Admiral Moffett of the conditions facing the Curtiss Aeroplane & Motor Company. On October 1, 1924, he wrote Patrick and Moffett to inform them that the Garden City plant was being dismantled, the employees involved in production were being laid off, and the engineering staff had been reduced to a small number in a more compact Engineering Division.[79] Keys gave three reasons why the Curtiss Board of Directors had decided to stop production at the Garden City plant: first, that contracts with the Army and Navy were placed through competitive bidding at prices that were below the plant's cost of production; second, that neither the Buffalo plant nor the Garden City plant had been able to obtain a continuous flow of work, leading to an inefficient process of hiring and then firing employees, which Keys believed was one of the main reasons for the high cost of production in the industry; and third, the duplication of costs in running two plants at less than 50% of their capacity.[80] "We cannot," he wrote, "take additional contracts at losses merely for the sake of keeping production facilities available, nor can we spend additional capital in carrying on the airplane industry."[81]

Though Keys saw his actions as indicative of the continual whittling away of the aircraft industry, Admiral Moffett chose to take a different tack, seeing the move as positive. "The Bureau notes with particular satisfaction," he told Keys, "that you expect this to materially reduce your overhead expenses and hopes that you will be enabled thereby to obtain contracts at a figure which will enable you to earn a profit, and to terminate the present unsatisfactory conditions under which you show yourselves to be operating."[82] Moffett said the Bureau saw this reduction in capacity, not as a problem, but as a constructive step that would benefit not only Curtiss, but the aircraft industry.

To Keys's satisfaction, the Curtiss name continued to break records. On June 23, 1924, the Air Service's Lt. Russell Maughan completed a successful "dawn to dusk" flight across the country in a Curtiss PW-8.[83] The outstanding performance of the year came from the Navy's R2C-2 Curtiss racer. The

Schneider Trophy races were canceled when Great Britain and Italy withdrew from the competition, but on September 27 Lt. David Rittenhouse, the winner of the 1923 Schneider Trophy race, broke the world's seaplane record in the R2C-2 with a speed of 227 mph, bettering his previous record by more than 30 miles an hour.[84] At the Pulitzer races in early October, an Army Air Service Curtiss R-6 racer took second place with a speed of 214 mph.[85]

At the end of the year, Keys wrote to former Secretary of War Newton D. Baker, who had returned to his law practice in Cleveland, to remind Baker of the promise Keys had given him when he took over the Curtiss company in 1920 that he would endeavor to bring back to the United States the world's aviation records. Keys informed Baker that the company had completed the program, which started in 1920, citing the speed records the Curtiss racers had set in 1922, 1923, and 1924. Keys said he thought that it might give Baker some pleasure to know the results that had been accomplished, and that Keys had fulfilled his promise. To drive home his concern about the current state of the aviation industry to one more possibly influential person, Keys told Baker that "these years of effort and responsibility have not resulted in any great material advantage either to the Curtiss Company or to me personally" and more pointedly, "we are now dismantling and closing, so far as production is concerned, the plant where these things have been done."[86]

In November, Keys testified before the General Board of the Navy. President Coolidge had directed Curtiss P. Wilbur, secretary of the Navy, to form a special board to consider recent developments in aviation and to make recommendations on the Navy's future needs for aircraft, battleships, and other equipment.[87] The Board called a large number of Air Service and Navy officers and representatives from the aviation industry to provide testimony. In addition to Keys, the Board asked Edgar Gott, president of the Boeing Company, Charles Lawrance, president of the Wright Aeronautical Corporation, and Chance Vought, president of the Vought Corporation, to give their views on what could be done to improve naval aviation.[88]

Keys appeared before the General Board for three hours on Friday morning, November 14. He took the opportunity to attack the factors he believed were causing the destruction of the aircraft industry and would, in his view, ultimately lead to a government monopoly of aircraft design, development, and production. Keys drew the Board's attention to three critical issues: the lack of continuity in orders for aircraft and airplane engines, which resulted in the lack of continuity in production, the system of competitive bidding, and what he claimed was a policy on the part of the Air Service and the Navy to buy aircraft at prices below their cost of production. The lack of continuity in orders, Keys argued, resulted in a lack of efficiency that increased costs to the Air Service and Navy. Keys produced a chart for the Board showing the Curtiss Company payroll from 1919 to 1924, pointing out how employment rose whenever the company obtained a contract and fell when the contract was completed.[89] A large fluctuation in employment ruined chances for efficiency. The solution was continuity in placing orders for aircraft. "Give them something steady to pay their overhead," he argued, "and your cost of airplanes will go down in five years."[90]

But the most destructive element facing the industry was the system of competitive bidding and the lack of recognition of proprietary design rights. By now Keys was well practiced in conveying his arguments. The system, he told the assembled admirals, was gradually wiping out the industry. As he had argued previously, Keys pointed out that under the competitive bidding system, companies that did not have to maintain an engineering division would always be in a position to underbid those companies that did. And with too little business to go around, too many firms would bid on contracts, even if it resulted in a loss, in the hope that it would keep them in business. "Keep that pressure on the industry month after month," he said, "what are you going to do to it? The answer is, just what you have done to it."[91] Keys listed the firms that had withdrawn from the industry and stated his belief that all the remaining firms had lost at least some of their capital. Capital was leaving the industry, and there was precious little incentive to attract new investment. At the rate the industry was going, there would be little chance of building up an industry that would be of any use in the event of a war.

The risk, in his view, was that ultimately the design and development of airplanes and engines would end up as a government monopoly, as the government would be the only organization that could afford to maintain a design bureau and design staff. And Keys was convinced that no one design organization would ever be able to obtain the results that several design organizations, working separately, could achieve. By all means insist on competition in design and creation, he said, but put an end to the destructive price competition. To Keys's apparent surprise, the discussion took a turn that he found to be the most constructive he had heard from the Navy. Rear Admiral Henry H. Hough, director of Naval Intelligence, asked Keys if placing orders directly with the Curtiss Company instead of putting them out for competitive bidding would solve the problem, to which Keys replied, "that would be the whole solution."[92]

Keys knew that the destructive policies that were hurting Curtiss and the other companies in the industry were not the fault of the contracting officers of the Air Service or the Navy, nor were they responsible for these policies. That, he knew, was the responsibility of the Congress, and it appeared that there might just be an opportunity to influence congressional thinking through the newly formed Special Committee of Inquiry into Operations of the United States Air Services. "I believe," Keys wrote Patrick after meeting with several of the Committee's members following his testimony to the Navy General Board, "that the Congressional Committee, if it were given the information, would become a strong advocate for almost any constructive plan. . . . It seems to me that the Congressional Committee, which includes at least three brilliant pleaders from the Floor of the House, might very well succeed in getting legislation which would enable both the Army and the Navy to use ordinary common sense in buying aircraft; but I don't think the Committee will do much in this line unless the subject is taken up constructively with the Committee."[93] Keys would do everything that he could to make sure it was.

The Lampert Committee Hearings

The Select Committee of Inquiry into Operations of the United States Air Services, which came to be called the Lampert Committee after its chairman, Representative Florian Lampert, began its investigation in the fall of 1924. The Select Committee began hearings against a background of continued expressions of concern over the state of American aviation from President Coolidge, the National Advisory Committee for Aeronautics, the military services, and the press.[94] To make recommendations for future government policies for aviation, the Select Committee intended to cover uses of aircraft in war, the organization of the military air services, the development of commercial aviation, the condition of the aviation industry, and what needed to be done to ensure that the United States had the right number and mix of modern aircraft, and a satisfactory basis for quantity production in time of war.[95] From the beginning of October 1924 until early March 1925, the Select Committee heard testimony from a broad range of military and government officials concerned with aviation and a broad range of representatives from the aviation industry.

Keys testified before the Select Committee during January 1925. In total, he testified at greater length than any other representative of the aviation industry. In the Select Committee's transcript of the hearings, Keys's testimony covered 121 pages.[96] Keys wanted to ensure that the Select Committee gained an understanding of just how damaging the government's approach to the aviation industry since the end of the war had been and why. Aware that none of the members of the Select Committee had much personal knowledge of aviation, he was willing to go into as much detail as the members of the Committee wanted to hear, patiently answering every question, though at times he allowed his frustration with the Army and Navy's policies to come through.

Keys's central argument was that there could not be an air service in the Army or the Navy adequate for the needs of national defense without a sound and creative aviation industry standing behind it.[97] But the money Congress had appropriated for aviation since the end of the war had, in Keys's view, not been spent with any consideration of maintaining a sound industry. And worse, he emphasized that the laws and policies the Army Air Service and the Navy had followed had been not only destructive to the industry, but if continued would also ultimately wipe it out.[98] Keys listed five major causes of the destruction of the aviation industry: the system of competitive bidding, the lack of continuity in government procurement, competition from the government facilities at McCook Field and the Naval Aircraft Factory, the withdrawal of private capital and capacity from the industry, and the destruction of the industry's morale.[99] He explained the impact of each of these causes in detail, describing how the Curtiss Aeroplane & Motor Company had struggled to cope in the face of these handicaps.

Keys based his arguments on two fundamental assumptions. For Keys, it was an article of faith that a manufacturer had a right to receive a fair return on the capital he had put at risk.[100] A second article of faith, stemming from Keys's philosophy of business that he had absorbed as a journalist, was his belief that the design, engineering, and production of aircraft were the proper functions of private industry, not the government.[101] Under the system of competitive bidding, a manufacturer was denied the opportunity to earn a fair return on capital, or even get back the capital he had invested in a design, because of the government's refusal to grant the manufacturer proprietary rights to his designs. The total amount of business the Curtiss Company had done with the Army and Navy over the previous five years amounted to $8,271,000, with net earnings to Curtiss of $94,536, which meant he barely broke even.[102] By denying the opportunity for a fair return on capital, the government made it "impossible for anyone except a philanthropist or a fool" to put capital into the aviation industry.[103]

Citing the experiences of the Curtiss Company and the companies that had exited the industry, Keys demonstrated at length how capital and capacity had been removed from the industry, despite the fact that some new entrants had come into the industry since the war. Keys explained how he had used the cash flow the company had received over the previous few years to pay down outstanding debt rather than reinvesting in the business. Using the same graph he had provided to the Navy General Board, Keys showed the Select Committee how disruptive the lack of continuity in orders had been to his company. For any highly skilled industry, the need to hire, train, fire, and rehire skilled workers was extremely damaging; such volatility would ultimately break any company. Keys told the Select Committee that even if the successful bids on government contracts were taken at fair prices, which he claimed they had not been, "I say that a combination of competitive bidding which makes prices below the cost of production in most cases, with an absolute lack of continuity so that no man can build up a pay roll and build the proper subversion for it, means destruction and can not help but mean destruction."[104]

What also concerned Keys about the system of competitive bidding was the damage it did to the creative design function, which was so vital to advancing the state of the art. Keys wanted to keep the creative engineering organizations in the industry in being, but the current system gave manufacturers little incentive to do so.[105] "Why go to the great grief and danger of carrying on experiments, and doing designing," he argued, "and then have the results of your labors taken by the Government and thrown open for anybody that builds airplanes and who is not making that expenditure? The company that has that expenditure must be crushed finally by the man who has not got it."[106] The end result, he believed, could only be a government monopoly on design. Summarizing his views on competitive bidding, Keys said that the policy was, in his judgment, "an element of destruction from any angle that you look at it and has absolutely nothing to recommend it."[107] Keys pointedly told the Select Committee that no other nation followed the same practice.

When asked for his recommendations for corrective actions to sustain the industry, Keys responded that one of the most important steps was for the government to adopt a policy that looked upon aviation "as a thing that has to be kept alive on an efficient low cost and continuous basis."[108] Keys argued strongly for abandoning the policy of competitive bidding and for allowing a manufacturer to retain proprietary rights to his designs. Retain sharp competition in design and performance, Keys said, but not in price.[109] Keys also argued for a "constructive, continuous and forward-looking policy" on the part of the Army and the Navy.[110] Congress, he said, had to decide what policies it wanted the Army and Navy to follow with regard to aviation.

Keys divided his specific recommendations into two categories: things that could be taken immediately and things that he believed should ultimately be done.[111] Keys recommended, first, destroying all obsolete and unfit equipment in the Army Air Service and the Navy; second, revising the laws governing procurement to remove the requirement for competitive bidding and making competitive bidding completely discretionary; third, ending the design and production of aircraft at Air Service and Navy facilities; fourth, strengthening the technical divisions of the two services so that they could better cooperate, not compete with industry; fifth, passing legislation to control civil and commercial aviation, including the licensing of pilots and inspection of aircraft; sixth, encouraging the expansion of the air mail service and support private air mail routes; and seventh, contracting military repair work to private firms. Over the longer term, Keys believed that all government aviation should come under a single department, which would have responsibility for all administrative functions, while the Air Service, the Navy, and the Post Office would retain operational control to ensure that their aviation assets were used in a manner appropriate to their respective functions. Keys also argued for a flexible, but continuous, procurement program for the Air Service and the Navy, which he believed would go far to remove many of the difficulties the aviation industry had encountered. He also supported a government role in laying out airways across the country and providing other aids to commercial aviation to stimulate its growth and development. Lastly, Keys strongly recommended the creation of standing committees in the House and Senate to follow developments in commercial and military aviation, thereby building up expertise within the Congress.[112]

Back to Work

Despite his own frustrations, Keys told the Select Committee during his testimony that "I do not want to get out of the airplane business, and I do not want to drop my engineering department. I do not want to quit winning any world records. I do not want to leave this aeronautical situation as it is."[113] Keys soon learned that the effort he had put into his testimony had not been wasted.

Frank Burke, president of the United States Shipping Board Emergency Fleet Corporation, heard that Representative Charles Faust of the Select

Committee had said that Keys and Howard Coffin, now a vice president with the Hudson Motor Car Company who also had testified before the committee, "had given the Committee more concrete information that could be considered as a basis for the Committee Report than all the other witnesses combined."[114] The Committee had found that the Army and Navy witnesses had given testimony from such conflicting angles, due to personal and departmental interests, that "had it not been for the testimony of yourself and Mr. Coffin the Committee would be more or less where they had started."[115] Thanking Burke for passing on Representative Faust's comments, Keys wrote that "personally, I think that the Committee has done its greatest work, regardless of what it may report, in that both the Army and Navy are stirring around every way possible to clean up many of the things that they know are wrong and that this Committee has looked into pretty thoroughly during this investigation."[116]

A number of benefits emerged from the Select Committee's hearings. First and foremost, the hearings stimulated discussion and debate across the country about the state of American aviation and aviation policy. In an address to the Society of Automotive Engineers' Aeronautic Dinner later that year, Keys called this outpouring of public interest in aviation "a miracle." For the first time, Keys said, a congressional investigating committee "really started something besides headlines."[117] During one 30-day period while the Select Committee's hearings were going on, Keys received 1625 editorials from all over the country voicing the sentiment that "we must have aviation, both military and commercial, in this country," and as Keys noted, "when the people of the United States want something, they intend to have it."[118] This growing awareness did not go unnoticed in Congress. As the *New York Times* commented, "Gradually, as a result of the House committee's hearings, it is dawning upon Washington that a major issue is up for settlement."[119]

The Select Committee's hearings continued until early March. During February, Keys provided a steady flow of information, comment, and suggestions to certain Committee members. He was keen to ensure that the issues the Select Committee raised were kept alive in the forum of public affairs. On his own initiative and at his own expense, Keys had several thousand copies of his complete testimony to the Select Committee printed up for distribution. He sent copies to every senator and representative in Congress, to all members of the Washington press corps, and to 258 newspapers around the country, a select list of government and military officials, members of the Aeronautical Chamber of Commerce, fellow businessmen, and the stockholders of the

Curtiss Aeroplane & Motor Company.[120] In total, Keys sent out more than 2000 copies of his testimony across the country. Most of the recipients were politely appreciative, but a few commented more directly. Edgar Gott, president of the Boeing Airplane Company, complemented Keys on his testimony and his determination "to bring order out of chaos" in the industry, while an investment banking friend told Keys that while it was difficult to see "how some of your stockholders can remain cheerful enough to continue to own your stock after reading your testimony . . . there never yet has failed to be method in your madness, and I think I sense it this time."[121]

While Keys had painted a glum picture of the state of the aviation industry, the Curtiss Aeroplane & Motor Company's financial condition had actually improved over the prior year. Commenting on the Curtiss Company's annual report, the aviation journal *Aero Digest* said that the company's results "are most gratifying, as a forecast of the earning possibilities for capital invested in the aircraft industry under wise and businesslike management. . . . Mr. Keys and his associates are to be congratulated."[122] Through disciplined control over expenses and a reduction in experimental engineering, the company managed to post a profit of $156,228 for 1924 while reducing debt. The working capital position was the strongest it had been since the end of the war; the company had a strong cash position, reduced inventories, and almost no short-term liabilities. The tone of Keys's report to the stockholders, issued on March 2, was in marked contrast to the blunt warnings on the outlook for the aviation industry expressed the year before.

For the first time, Keys allowed himself a note of cautious optimism. While competition from government plants and the policy of competitive bidding on government contracts had continued through much of 1924, toward the end of the year this had begun to change, particularly with regard to contracts with the Army Air Service. Keys said that the leaders of the Air Service had "made an intelligent study of the matter and began, of their own volition, to shift their policy with regard to the placing of contracts."[123] Keys also mentioned the work of the Select Committee of Inquiry, saying that the work of the committee had been constructive and that its preliminary findings "seem to indicate that Congress will take a stand against the destructive tendencies that have grown up in the Air Services since the close of the War."[124] Keys did not, however, give a clear outline of the company's policies for the coming year. The aviation industry, he said, was in "a curiously uncertain state." Keys explained that after five years, "during which period the industry as a whole has dwindled as a result of policies or lack of policies of the Government Departments, public opinion appears at this time to have taken a definite constructive turn. The policies of your company will be determined when that public opinion has crystallized into definite Government policies.... Your company is ready to move forward or to maintain a cautious attitude, as conditions may warrant."[125] Keys and the other directors had sufficient confidence in the company's financial condition and future prospects to approve a 6 percent dividend on the preferred stock.

The good news underlying Keys's comments was that after a dearth of business, the Curtiss Company was finally getting new government contracts, and without the threat of competitive bidding. In January, the Navy placed a direct order with Curtiss for nine pursuits based on the revised XPW-8B prototype, designated the F6C-1.[126] A few weeks later, the Army Air Service awarded Curtiss a contract for ten examples of the Curtiss entry in the Corps Observation airplane contest, the O-1, and in early March it gave Curtiss a contract for fifteen production versions of the XPW-8B, redesignated the P-1, without resorting to competitive bidding.[127] In addition, the Army and the Navy had agreed to place a joint contract for three Curtiss racers, designated R3C, for the 1925 Pulitzer and Schneider Trophy races in the fall. With contracts for additional D-12 engines for the Boeing and Curtiss pursuits on order, by March the Curtiss Company had over a million dollars in new government business on its books, with the prospect of more to come.[128] In early April, Keys wrote to W. L. Gilmore, chief engineer at the Curtiss Company, that Major General Patrick had issued an order to the Air Service depots to stop major overhauls on older planes with the ultimate goal of having them condemned. Keys exclaimed: "Many things are happening in aviation!"[129]

Gaining Momentum: The Morrow Board

The debate on aviation policy took a dramatic turn in September. On September 3, the Navy dirigible *Shenandoah* crashed in a storm in Ohio with the loss of fourteen crew members, including the commander, which led Air Service Colonel William Mitchell to condemn "the incompetency, criminal negligence, and almost treasonable administration of the National Defense by the Navy and War Departments."[130] With the entire range of aviation issues gaining public interest and political support in Congress, President Coolidge responded to the *Shenandoah* disaster and Mitchell's attacks by creating, ostensibly at the request of the secretaries of War and the Navy, the President's Aircraft Board, to study the development and use of aircraft in national defense.[131] In fact, Coolidge wanted to counter what recommendations might come out of the more aviation-friendly Lampert Committee.[132] Coolidge appointed an old and trusted friend, Dwight W. Morrow, a lawyer and senior partner of the Morgan Bank, to be chairman of the Aircraft Board.[133]

The Morrow Board, as it came to be called, set to work at once. It called its first witnesses on September 21. Over the next month, the Board heard from ninety-nine witnesses, primarily from the Air Service and the Navy, the Post Office, the Department of Commerce, the National Advisory Committee for Aeronautics, and the aviation industry.[134] The Morrow Board completed its hearings on October 16, 1925, and then began working on a full report of its conclusions and recommendations to the president. Members of the aviation industry appeared before the Aircraft Board on October 15, beginning with Charles Lawrance, who spoke in his capacity as president of the Wright Aeronautical Corporation and as president of the Aeronautical Chamber of Commerce.

Lawrance presented the Aircraft Board with a statement representing the unanimous views of the leading companies in the aviation industry. Keys was in complete agreement with the industry's statement and was one of twenty-five executives who signed it. With regard to the conditions a peacetime aviation industry would need in order to be available for a national emergency, the industry statement contained four recommendations:

1. Secure continuity of production.
2. Stop direct competition from government-owned facilities.
3. Eliminate destructive price policies in purchasing aircraft, especially the policy of competitive bidding, which did not allow for a fair profit commensurate with the risk.
4. Recognize and honor proprietary rights to a design.[135]

Keys had argued for these recommendations on numerous occasions. He was no doubt pleased to see that the industry statement also recommended that the government place orders only with firms that had "created good and adequate facilities for design and engineering development." He had stressed this point repeatedly in his discussions of the damaging impact of competitive bidding.[136]

Keys gave a short, twenty-minute testimony to the Morrow Board on October 14, two days after the Curtiss R3C-1 racers placed first and second in the 1925 Pulitzer air races.[137] Drawing on the arguments he had presented

Clement Keys, *lower right,* stands near the Army Air Service's Curtiss R3C-2 Racer at the October 1925 Schneider Trophy race held at Baltimore, Maryland. On October 26, Army Lt. James H. Doolittle won the race in this airplane. The next day Doolittle set a new seaplane speed record, pushing the R3C-2 to 245 mph. (459836, RG80, NARA)

to the Lampert Committee of Inquiry in January, in his brief testimony to the Morrow Board Keys used his time to elaborate on what he called "the virtual destruction of the airplane industry as it existed in 1919."[138] Keys again stressed the lack of continuity in production, competitive bidding, and competition from the government as the three main causes of the industry's decline.[139] As if to illustrate to the Morrow Board what could be done through a successful partnership between government and the aviation industry, two weeks later Army Air Service First Lt. James Doolittle captured the 1925 Schneider Trophy Race in a Curtiss R3C-2 racing seaplane, setting a new world seaplane speed record.[140]

The Morrow Board issued a report of its findings on December 2, 1925. The Board's report was supportive of the aviation industry, accepting many of the arguments that Keys, Lawrance, and others had presented in their testimony. The Board recognized the importance of a sound aviation industry to the nation's defense, but also noted that for some time to come, at least until the expansion of commercial aviation and foreign markets, the military air services would be the industry's main source of orders. The Board grasped the importance of maintaining competent design staffs, an issue that Keys had stressed repeatedly. The Board's recommendations addressed the main defects in the government's approach to the industry since the end of World War I, specifically, the lack of continuity in orders, the policy of competitive bidding, and government competition with private industry—all the points Keys had raised in his comments before the Board. The Board issued seven recommendations for the industry:

1. Adopt a policy of continuity in orders and of a standard rate of replacement.
2. Grant production orders only to companies that maintain design staffs of reasonable size and keep them active.
3. Allow full recognition of proprietary rights.
4. Eliminate governmental competition with the civil industry in production activity except in those projects impracticable of realization by the civil industry.
5. Grant small orders for experimental designs to be distributed among firms having design and production staffs of proven competence at prices sufficient to cover the overhead expenses of maintaining design and experimental departments.
6. Modify the existing statutes requiring competitive bidding to allow these recommendations to be put into effect.
7. Provide governmental aeronautical research and its results to the civil industry.[141]

The Morrow Board further recommended that assistant secretary positions be created within the War and Navy Departments to oversee military aviation matters. The Board saw no need to build a large air arm, as America's geographic position protected it from air attack. However, the Board did recommend special appropriations from Congress, for a period of five years, to allow several of the newer types of aircraft to be purchased in quantity for the Air Service and to facilitate replacement of remaining war stocks.[142] Stating that the report of the Morrow Board "ought to be reassuring to the country, gratifying to the service and satisfactory to the Congress," President Coolidge announced that he approved of the Board's report and would ask the Congress to enact legislation based on the Board's recommendations.[143]

Congress Acts, At Last

The Lampert Committee issued its report twelve days after the Morrow Board, on December 14, 1925. Where the Morrow Board had been reassuring about the state of American aviation, the Lampert Committee report concluded that "an alarming situation still exists in the Army and Navy air services."[144] The Lampert Committee concurred with the Morrow Board's view that the aviation industry was essential to national defense and had to be sustained. The committee's report identified the lack of continuity in orders, the failure to recognize a manufacturer's rights to a design, and the system of competitive bidding on government contracts as central causes of the aviation industry's decline. The Committee recommended an end to the system of competitive bidding, recognition of proprietary design rights, and government competition with the aviation industry, and a continuing, five-year program of aircraft procurement.[145]

The reports of the Morrow Board and the Lampert Committee were generally well received across the country, but there was pressure on the Congress to actually get something done.[146] In particular, the Morrow Board report has been called "a milestone in the history of American aviation" as it "came from a blue-ribbon presidential panel and carried with it the weight and prestige of the chief executive."[147] Keys adopted a decidedly positive tone in his commentary for the Curtiss Aeroplane & Motor Company annual report, which was released on March 4, 1926. The conservative financial policies and strict cost control that Keys had imposed on the company, together with an increase in government contracts, resulted in a profit of $156,780. Keys announced the payment of a dividend on the preferred stock.

Terming the year "exciting and interesting," Keys referred to what he called the public awakening concerning the uses of aviation for national defense and commerce and the results of the investigations of the industry by the Morrow Board, the Lampert Committee, and the Joint Committee on Civil Aviation, which would likely lead to legislation for a stronger air service for the Army and the Navy and for the promotion of commercial aviation. Whether or not such legislation did emerge from the Congress, Keys affirmed that "a great deal of progress has been made and probably will continue to be made for some time to come."[148] The most critical development during the year, Keys argued, was the change in the conditions in the contractual relationship between the aviation industry and the Army and Navy. "There has been a clear recognition in Government circles," Keys said, "that Government competition against the industry is not in the interests of the Government or the people; that the contractor who maintains an engineering department and creates thereby new designs and new advances in the art is entitled to receive a return for this work and not to have it taken from him without reward by the Government; that destructive competitive bidding tends to destroy rather than to benefit both the industry and the Air Services; and that a sound industry is an essential part of the machinery for national defense."[149]

Following publication of the Morrow Board and Lambert Committee reports, proposed legislation based on their recommendations quickly emerged in the Senate and the House. On June 24, 1926, President Coolidge signed the Naval Aircraft Expansion Act, which authorized the Navy to purchase 1614 aircraft over five years beginning in 1927 at a cost of $85,078,750.[150] The Air Corps Act that Coolidge signed on July 2 changed the name of the Air Service to the Air Corps and authorized it to acquire 1800 new aircraft over a five-year period. The two military programs planned for an Air Corps of 2200 aircraft and a naval air service of 1600 aircraft by the end of the five-year period. The acts created Assistant Secretaries for Aeronautics in the Navy and War Departments, giving aviation a more prominent voice in the military.[151]

The aviation industry was pleased with the legislation that Congress had produced. Referring to the Army Air Corps Act, *Aviation* commented that "those who testified before the Lampert Committee and labored over the recommendations of the Morrow Board must feel a considerable sense of gratification. . . . The entire situation can only be considered favorable."[152] In its annual *Aircraft Yearbook*, the Aeronautical Chamber of Commerce wrote that "the year 1926 will go down in history as the year when aviation became permanently established in the United States, upon a sound program of constructive legislation, giving civil aviation a legal status, and the Air Services of the Army and Navy legal authority for greatly improved conditions upon a basis of permanent progressive growth."[153] The new system of procurement came in for particular praise, *Aviation* claiming that the Congress had made "some very radical concessions to the peculiar needs of the aeronautical industry. The system of competitive bidding, which is necessary for the purchase of all other commodities, is waived in the procurement of aircraft."[154]

The reality was somewhat different, particularly in regard to the military air services. Few industry commentators noted that the Congress had authorized new spending only for procurement and had not appropriated the funds requested. There was a clear policy toward the expansion of military aviation, but the money would still have to be subject to the executive budget process and congressional approval on a yearly basis. Nor was the question of the procurement process as clearcut as *Aviation* made it out to be. Keys and others in the industry and the air services had hoped that the recommendations of the Morrow Board and the Lampert Committee would be adopted in full and that the practice of competitive bidding would be eliminated once and for all.

Instead, the House Military Affairs Committee and the Naval Affairs Committee set up a special joint committee to work out a compromise between competing bills governing procurement. Representative Carl Vinson (D. Georgia), a supporter of the aviation industry and a member of the House Committee on Naval Affairs, and Representative John McSwain (D. South Carolina), a member of the House Committee on Military Affairs and a leading opponent of changes in procurement practices, introduced diametrically opposing procurement bills the day after the House had passed bills for the Navy and Army air services.[155] Where Vinson's bill called for an end to the practice of competitive bidding, McSwain wanted to throw open competition to anyone and everyone. McSwain, who appears to have had little understanding of the growing technological complexity of aviation, "refused to acknowledge the necessity of preserving specifically the established companies in order to assure an efficient and competent supply of equipment. On the contrary, he attacked these manufacturers without reservation."[156]

The bill that emerged from the special joint committee and that was added as Section 10 to the Army Air Corps bill was a compromise between those who advocated reforms in procurement and those who were opposed.[157] The new bill was open to a number of interpretations, depending on the reader's point of view. Reflecting Representative McSwain's suspicion of the aviation industry and his desire to harness the inventive genius of America, Section 10 called for the broadest possible design competition, with public announcements inviting bids. Section 10 (k) allowed the secretary of War or Navy to negotiate contracts for experimental designs, and if the resulting aircraft proved to be dramatically superior, then to negotiate a contract for quantity production without the benefit of a competition. But Section 10 (t) stated that whenever the secretary entered a contract for aircraft, the secretary was authorized "to award such contract to the lowest responsible bidder that can satisfactorily perform the work to the best advantage of the Government."[158]

As Keys noted in a letter to Representative Vinson some six months later, "it would take an extraordinary amount of courage" for the secretary to rule for any other company, thus "perpetuating in its very worst form the main production evil against which this law was aimed, viz. , competition on price alone."[159] In practice, however, over the next five years, the Air Corps and the Navy tended to award contracts for new aircraft without resorting to competitive bidding. As Keys told F. Trubee Davison, the newly appointed assistant secretary of War for Aviation, "I find in a general way definite opinion that with the new legislation and the new machinery to enforce it . . . the dark days of American aviation seem to be definitely at an end."[160]

Keys waged a long, hard fight against the system of competitive bidding. While the Air Corps and Naval Aircraft Expansion Acts may not have given him everything he had hoped for, the results for Curtiss Aeroplane & Motor Company were positive. "The procurement provisions," he said in his remarks to stockholders in the Curtiss Aeroplane & Motor Company's 1926 annual report, "while still leaving many things to be desired, are a very great improvement on all previous laws in that they provide an incentive to the engineering firms to strive for excellence in product and thereby secure a steady and continuous share of Government orders."[161]

Through the Air Corps and Naval Aviation Acts, Keys had obtained his most important goal: a chance for the Curtiss Aeroplane & Motor Company to survive and prosper. The new laws, he said, "provide that firms, which by their own efforts, produce outstanding designs and high performance for the Army and Navy, shall be able to go forward in the manufacturing of their own products at reasonable prices and shall be, to a certain extent, immune from competition of firms which produce nothing new but merely bid on the manufacture of other people's products."[162] Keys stated that the five-year program under the Acts "assures a fairly continuous flow of business in moderate volume over the next five years, provides a great incentive toward maintenance of quality and high performance, and in general makes this branch of the aeroplane industry a business instead of an adventure."[163]

4

A FRAMEWORK FOR COMMERCIAL AVIATION AND THE FIRST AIR TRANSPORT COMPANIES

———■———

The Advance Agent of Prosperity

KEYS UNDERSTOOD HOW IMPORTANT TRANSPORTATION SYSTEMS WERE TO the economy. In an article he wrote for the January 1909 issue of *The World's Work*, titled "The Advance Agent of Prosperity," he argued that the railroads had opened up the land to "the mighty streams of commerce."[1] He believed the airplane could potentially do the same. What the airplane offered was speed, providing a savings in time over other forms of transportation, as well as freedom from the constraints of geography. For a country like the United States, with great distances between major cities, the airplane could provide a valuable service, reducing travel times from days to hours.

Keys had no sentimental or romantic ideas about air transportation. To him, air transport was a business, just like the railroads, and it was subject to the same disciplines and objectives common to all other businesses. Air transport companies would, he believed, be required to show a sufficient margin of profit to provide an adequate return on capital invested in the enterprise.[2] Admittedly, at the time the airplane was not a practical means of transportation; all it could offer was a promise of things to come. Yet, while the challenges were enormous and the timing of success was uncertain, here was an opportunity for a new generation of pioneers, the "builders" that Keys so admired, men who "wander in a wilderness, seeing things that are not there" and then go out and build them.[3]

In his 1909 article, Keys had asked, rhetorically, but perhaps more to himself than to a larger audience, if the mightiest feats had all been done or if there were "yet some battles to be won, some laurels of fame and fortune to be gathered."[4] Air transportation was a new field that would demand men of vision, ability, and perseverance. There would be no guarantee of success and no certainty of profits. He was an expert on the railroads with a firm grasp of the economics of transportation systems. Perhaps it crossed his mind that as a pioneer in this new field, he could achieve with airplanes what Hill and Harriman had done with the railroads. When commenting on the outlook for commercial aviation during his testimony to the Lampert Committee in 1925, Keys noted his strong inclination to go into it. "My own feeling," he said, "was very strong that I should like to embark as a pioneer in this line of commercial aviation, when the time is right."[5]

In the early years of the 1920s, the time was decidedly not right. With the exception of the Post Office air mail service, "organized civil aviation was practically nonexistent in the years after the armistice."[6] The war-surplus airplanes available were expensive to operate, inefficient, and not especially safe, there was no infrastructure to support commercial air transport, the public had ambivalent feelings toward flying, capital for investment was limited, and, as with military aviation, there was no government policy on commercial aviation.[7] The framework for successful commercial air transport had yet to be built.

The Drive for Federal Regulation of Commercial Aviation

Keys had his first exposure to commercial air transport in 1919 with the American Aviation Mission. The Mission's report noted the airplane's potential for offering "the fastest and most direct means of transportation for persons, mail, and light freight known to civilization."[8] To stimulate the development of commercial aviation, the report recommended that the federal government should support building the infrastructure necessary for commercial air transport. The report suggested that the government help provide flying fields with aids for navigation, a meteorological service for aviation, a body of law governing the airworthiness of aircraft and the use of aircraft for interstate commerce, and the encouragement of private enterprises to carry the air mail.[9] The Mission proposed that the government should set up air transport services on its own and maintain these services until they were proven to be commercially successful, at which point they could be sold to private enterprise.[10] Commercial aviation, as well as the military air services of the Army and Navy, should be concentrated in a single government agency with representatives from the Department of Commerce, the Navy, and the War Department.[11]

In 1920, the National Advisory Committee for Aeronautics issued recommendations for a national aviation policy in its annual report. NACA's first recommendation was that Congress should enact legislation to give the federal government authority over air navigation, specifically licensing pilots and aircraft, and provide aid to the states to build landing fields.[12] In contrast to the American Aviation Mission's report, NACA did not agree with concentrating all air activities, both civil and military, within one government department but argued instead that control and regulation of commercial aviation should be placed in the Department of Commerce.[13]

The form that federal regulation and administrative organization of control over aviation should take became a central theme in the debates that followed. Between 1919 and 1925, the question of a national policy for aviation would be "one of the most intensely discussed and debated questions on the national scene."[14] On one side were proponents of all-encompassing, detailed regulation through a bureau in the Commerce Department or a Department of Aviation, as the Aviation mission had proposed. The alternative was a broader approach that would give the Commerce Department general authority for regulation but would allow the Department to develop specific regulations on its own. Although many bills were put forward, no legislation governing the regulation of commercial aviation would emerge from Congress until 1926.[15] Within Congress there was a lack of concern about aviation, and without broad public interest Congress was under little pressure to act.[16] There was also a "lack of leadership" in the executive branch.[17] Neither President Harding nor President Coolidge pushed for aviation legislation. What did emerge through this debate was a general consensus among the aviation industry, NACA, government policymakers, certain members of Congress, and business interests of the need for the federal government to support development of commercial aviation.[18] The form this support should take would be through regulation to support safe operations, federal government funding to build the infrastructure needed for commercial aviation, and indirect subsidies provided through air mail contracts to support commercial airlines.[19]

In contrast to his more forthright attack on the military's policy of competitive bidding, in the debates on commercial aviation Keys played a more supportive role until the opportunities for concrete action emerged. At least until 1923 Keys's preoccupation was ensuring that Curtiss Aeroplane avoided bankruptcy. As he told the Lampert Committee, "I have some four thousand stockholders who invested their money in the Curtiss Company, and I have to take care of them as well as I can."[20] He was, however, well-connected to several of the principal participants pushing for federal regulation of commercial aviation, particularly Howard E. Coffin and Samuel S. Bradley, who he knew from the American Aviation Mission. They would, from time to time, call on Keys for information and observations on aviation.

In March 1921, the aviation industry gained a significant supporter when Herbert Hoover became secretary of Commerce in the Harding Administration.[21] While Hoover believed in "decentralized and self-regulatory" economic units with a restrained role for government, he realized that commercial aviation "demanded mechanisms uniquely geared to the needs of a fledgling enterprise."[22] Hoover saw parallels with government regulation of water transport and shipping, as well as the railroads, and thought that the government could play a formative role in developing commercial aviation. Over the next several years, Hoover pushed for legislation in Congress that would give the Department of Commerce authority to regulate commercial aviation and build the infrastructure air transport companies would need. His chosen method was to work closely with the aviation industry, arranging conferences between government officials and representatives of the industry to build consensus.[23]

On April 1, 1921, President Harding wrote to Dr. Charles Walcott, NACA chairman, asking him to set up a NACA subcommittee with representatives from the Army, Navy, Post Office, Commerce Department, and civilian life to make recommendations on legislative action that would provide for federal regulation of air navigation.[24] Walcott quickly brought together the subcommittee, which included Frank Russell, vice president of Curtiss Aeroplane and a director and treasurer of the Manufacturers' Aircraft Association.[25] In all likelihood, through his participation Russell reflected Keys's concerns, and he reported back to Keys on his work with the Subcommittee.

The NACA Subcommittee submitted its report to the president on April 9, concluding that it was "a pressing duty of the Federal Government to regulate air navigation," as independent state regulation would interfere with the development of aviation, and recommending both legislation and that a Bureau of Aeronautics be established in the Department of Commerce.[26] President Harding mentioned this recommendation in his address to Congress three days later. In June, a Commerce Department aide to Hoover happened to meet with Howard Coffin and Samuel Bradley, who were also observing the Army Air Service bombing tests off Virginia.[27] The aide arranged for Coffin and Bradley and other representatives of the aviation industry to meet with Hoover on July 18.[28] Hoover asked Coffin to chair a committee to revise a draft bill for a law governing regulation of air navigation that NACA had submitted to Congress.[29] In addition to Bradley, Coffin brought in Benedict Crowell, who had returned to his business career, George Houston of Wright Aeronautical Corporation, and Clement Keys, all of whom were fellow members of the American Aviation Mission.[30]

Although the Aviation Mission's report had recommended a unified department to deal with all government aviation, at the advice of Wright Aeronautical Corporation's counsel, the committee adopted "a broad, flexible approach to legislation as the best means to establish safety requirements necessary to promote civil aviation" through a bureau of the Commerce Department.[31] In August, Senator James W. Wadsworth (R-NY) introduced the bill in the Senate where there was opposition to "an all-encompassing interpretation of federal authority under the commerce clause at the expense of state sovereignty."[32] Proponents of states' rights believed the government had no authority to regulate intrastate aviation.[33] Testifying before a Senate subcommittee holding hearings on the bill in December 1921, Frank Russell, no doubt echoing Keys's views, argued for regulation of aviation to include inspection of aircraft and licensing of pilots.[34] Russell went on to add that "the development of an industry depends primarily on the demand and the possibility of interesting capital."[35] When businessmen asked the Curtiss Company about getting into commercial aviation, Russell replied: "we are forced to advise them that the time is not ripe, that there isn't the control in this country sufficient to make it a stable and safe business in which to invest their money."[36]

A more restricted version of the bill, which limited federal regulation to interstate and foreign commerce, passed the Senate on February 12, 1922.[37] But the aviation industry and supporting groups did not want regulation of intrastate flying eliminated from the bill.[38] To get a similar bill through the House allowing greater federal control of aviation, the Commerce Department's solicitor, William B. Lamb, began working on a revised draft with comments from Major General Mason Patrick, Keys, and others from the aircraft industry.[39] Congressman Samuel Winslow finally submitted a bill with stronger regulatory power in January 1923. Neither the Wadsworth bill nor the Winslow bill survived the 67th Congress, which adjourned in March 1923.[40]

The aviation industry and its supporters would have to wait until the 68th Congress convened in December 1923.[41]

With no legislative support for federal regulation, commercial aviation made little progress. This was the era of the barnstormer, in which a young pilot took a surplus Curtiss JN-4 to small towns around the country giving rides to the daring few.[42] Inexperienced pilots flying older and poorly maintained surplus aircraft too often turned these "joy rides" into something far from joyful. During 1922, there were 126 accidents involving itinerant flyers, resulting in 62 fatalities and 100 injuries.[43] Sadly, most knowledgeable observers believed that the majority of these accidents and fatalities could have been avoided had there been effective regulation.[44] As the Aeronautical Chamber of Commerce's *Aircraft Yearbook 1923* noted:

> [The] explanation of the hesitant attitude shown by the public and capital toward aircraft as common carriers is to be found primarily in the increasing number of accidents among itinerant and irresponsible pilots, and secondarily in the manner in which the press features all accounts of disaster. Until we have Federal Air Law, which will encourage conservative flying and discourage recklessness, and until the newspapers discern that real news is constructive instead of destructive, public and capital will justifiably regard aviation as an experiment.[45]

During 1922, however, there were several positive developments. The Post Office's air mail service continued to improve. This service carried over 60 million pieces of mail in 1922, with a 95 percent completion rate, while reducing forced landings by nearly two-thirds.[46] Paul Henderson, Second Assistant Post Master General in charge of the air mail service, began planning for night flights over the Chicago to Cheyenne, Wyoming sector.[47] If successful, this would cut the flying time along the transcontinental air mail route from New York to San Francisco to around 30 hours, one-third the amount of time by train.[48] The Wright Aeronautical Corporation conducted a detailed study of a proposed contract air mail route from New York to Chicago, calculating the cost of operating the service and the postal rate that would cover these costs.[49] The study appeared in the Aeronautical Chamber of Commerce's *Aircraft Yearbook 1923* and in the aviation magazine *Aerial Age Weekly*.[50] It is highly likely that Keys read the study.

More important was a move in Congress toward transferring the air mail service from the Post Office to private contractors. The accepted theory behind the air mail service was that it would serve as a laboratory for developing commercial aviation, proving the feasibility of carrying mail by air.[51] Representatives Halvor Steenerson (R-MN) and M. Clyde Kelly (R-PA) introduced bills in Congress to authorize the Post Office to contract for the air mail with individuals or companies.[52] While neither Kelly's bill (H.R. 10717, March 3, 1922) nor Steenerson's bill (H.R.11193, April 6, 1922) advanced through Congress, it was a beginning.[53]

In late 1923, Keys returned to the challenges of developing commercial air transport. With the Curtiss reorganization behind him, he had more time available. The figures for the air mail service for the fiscal year 1923 (July 1, 1922–June 30, 1923) came across Keys's desk in early October 1923. As Keys reviewed the figures, what struck him was the sheer volume of mail that the service had built up without any advertising or special promotion of the air mail. During fiscal year 1923, the air mail service had carried 68 million letters, a 13 percent increase over the prior year. This was a remarkable number for a service that few people sought out. Few businessmen Keys knew insisted their correspondence go by air mail. There was, he thought, the potential for a greatly expanded volume of business if it was actively promoted. Looking at rough estimates of costs and loads, Keys realized that with more modern aircraft than the Post Office's Army surplus DH-4s, it would be possible to double the useful load at no increase in cost per mile. If, as he suspected, people would be willing to pay more than the standard 2¢ rate for the rapid transcontinental air mail service then in effect, then the possibility of a reasonable business offering sustained profits existed. And that could attract capital for investment.[54]

Keys sent his preliminary analysis of the air mail service figures along with his conclusions to Howard Coffin. He told Coffin that he was "wondering if this experiment has not produced sufficient data and sufficient information to justify a very close study of it by business men contemplating business adventures."[55] There was a lot more work to do, but his overall conclusion was straightforward. "It may be time," he said, "to contemplate establishing mail and packet freight service on a limited scale, charging what the traffic will bear. This is the way every transportation industry in the world has been established, and is, I think, the only known method of establishing such service except the one alternative, subsidy by taxpayers. I do not think anyone in business circles would be interested in a subsidized industry; but I, for one, would be interested, possibly, in a limited adventure in traffic, provided it did not contemplate passenger traffic."[56]

CHAPTER FOUR

At Coffin's suggestion, Keys wrote a letter to Paul Henderson saying that Keys wanted to talk with him about the air mail service. Keys included a copy of his memo to Coffin, telling Henderson that "there is a good deal of interest, at the present time, in transportation by air; but nothing of a definitely constructive sort can be done, or, in my judgment, should be done, until we are able to talk intelligently to business interests concerning the figures of the Air Mail, and the facts that lie back of these figures."[57] Henderson responded to Keys's inquiry, sending Keys more detailed information on the Post Office's transcontinental air mail route and a report he had prepared for the postmaster general on the air mail service's night-flying experiments over the transcontinental route the previous August.

Keys replied a few days later, saying his letter was "a very frank laying of my cards on the table."[58] For the previous four years he had been carefully watching the development of air transportation in Europe and the United States, but up until the present time he had declined to take an interest in these ventures and had warned others not to get involved. His attitude had now changed. He was having engineers at Curtiss Aeroplane analyze the data he had received from Henderson. "At the present time," he said, "I am frankly interested in the possibility of establishing, on a sound basis, some main transportation lines in the United States. If I can become persuaded that tonnage would be available in a reasonable number of years in sufficient quantity to justify the capital expenditure necessary, and at rates that would pay a reasonable return on such an investment, capital will answer a call forthwith."[59]

Keys met with Henderson on November 9, 1923.[60] That afternoon Keys returned to his office where he wrote a letter to the magazine *Aeronautical Digest*, responding to a request for a comment on an article by Charles Babson, the noted business journalist, on the possibilities for commercial aviation. With his conversation with Henderson fresh in his mind, Keys outlined some of his main concerns. He thought, first of all, that the growth of air transportation would be slow, due to the high cost of operating the equipment then available. The great challenge was to increase the payload of commercial airplanes to a point where, at the maximum possible rates, there would be a sufficient margin to cover not only operating costs, but also depreciation and capital costs, leaving a profit margin appropriate for a new enterprise. In Keys's judgment, that point had not yet been reached, though he was hopeful that it could be realized in the near future.

As a businessman, Keys believed the safest approach at that time was to adopt "a willingness to study in detail not only efficiency in manufacturing and operations, but also tonnage available, rates that this tonnage would pay, and overhead costs in obtaining it."[61] Keys also sounded a note of caution for potential investors. "Personally," he said, "I should regret very much to see the public plunging blindly into the purchase of airplane transportation stocks at this time, and would expect that such ventures would not, on the whole, prove profitable in the near future, to the stockholders who buy them on a sentimental basis."[62] Keys knew that investing in a new industry, in its early stages, was not for the fainthearted. Capital, when it decided to enter the field of aviation, would have to come from practical, hardheaded businessmen who clearly understood the risks they were taking.

Before moving forward on any venture, Keys wanted to identify and understand the risks and the potential rewards. Keys decided to have Curtiss Aeroplane undertake a survey of commercial aviation covering potential tonnage and rates, laws governing air commerce, equipment, operating costs, and capital available.[63] The studies on operating costs that Keys had seen had all been done on equipment that Keys knew was obsolete. He wanted an analysis of what would be required to get the lowest level of operating costs possible. The amount of capital that might be available for investment in such a venture was the last question to consider, as it would, as Keys knew, be entirely dependent on the results of the complete analysis. "It is fairly obvious from my own knowledge," Keys wrote, "that when a group of sensible business men decide, not on sentiment, but on business, that air transportation will pay the kind of a profit that a pioneer industry must pay, capital will be available forthwith."[64]

Keys wrote to Henderson in late November 1923 to tell him that the Curtiss Company had decided to undertake a detailed study of commercial aviation. The study would take several months and would be as comprehensive as possible, looking at "all the known factors that are capable of analysis," and based on that analysis "would rest a decision whether or not other interests will embark large amounts of capital in this field."[65] Keys made reference to "a new group of serious men who are studying aviation," and he hoped that Henderson could get up to New York to meet them.[66] Keys had been sounding out other businessmen he knew in New York who he thought might be potential investors in any new venture. Henderson had offered to cooperate with Keys on the Curtiss study and had given him a tantalizing hint that some of the legal issues might soon be dealt with. Henderson informed Keys that he had talked with several members of Congress about Keys's plans, without mentioning Keys by name. Henderson thought that with the right sort of campaign it would be possible to get the legislation necessary to allow for private companies to carry the air mail.[67]

The study took four months to complete. Keys paid to have his men travel the route of a potential transcontinental line from New York to San Francisco, observing the terrain, the facilities en route, and talking with anyone who could comment on potential traffic volumes.[68] The report concluded that in the spring of 1924 there was no commercial basis for setting up an air transportation service. With the equipment available the costs were higher than comparable means of transport, and the service would not be sufficiently competitive to make potential users pay a premium over the costs. Of primary importance was the almost complete lack of landing fields along the proposed route. Establishing these facilities at the required intervals would require an enormous amount of capital and were more properly the responsibility of either the municipalities along the way or the federal government.[69] There were still no laws governing air commerce, or the licensing of pilots and the inspection of aircraft, nor was there any apparent movement within the government or Congress to enact the relevant legislation.

The report's conclusions were a disappointment. Keys believed that over the long term commercial aviation would develop, but based on the report, that would not be any time soon. In July he told an interviewer that "this scattered unorganized industry of flying has not much capital. It has courage and brains, but it lacks the power or the character to carry on organized industry; and only by organization can an industry live in America."[70] There was, however, a small ray of hope for the future, as he explained:

> The reason why I personally am in aviation—although greatly discouraged and almost beaten to a point of retirement from it—is that I have believed and still believe, that the fastest safe method of carrying anything, big or little, from any one place to any other place cannot help but win; because it cannot help but serve. Here, then, is a proper servant of the world; of banking, of the mails, of express, of human commerce on the whole. Today it is a beggar in rags. So, once, was the telephone, the railroad, the steamship. Someday, some men, not I nor those who have fought the early battles of aviation, but other men, bigger, more courageous, perhaps, more deserving of success, will take this most modern and most efficient of transportation agents and win with it victories of commerce equal to those of Hill, Huntington—of all the transportation

pioneers. I wish them, from my heart, all luck, and speed and safety when their day shall come.[71]

A Framework for Commercial Aviation Emerges

On a visit to the United States in January 1925, C. G. Gray, the editor of the leading English aviation journal *Flight*, gave his impressions of the state of American aviation. "The general atmosphere of Aviation in America," he said, "impressed one as being in that state when something is just going to happen. Not so much the calm before the storm but rather the slump before the boom."[72] By the end of 1924, the framework for a commercial air transportation system was in process. On July 1, 1924, the Post Office had introduced a transcontinental service that flew by day and by night, carrying a letter from New York to San Francisco in 34 hours and 45 minutes.[73] The Post Office had demonstrated that it was possible to conduct a regularly scheduled air mail service across the country. The Post Office planned a night air mail service between New York and Chicago for the spring of 1925 after building a lighted airway between the two cities. This new service would cut delivery time from 20 hours by train to 9 hours by mail plane; a letter mailed in New York in the late afternoon would arrive in Chicago the next morning.[74] Referring to the enormous amount of work the Post Office had done to get the transcontinental service up and running, Gray commented that "the real work was done on the ground, not in the air," a theme that Keys would constantly emphasize in his own ventures.[75]

The most important event in the first part of the year was passage of the Contract Air Mail Act on February 2, 1925. Representative Clyde Kelly had resubmitted his bill to transfer the air mail to private contractors as H.R. 7064. The House Committee on the Post Office and Post Roads held hearings on the proposed bill on April 29, 1924. The bill passed the House in December 1924 and the Senate in January 1925. President Coolidge signed the bill, known formally as the Air Mail Act of 1925 but informally as the Kelly Act, into law on February 2, 1925.[76]

As one historian has argued, "no event was more pivotal in the eventual assumption of civil air regulation by the Federal Government than the enactment of the Air Mail Act of 1925."[77] Federal regulation now seemed a certainty as contracting the air mail with private parties "would be doomed to failure without Federal airway development and regulation."[78] As the year went on, the momentum for federal regulation continued to build. The Lampert

Committee heard extensive testimony from Paul Henderson on the postal air mail. In his testimony, Keys stressed the need for regulation to license pilots, inspect aircraft, and build the infrastructure necessary for commercial air transport.[79] In May Howard Coffin, acting on behalf of the Commerce Department, asked the American Engineering Council to undertake a survey of commercial aviation; Coffin and Keys each contributed $1000 to fund the survey.[80] In the fall the Morrow Board continued the inquiry, calling Postmaster General Harry S. New, W. Irving Glover who had replaced Paul Henderson as second assistant postmaster general, as well as Henderson.[81]

The Morrow Board report came out on December 2, 1925, with the Lampert Committee report following twelve days later.[82] The American Engineering Council's study was issued shortly thereafter. The recommendations of the three reports were along sufficiently similar lines that provided support for Coolidge to ask for legislation to implement the recommendations. Senator Hiram Bingham (R-CT) introduced a bill in the Senate on December 8, 1925, with Representative James Parker introducing a similar bill in the House two days later. The bill limited federal regulations to airplanes and pilots engaged in interstate and foreign commerce, created a new assistant secretary position for aviation in the Commerce Department, and gave the federal government authority to install aids to navigation.[83] Airports became the responsibility of the states and municipalities.[84] The bills made their way through the Senate and the House to become the Air Commerce Act. The act became law on May 20, 1926, with President Coolidge's signature.[85]

The framework for commercial aviation, and particularly commercial air transport, was now in place after nearly eight years of debate. The process of transferring the air mail service to private contractors had begun, and the uniform regulations necessary for air commerce would soon be put in place. Keys had not been a leader in this debate, but he had been keenly interested in and supportive of commercial aviation. Even before the Air Commerce Act had become law, he was deeply engaged in his first venture in air transportation.

Building National Air Transport

The Kelly Air Mail Act was the incentive Keys and Coffin needed to act on the plans they had been discussing for over a year. They decided not to wait for the federal regulation of commercial aviation that was likely to follow. As Henry Ford, another early entrant into air transportation, said at the time, "I feel it is now or never to get hold of commercial flying and make a success of it."[86] In the two months following passage of the Kelly Act, more than 2000 inquiries came in to the postmaster general.[87]

The planned New York to Chicago air mail route had the greatest business potential.[88] Keys and Coffin wanted to set up a company to operate a mail and express service between the two cities, using the new lighted airway that the Post Office was building along the route.[89] There would be risks to this venture, and there was no certainty of profit. The proposed company would need sound organization and sound financing, with an amount of capital that would be sufficient to see the enterprise through the early stages of operations, which would likely be at a loss. Keys knew that the providers of this capital had to be sensible businessmen who saw aviation as a business and not some romantic ideal. Keys had already established contacts with several influential businessmen in New York who were interested in aviation. Coffin had contacts with Detroit businessmen who were keen to become involved in commercial aviation and make that city a center of the aeronautical industry.[90] Among Coffin's contacts in Detroit was Edsel Ford who, with his father Henry, had become interested in commercial air transportation. New York and Detroit would provide a good base, but Keys and Coffin needed to add Chicago businessmen to their group of potential supporters.

In March, following passage of the Kelly Air Mail Act, Howard Coffin and Edsel Ford invited men from industry, banking, aviation, and the government to meet in Detroit to discuss forming a commercial air transport company.[91] Keys and Paul Henderson were at the March 28, 1925, meeting in Detroit with over seventy other attendees. The group appointed a General Committee, with Coffin as Chairman, to take up the question of forming an air transportation company. The General Committee authorized Coffin to appoint an active committee of five members, designated the Air Transportation Committee, with "carte blanche authority to proceed with the organization of a private commercial air transportation corporation with a suggested capital of not less than three million dollars."[92] The members of the Air Transportation Committee were Coffin, Edsel Ford, William B. Mayo, chief engineer at the Ford Motor Company, C. M. Keys, and Leonard Kennedy from the Curtiss Aeroplane & Motor Company.[93]

During April and early May, the members of the Air Transportation Committee traveled to New York, Washington, DC, Philadelphia, Dayton, Detroit, Chicago, and St. Louis meeting with prominent businessmen and bankers to encourage them to join the new company.[94] The Committee benefited from the announcement that the Fords were entering aviation. On April 13,

Clement Keys stands nineth from the right, just to the left of the radiator on this Stout 2-AT Pullman airplane, with a group of businessmen gathered in Detroit for one of the meetings to organize National Air Transport. Paul Henderson is seventh from the right, standing next to a hatless Henry Ford. Courtesy of Smithsonian National Air and Space Museum (NASM 2006-27070).

1925, the Fords started an air service between Detroit and Chicago to carry machined auto parts between Ford auto plants.[95] Public reaction to the Fords' commercial venture was "astonishing" according to one reporter; people said that if Henry Ford had enough faith in the airplane to enter the business, "then flying must be practicable."[96] It was a psychologically important moment for commercial aviation. Ford gave commercial aviation a new credibility and spurred others to accelerate their own ventures.[97]

The Air Transportation Committee found wide support for the proposed air mail company. On the last day of April, Keys gave an address in New York to the Exposition of Inventions in which he said that a sufficient number of men of wealth interested in aviation had given their support to insure the organization of a night mail and express service between New York and Chicago. The personnel and financial structure of the new company would be announced in two weeks. Keys said the entry of the Fords into commercial aviation was a principal factor in insuring the project went forward. The backers of the venture, Keys said, were prepared to risk substantial amounts of capital on what was a gamble. "The commercial end has been thoroughly studied, and it is believed that enough high-grade freight demanding speedy transportation can be obtained to make the venture a success. But, as in other pioneering enterprises, only the future can tell what will be made or lost."[98]

On May 21, 1925, thirty-five men, prominent in business and finance, met at the Drake Hotel in Chicago to approve the Certificate of Incorporation for a new company, National Air Transport, Inc. (NAT). The group elected Howard Coffin president and Keys as chairman of the Executive Committee. At the meeting, Colonel Paul Henderson announced that he was resigning from his position as second assistant postmaster general to become vice president and general manager of NAT. The list of the other executives, directors, and stockholders in National Air Transport read like a 'who's who" of American business. Charles Lawrance, president of Wright Aeronautical Corporation, was first vice president and a director; Lester Armour, of the Armour meatpacking family of Chicago, was named a director, as was Philip K. Wrigley, of the wealthy Wrigley family; William A. Rockefeller of New York, Charles F. Kettering, head of the General Motors Research Laboratory, and Roy Chapin, president of the Hudson Motor Car Company were directors. Among the prominent stockholders were Marshal Field III of Chicago, Glenn Curtiss, E. G. Wilmer, president of Goodyear Tire & Rubber, and Richard Hoyt, Chairman of Wright Aeronautical and a partner in the New York investment banking firm Hayden, Stone & Co.[99]

The group decided that stock in NAT would not be offered to the public, believing it unwise when the future of commercial aviation was uncertain. As Coffin explained, "the men behind the National Air Transport are willing to spend their money and we do not intend to sell stock to widows and orphans."[100] The new company was capitalized at $10 million, which was a powerful statement of the group's commitment to air transportation. The delegates to the meeting offered to put in $5 million, but only $2 million was accepted, representing 20,000 shares at $100 par value, as this was deemed sufficient to set up the New York–Chicago mail service. The delegates decided to allocate the majority of the capital in roughly equal shares between Chicago, Detroit, and New York, with smaller allocations to investors from other cities. Keys subscribed to an allocation of 1000 shares.[101]

The announcement of the formation of National Air Transport aroused wide public interest. *The New York Times* noted that "nothing so ambitious has yet been undertaken in America."[102] Coffin explained why the new company was being formed. "It is a very simple story," he said, "A group of men have associated themselves to do a much needed job with their own money. The establishment of an overnight air express service between Chicago and

Clement Keys and Howard Coffin, *right*, at the White
House on September 12, 1925, to tell President Calvin
Coolidge about forming National Air Transport Company.
(LC-DIG-npcc-14431, Library of Congress)

New York is a practical engineering task."[103] Coffin said the new company
intended to transport mail and express packages; there was no intention to
carry passengers. And, unlike airlines in Europe, there would be no direct
government subsidy. NAT, Coffin insisted, "is proceeding on the theory that
the New York–Chicago route, for instance, can, will and must pay its own
way . . . it purposes to meet an existing demand for faster transportation service between America's two largest industrial centers at a price that will render
the enterprise profitable."[104]

Framework for Commercial Aviation

The magazine *Aviation* called the announcement "a milestone in the history of American aviation."[105] The new airline, the magazine's editorial said, would have the three requisites for success: ample financial backing, experienced management, and efficient aircraft. "With these elements available, the possibility of airplanes making money in transportation will have its first real test in the world."[106] The magazine also commented on the new company's backers: "[T]he impressive list of capitalists and aviation experts who are interested in the new transport company will command the greatest respect from the business world. Confidence of this kind is needed when industry has to decide whether or not it will entrust its valuable papers and packages to the air. In securing traffic this element will be of the greatest assistance."[107]

Keys featured prominently in articles about National Air Transport, which pointed to his experience and acumen in business. *Aviation* said that Keys "has been probably the closest student in this country of aeronautical development from the business point of view" and noted that as a financial writer, Keys had "won a reputation for keen analysis and observation that has been enhanced by the application of his talents to financial problems of large magnitude, one of which has been the direction of the policies of the Curtiss Company."[108] An article in the magazine *The Independent* stated that Keys "is primarily a sound financier and economist. As such, he has no desire to manufacture a product that lacks a profitable market. . . . Mr. Keys is neither an optimist nor a pessimist. He is a businessman. The fact that he has associated himself prominently with this new flying project indicates that events of the last few months have brightened the horizon for commercial aviation."[109]

NAT's directors planned an overnight mail and express package service between New York and Chicago. They now had to turn that plan into a working airline, a real business, with routes, schedules, airplanes, ground facilities, staff, and other equipment necessary for the smooth operation of the service. After considering the available alternatives, the Executive Committee chose the Curtiss Carrier Pigeon, an airplane the Curtiss Aeroplane & Motor Company had designed in response to a Post Office competition for a more modern single-seat mail plane to replace the air mail service's aging DH-4s. On July 2, Howard Coffin announced that NAT had placed an order with Curtiss for ten Carrier Pigeons for delivery on October 1, 1925.[110]

On July 15, 1925, Postmaster General Harry New announced the opening of bids on eight Contract Air Mail (CAM) routes.[111] Unfortunately for NAT, the New York–Chicago route was not among them, but Paul Henderson, who

The Curtiss Carrier Pigeon, National Air Transport's first airplane. Built for the 1925 U.S. Post Office's competition for a new mail plane, Curtiss Aeroplane built ten improved models for National Air Transport. (*Division of Air Mail Service*: Correspondence of National Air Transport, 1926, Box 1, Entry 161, Records of the United States Postal Service, RG 28, NARA)

resigned from the Post Office two weeks later to become NAT's general manager, recommended to NAT's Executive Committee that the company bid on CAM 3, the Chicago–Dallas route, as this would be a logical extension of the New York–Chicago route should the company win that route when it was ultimately put out to bid. Postmaster General New announced the award of the air mail contracts on October 7, 1925. NAT won Contract Air Mail 3, the Chicago–Dallas route.[112] Then came the task of building up an organization on the ground to support the air mail service between Dallas and Chicago.

It took nearly a year for NAT to begin operations, but at 6:05 a.m. on May 12, 1926, an NAT Carrier Pigeon took off under a cloudless sky from Chicago's Maywood air field to begin the Chicago–Dallas air mail service. Two hours and ten minutes after the first Carrier Pigeon had left Chicago, two Carrier Pigeons took off from Love Field in Dallas to inaugurate the return leg of the route. During its first three months of operations, 97 percent of the NAT flights arrived on time, and out of 368 flights, there had only been ten forced landings, all but one of which was caused by severe weather.[113]

Beginning in April 1926, Keys convened the meetings of the NAT Executive Committee at the offices of C. M. Keys & Co. at 60 Broadway in New York City.[114] As chairman, Keys directed the issues to be discussed and the resolutions that came out of the discussion. The Executive Committee received reports from Paul Henderson on NAT's operations, but it dealt primarily with policy issues, particularly the upcoming bid for the New York–Chicago air mail route. Keys was actively involved in these discussions.

On November 15, 1926, more than a year after NAT was organized, the Post Office advertised bids for the New York–Chicago route. NAT was not the only company interested in the route. Colonial Air Transport, which had been awarded CAM 1, the Boston–New York route, also wanted the New York–Chicago route, as did several other bidders. In response to the Post Office's advertisement, NAT and Colonial submitted separate bids, but on January 29, 1927, the postmaster general rejected all the bids submitted and announced that the bidding on the New York–Chicago route would be reopened.[115] Keys and Coffin held discussions with Colonial Air Transport about setting up a joint venture company to submit a new bid for the New York–Chicago service, but nothing came from this idea.[116] On March 8, 1927, the Post Office put out a new advertisement for bids on the New York–Chicago route, which became CAM 17. Keys and the other members of the Executive Committee faced a dilemma. NAT's previous bid for the route had not approached the estimated cost per pound, and it was likely that the second round of bids would be even more competitive.

Keys presented two choices to the Executive Committee:

1. Shall we adhere to the pioneering courage which was responsible originally for the inception and organization of the NAT, anticipate an increase in poundage and bid an extremely low rate in order to secure the contract, regardless of current losses that stare us in the face, or;
2. Shall we bid on the basis of present existing poundage, adhering to the usual stereotyped attitude which prevails in established sound business practice.[117]

The Executive Committee unanimously decided to make a calculated gamble and bid on the basis of the probable volume of mail. The New York–Chicago contract was too valuable to NAT's future prospects to risk losing it, if there was a way to make a reasonable bid. In making this decision, the Executive Committee benefited from more recent information on the volume of mail

over the route and a harder look at actual costs. By reducing NAT's estimated costs to narrow the gap between costs and revenues per pound, and by
making a bet that the volume of mail would continue to grow, the Executive
Committee determined that NAT could bid at a rate of $1.24 a pound, based
on an assumed volume of 1,500 pounds of mail within six months of the award
of the contract and continued growth in volume beyond that.[118]

The gamble worked. On April 2, 1927, the postmaster general rejected two
lower bids and awarded the contract for the New York–Chicago route to NAT
as the most responsible bidder.[119] After just two years, NAT had a route system totaling 1700 miles, longer than any airline in Europe. Both the Chicago–
Dallas route and the New York–Chicago route would be fully lighted and
equipped with beacons along the way, at government expense, so that pilots
could fly the route at night. Best of all, the cost of setting up the company
was less than expected. The organizers had estimated that the company would
need $2 million just to set up the New York–Chicago service. The actual cost
of operating the Chicago–Dallas route and adding equipment and personnel
for the New York–Chicago line was less than $1.5 million.[120]

The success of National Air Transport demonstrated what Keys had
been saying since 1924: to succeed, an industry required organization. For
more than two decades, Keys had been a keen student of transportation systems; as chairman of the Executive Committee, he translated his theoretical
knowledge into practical experience, helping to build a transportation system from the ground up. In its first year of operation, NAT's airplanes had
flown 700,518 miles, completing 95.8 percent of scheduled flights; NAT lost
no airplanes, though bad weather caused 101 forced landings.[121] This record
was due to good organization, careful planning, and attention to the myriad
tasks needed to ensure the service continued on schedule. The effort had not
merely gone into the flying aspects of the business; NAT had spent a considerable amount of time and money, perhaps more than other air mail contractors, on promoting the air mail.[122]

Organization and administration were themes that Keys would return to
again and again as his aviation interests grew. He would become famous for
one particular phrase that captured his philosophy. In May 1927, Keys wrote
a letter to the British writer H. G. Wells, who had written an article for the
Sunday edition of *The New York Times* claiming that the business and administrative side of aviation in Europe was not up to the mechanical.[123] Keys wrote
to Wells saying that "the point of my approval of your criticism is that both
the Curtiss Company and the National Air Transport are operated on the

basis that nine-tenths of every flying operation is on the ground and not in the air."[124] Keys went on to articulate his philosophy, based on the lessons he had learned from his work with NAT:

> We spend more money in ground work than we do in flying. We regard the executive positions as far more important than the flying positions, and the flying personnel and equipment are under complete control of the executive branches of the business. For that reason I can hardly help but endorse your theory that what is needed in the air is not so much equipment and strongarmed young men, as it is business brains and training. It is not the best steamships or the best locomotives that make the most money and render the best service. In no form of transportation is the equipment or the operating personnel so important as the executive management. To a certain extent this principle seems to be forgotten in the air business. I suppose there is too much romance about it, particularly for those who have come recently into it. To us in the Curtiss Company and in the N.A.T. it all seems like good hard work, just the same as any other job of the same size.[125]

Keys's participation in NAT was formative in another way; it was his first experience as an entrepreneur, a promoter. He and Howard Coffin had started something completely new, based on a shared vision of the possibilities for commercial aviation. Together they had won the support of interested investors, raised capital to finance the venture, incorporated a company, and established a management structure. These were new skills for Keys. Up to that point in his business career, he had worked with established companies; he had steered Curtiss Aeroplane through desperate straits, but in that case he had been working with a structure and a business that had been in existence for several years. Helping to build NAT proved to be Keys's apprenticeship in creating new ventures in aviation. As a learning experience, his timing could not have been better. In the next two years, he would take advantage of more opportunities than he might have dreamed possible in the spring of 1927.

Organizing Transcontinental Air Transport

Keys believed that air transportation was not limited to carrying mail and express, but it was bound to include passengers. The private air mail lines, like

National Air Transport, had been a " . . . wise and excellent means of pushing forward the development of air transport as a whole . . . ," but he believed that " . . . no trunk line can be built up on air mail alone . . . all systems must be complete systems operating every branch of transport aviation . . . ," carrying mail, express, and passengers.[126] Moreover, he believed that passenger services, once initiated, should be developed fully. "I have never had the idea," he later wrote, "that these passenger and other services should be developed to a minimum possible extent. I have always held the opinion that they can be made profitable in themselves and that they should be developed to the fullest extent possible with the range of business common sense."[127]

Keys was initially cautious; in his discussions with Howard Coffin on the possibilities for air transportation, he made clear that he was not interested in passenger services, but only in air mail and express.[128] NAT's initial operations provided more data as well as practical experience, but Keys remained dubious about the potential for a profitable passenger service. Responding to comments from Postmaster General Harry New in August 1926 to the effect that air mail contractors operating at a loss should start carrying passengers in order to make a profit, Keys wrote to New saying "I am rather definitely of the opinion that organized passenger transportation on the present basis of equipment, airways and landing fields cannot pay."[129] Nevertheless, despite his reservations, that same year Keys and several of his colleagues among the NAT directors, notably Paul Henderson and Howard Coffin, began to study a possible passenger service between New York and Chicago, and between New York and Los Angeles.[130] This statement followed a pattern. Keys would often publicly express his reservations about a plan, but would conduct a study in private to ensure that his reservations could be addressed before proceeding.

Spanning the continent had the potential for exploiting the airplane's growing advantage in speed over other forms of transportation. If people could be persuaded that the savings in time outweighed the disadvantages of flying, namely, safety and comfort, air travel could offer a significant improvement over rail travel, its main competitor. In 1926 it took eighty-four hours to travel by rail from New York to Los Angeles; the U.S. Post Office's DH-4s were flying the transcontinental air mail route from New York to San Francisco in just thirty-four hours. It seemed more than possible that a transcontinental passenger service could make the trip in forty-eight hours or less.

A flight from New York to the West Coast, however, had to cross three mountain ranges: the Alleghenies, the Rocky Mountains, and the Sierra Nevada Mountains. To save time, some part of the route would have to be done at night. But these mountain ranges were difficult to fly over in the air-craft then available and posed danger in bad weather or at night. During 1926, the Pennsylvania Railroad contacted Keys and National Air Transport to pro-pose a solution to the problem in the form of a combined air–rail service.[131] The Pennsylvania Railroad would take passengers from New York by train overnight to Columbus, Ohio, a point that Paul Henderson believed was far enough west of the Allegheny Mountains and far enough south to avoid the weather around the Great Lakes. From Columbus passengers would con-tinue their journey west by airplane, connecting with another railroad that night and flying to Los Angeles the next day. This combined air–rail service was estimated to take forty-eight hours, a savings of more than a day over the railroads.[132]

Nothing came of these discussions at the time, but the idea of a trans-continental passenger airline began to grow firmer in Keys's mind. Clearly, a route from the East to the West Coast would be one of the main air transpor-tation lines in the United States that he was contemplating. Keys knew well that the business capital of America was New York, but the industries con-trolled from the streets of Manhattan were located all across the country. A transcontinental passenger system, with feeder lines to the north and south such as NAT could provide, would bring together the component parts of an industry, with a considerable savings in time.[133] And it made good business sense to offer a single through service across the country rather than have a sequence of regional airlines passing passengers off one to another as the railroads had done. James J. Hill's Great Northern Railroad and Edward H. Harriman's Union Pacific and their associated lines extended from the West Coast to Chicago but not beyond. The airplane would achieve what Harriman and Hill had not, a point that would not have been lost on Keys.[134]

On June 14, 1927, a few weeks after Charles Lindbergh's successful flight to Paris, Keys and Coffin wrote to Lindbergh telling him that they had been "seeking to establish a truly national air transport service," and while they had focused on a private air mail service, they "were looking forward to the exten-sion in the very near future of our service to that of carrying passengers."[135] They thought that after his Paris flight Lindbergh might want to play a lead-ing role in this development. Keys and Coffin were ready to form a new air-line company, to be known as the Lindbergh National Line, or some other name that would suit him, and they were prepared to underwrite all the capi-tal required for this new enterprise. They offered Lindbergh a one-fifth inter-est in the company, a salary sufficient to cover his personal needs, and hoped that he would accept a position as president or director of flight operations for what would be "the first great passenger trunk line in the country."[136]

Ten days later Lindbergh met in Washington and New York with several representatives of the groups involved in the project to work out basic principles and objectives. They proposed that Lindbergh would be president of the new company. The initial route of the company, which would be planned and operated in conjunction with a major railroad, was tentatively set as New York to St. Louis, with an overnight train service from New York through the Alleghenies to Columbus, Ohio, then after breakfast a plane ride to St. Louis, arriving around noon. This route would be the first link of what would ultimately be a transcontinental system.[137] Lindbergh turned down the offer; he had little interest in taking on a position where he would be running the daily operations of an airline.[138] Though nothing emerged from this flurry of activity around Lindbergh over the summer of 1927, the plans outlined, the interest expressed among the various parties, the capital made available, and the commitment to forming a transcontinental passenger airline would bear fruit early the following year.

The day after sending his joint letter to Charles Lindbergh, Keys and his wife left for Europe to attend the International Chamber of Commerce meeting in Stockholm. Keys was part of the American delegation, representing the American aviation industry. The trip gave Keys an opportunity to observe the latest developments in European air transport, one of which was the growing readiness of Europeans to use air travel and to talk with the principal men involved. In Stockholm, Keys met with Martin Wronsky, managing director of Deutsche Lufthansa and the heads of the Danish and Swedish airlines. In the Hague Keys met with Dr. Albert Plesman, director of KLM, the Dutch airline.

Keys later recalled how his meeting with Dr. Plesman at KLM's headquarters had made a very deep impression on him. Plesman kindly allowed Keys to go over KLM's accounting system in detail with the company's accountants. Keys came away amazed at the amount of money that KLM allocated for the depreciation, obsolescence, and replacement of airplanes and other equipment. When he questioned Plesman about this, Plesman told him firmly that the operation of any airline must include a sufficient amount of money to replace its fleet of airplanes due to being worn out or obsolete, and that this replacement could happen much faster than most airlines estimated. Plesman told Keys that on some of KLM's divisions he replaced the aircraft even though they had used up only one-sixth of their useful lives. And, Plesman added, there was the ongoing cost of experiments in improving radio

During the 1920s, Clement Keys made several trips to Europe where he had the opportunity to experience and study European commercial air transport. Here Keys, *third from right*, stands next to the Imperial Airways Armstrong Whitworth Argosy II *City of Liverpool*. (Courtesy Donald Furin)

communications and adding safety devices to the airplanes, among other areas. This effort had to be maintained, and often the expenditures had to be completely written off. Keys took away from his talk with Plesman the lesson that any airline venture would likely need greater capital than he had originally estimated, but also the lesson that a substantial portion of this initial capital should be set aside in a reserve for these unforeseen expenditures.[139]

But the one big lesson that Keys took away from his European trip was that "people are not afraid to fly, even in this generation, on well-organized lines between good terminals in first-class ships."[140] Keys thought that the movement of mail and express would be more valuable as a source of revenue and to the commerce of the nation than passenger traffic, which he believed would develop more as a luxury than a necessity, but he came back from Europe more confident than before that it would develop. "Confidence in the organization and equipment is, of course, absolutely essential," he said, "but the fact that the confidence has been gained and held by four or five great organizations in Europe is clear proof that it can also be gained and held by many organizations in America. Therefore the matter of passenger traffic is the basis of an industry in America—a new industry hardly yet born, but something that it may be worthwhile to develop in the fullness of time in a typically American way."[141]

After completing an extensive tour with the *Spirit of St. Louis* around the United States, Mexico, and Central and South America, Charles Lindbergh returned to St. Louis in February 1928 determined to devote his time and energy to the development of commercial aviation.[142] He wanted to start an airline that would fly not just from New York to St. Louis as Keys and Coffin had proposed to him the previous June, but on to California.[143] Lindbergh took his ideas to his St. Louis friends William Robertson, Harold Bixby, and Harry Knight. An initial approach to Henry Ford failed to gain Ford's support. Rebuffed by Ford, the four men decided that it might be better to approach a major railroad with their idea for a transcontinental airline, for a railroad would likely be part of the service and be more interested in transportation than Ford had been. Harold Bixby had previously met with some railroads in New York to gauge their interest in participating in an airline venture, but without result. Bixby proposed they meet with Keys. Bixby knew Keys from his work with a subsidiary of the Curtiss Aeroplane & Motor Company. He had a favorable impression of Keys's abilities and his commitment to aviation. Bixby also knew of Keys's financial connections on Wall Street.[144]

Keys met with Lindbergh and his colleagues, probably sometime in late February, and went over their plans with them. He told them of his own interest in setting up a transcontinental airline. Keys believed they were a year too soon, but since there was a strong argument that the passenger business was inevitable, and since there was capital available to back the venture, it made sense to start then.[145] Keys proposed that they consider forming a company with an initial capital that was bigger than what Lindbergh and his colleagues were considering, but likely reflecting what Keys had learned from his conversations with Plesman of KLM. Keys pointed out that the airline would probably need substantial reserves to carry it through the initial years of operating losses. Keys also agreed that it made sense to involve one of the major railroads. He recommended that Lindbergh and his group approach General William Atterbury, president of the Pennsylvania Railroad, who had discussed the air–rail service idea with National Air Transport a few years before.[146]

Atterbury was interested. Several years earlier he had decided that the Pennsylvania Railroad would adopt a policy of cooperation with this latest form of transportation, when the time was right. After Lindbergh's flight to Paris Atterbury decided it was time for the Pennsylvania Railroad to get involved. He did not see aviation as a threat to the railroads, but as potentially

Framework for Commercial Aviation

complementary through a combined air–rail service that would stimulate traffic for both. When Bixby approached Atterbury with Lindbergh's idea of building a transcontinental service, Atterbury was receptive, but he wanted Keys to be involved. Atterbury recognized that the venture would need a large amount of capital, and he wanted the financiers who had successfully underwritten National Air Transport to participate in whatever came out of the discussions.[147] Atterbury assigned Daniel Sheaffer to represent the Pennsylvania Railroad, and Keys called in Paul Henderson from Chicago, who was also keen to set up a transcontinental passenger airline, to participate in the talks.

Early in March Keys, Sheaffer, Henderson, Lindbergh, and his colleagues held talks to discuss their ideas for the transcontinental airline, working out a tentative route and schedule, organizational structure, possible participants, capital needs, and sources of financing.[148] Later that month Keys, on his own, wrote up a tentative plan incorporating all the information he had available up to that point. He suggested organizing an airline to operate three-engine passenger planes, built by Boeing, Fokker, or Ford, on a combined air–rail service from New York to Los Angeles. The plan envisioned the Pennsylvania Railroad taking passengers from New York overnight to Columbus, Ohio, where they would transfer to an airplane and fly in easy stages to Wichita, Kansas by the end of the day. Here the passengers would take another overnight rail trip on the Santa Fe railroad arriving at Clovis, New Mexico, in the morning where they would again board an airplane for the last leg to Los Angeles, arriving in the late afternoon. This schedule limited the flying portions of the route to sections of the country that were considered to be the most favorable for flying and avoided all night flying, as it was generally agreed that neither the airplanes likely to be available nor the aids to navigation were adequate for carrying passengers by night.[149]

Keys sought broad geographic participation in the proposed transcontinental airline. He hoped to interest the Boeing Company, Western Air Express, and the Southern Pacific Railroad in participating, thereby linking aviation, financial, and transportation interests across the country. Keys led the negotiations with the different potential participants in the transcontinental airline venture. He was sensitive to the fact that the transcontinental airline would potentially threaten the operations and plans of Western Air Transport and the Boeing Airplane Company. This was, in part, the reason why he wanted the venture to be a passenger line with no air mail contracts. The Southern Pacific declined to participate, as did Harris Hanshue, president of Western Air Express, who had his own plans for expansion.[150]

The Boeing Airplane Company had similar reservations about joining the venture. William Boeing had already invested a considerable amount of capital in Boeing Air Transport for its planned San Francisco–Chicago passenger service. He, too, saw the transcontinental line as creating competition for his own operation and saw no advantage in helping to develop traffic from the East Coast, particularly as Keys's planned transcontinental airline would have no link to the Boeing line. Keys thought that a strong argument could be made to Boeing that the eastern traffic the transcontinental line would develop would benefit his own service. In any event, he wanted to convey to Boeing that he and his colleagues in the transcontinental venture would be happy to cooperate with Boeing Air Transport in terms of schedules and rates. Keys told Henderson: "Frankly, I want to cooperate with Boeing very closely in every way and I think that he will see that even if we do run a Los Angeles line in cooperation with his passenger line to Chicago it will be more helpful to him than the other will be harmful."[151]

At the end of April, Keys took stock of the situation as it then stood. Boeing and Western Air Express had declined to participate. The Pennsylvania Railroad had refined its own intentions; the Railroad wanted to be an investor in an airline but did not want to operate one. Keys noted in a memorandum that the Pennsylvania's management "[does] not want to control operations in any way, but desire[s] to back a group in whom they have complete confidence." Keys noted that "this position is logical, clearcut and we can meet it fully."[152] Keys planned to organize a single company that would control passenger services directly.[153] The stock would not be sold directly to the public but would instead be placed with a key group of investors. Keys wanted to have as shareholders banks throughout the United States, the leading airline operators or representatives of their management, and the railroads which would be operating part of the service. He also proposed the formation of a large board of directors representing these interests, as well as leading financiers involved in the underwriting and some of the airplane and engine manufacturing companies. Keys wanted General William Atterbury to be chairman of the Board, and he took the role of president and chairman of the Executive Committee for himself, perhaps because as he wrote to Paul Henderson, "I seem to have the complete confidence of the bankers, who are sold on the idea that my experience and your experience and the experience of the NAT group is the safest rock upon which they can found their hopes and expectations."[154]

The proposed combined air–rail service would cross the country in approximately forty-eight hours, leaving New York at 6:05 p.m. and arriving in Los Angeles at 5:30 p.m. on the third day. This schedule would require a total flying time of twenty-one hours and approximately twenty-seven hours of train travel. Keys believed, based on the experience of other transportation companies, that the travelling public would patronize this new service even at a higher price to gain the savings in time. He calculated that with an average of eighteen passengers per day, a load factor of 60 percent, paying a fare of $375 ($6499 today), the airline operation would be in the black. With the number of first-class railroad passengers traveling east from Los Angeles varying from 1500 per day in the summer to 500 per day in the winter, the target of eighteen passengers on the new service did not appear to be unreasonable. To calculate the breakeven point, Keys added 25 percent to the estimated cost figures based on the experiences of National Air Transport. He expected that the municipalities along the route would provide landing fields and that the owners of these fields would provide restaurants and other passenger facilities. Keys estimated that the airline's main expense would be for equipment, which he calculated would run between $753,000 and $1,057,000. The planned capital of $5 million would then be more than adequate to cover contingencies.[155]

Transcontinental Air Transport, Inc. was incorporated on May 14, 1928, with $5 million in capital, C. M. Keys as president, and Colonel Paul Henderson as vice president. Keys obtained most of the directors he wanted, an impressive list of names well-known in business and financial circles.[156] Keys issued a statement on May 15 stating that the New York to Los Angeles route had been selected to permit the maximum comfort and safety to the passengers. "No night flying is involved," he said, "and the country over which most of the flights will take place is the most suitable in the country, according to many experts. The company plans to commence operations with the most modern American aircraft. These will be tri-motored planes carrying fourteen passengers each.... The planes will be equipped with every known safety device, radio communication and steward service.... Rates for the new air–rail trip will be at a reasonable advance over present transcontinental railroad rates."[157]

The announcement of Transcontinental Air Transport (TAT) made the front pages of the nation's newspapers. After outlining the basic features of the service and its backers, *The New York Times* went on to say in its front page article: "It is part of the new miracle wrought by the union of science and finance, with planes and railroads acting as auxiliaries instead of rivals.... To meet the spread of this broad land the new combination has been devised by

Getting Charles Lindbergh to join Transcontinental Air Transport as technical advisor was a coup for Keys. Lindbergh, *second on the right,* stands next to the Ford 5-AT Trimotor *City of Columbus.* Lindbergh used this aircraft to determine TAT's air routes across the country. (2008-3-31_image_655_01, Peter M. Bowers Collection, Museum of Flight)

men equally responsible for management and transportation. It may be only the beginning of an entirely new system of general travel in this country."[158]

Keys finally managed to persuade Lindbergh to join the transcontinental airline venture. He created a position for Lindbergh that took advantage of Lindbergh's talents, interests, and keen desire to promote the development of commercial aviation. Keys asked Lindbergh to become chairman of the Transcontinental Air Transport's Technical Committee. He outlined the terms of his agreement with Lindbergh in a letter following a meeting the two men had in New York. "Your work for the company," the letter stated, "will be of a technical and advisory character. You will act as Chairman of the Technical Committee, which will have under its immediate authority all matters concerning the choice of equipment, the layout and equipment of fields, all services pertaining to flying, safety appliances, etc."[159] Lindbergh would be neither a director nor an officer of the Transcontinental Air Transport, but in return for his services, Transcontinental Air Transport agreed to pay Lindbergh $250,000 to purchase 25,000 shares of TAT stock at the underwriter's price of $10 a share. In addition, Lindbergh was given an option to purchase an additional 25,000 shares at the same price, and an annual salary of $10,000.[160]

Getting Lindbergh to accept a position with TAT was a coup for Keys, adding luster to what was already considered a major advance in aviation. His announcement that Lindbergh would become Chairman of the Technical Committee was once again national news, giving Keys the title of "Lindbergh's boss." In a short statement, Keys described Lindbergh's role: "all matters concerning choice of equipment, fields, general service, flying routes, safety appliances etc.," would come under his authority and that of his Technical Committee which would "be made up entirely of practical and experienced men."[161] In an editorial entitled "Lindy Landed," the *New York Telegraph* said that "at last a commercial enterprise has grabbed Colonel Lindbergh. After a year of repeated efforts to obtain the one name that would ensure instantaneous success in any venture, America's leading financiers and business executives have given up hopes and surrendered to the victors in the mad rush for the Colonel's exclusive services.... The Transcontinental Air Transport, Inc., of which C. M. Keys, president of the Curtiss Aeroplane & Motor Corporation, is the head, landed Lindy—and it is but natural that this organization should accomplish the apparently impossible."[162]

Building Transcontinental Air Transport

At their first meeting to discuss his new job as chairman of the Technical Committee of TAT, Lindbergh asked Keys a simple question: "When shall we start?" Keys had to admit that he had no idea. "I did not know and did not pretend to know when it would be possible to start," he told Lindbergh, "None of us knew."[163] They thought, naively, that the job would take at most a few months and that Transcontinental Air Transport might be able to start in the fall or the winter at the latest. It would actually take one year and three months to get TAT into the air, what Keys later described as "a long year of hard driving, of thousands of disappointments and delays, of little failures and big triumphs."[164]

In that first discussion, Keys and Lindbergh laid down a number of things that had to be done before the airline could start service. These were, in Keys's words:

1. There had to be at every stop an adequate modern and complete field capable of taking care of the takeoff and landing of heavily laden transport planes carrying ten or twenty people.
2. There had to be, all along the line, a complete system of communication and weather reporting.
3. There had to be a sufficient amount of the best equipment we could find in the country, no matter who built it or how much it cost.
4. There had to be an adequate group of trained and experienced pilots who had had a specified number of hours of experience on big ships with passengers and who had flown the route a sufficient number of times to be familiar with the landscape, with radio communication, with the approaches and takeoff of the fields and the general flying conditions.
5. There had to be adequate station and comfort facilities at all landing fields and also the latest known appliances and fittings on the planes for the comport of passengers.
6. There had to be a system of emergency landing fields along the entire air route.
7. The combined schedule of the air and rail service had to be worked out and coordinated.[165]

Little of this infrastructure existed. All of it would have to be built up and put in place before the first passenger stepped onto an airplane.

Keys wasted little time, calling on Lindbergh's technical committee and a network of associates at the other companies in his group to begin tackling these challenges. He had Lindbergh conduct a survey of the proposed route to the West Coast, examining landing fields and terminal facilities for their suitability for the transcontinental service.[166] He sent representatives to Europe to study European passenger operations and equipment, contacted the Department of Commerce to get meteorological data on the two alternate routes across the Southwest, and where he needed to, sought the advice of outside experts to work on a radio communications system as well as on necessary improvements to airports along the route.[167]

With Lindbergh and others addressing the myriad technical problems TAT faced, Keys took on the more delicate issue of competition with the West Coast airlines and ensuring the cooperation of Western interests with TAT. Keys wanted to avoid the destructive competition that had been so damaging to the railroads in the late 19th century; he hoped instead to engender cooperation within the air transport industry.[168] Writing to one competing airline, Keys stated: "there will, of course, be lots of competition in the air transport lines and at first no doubt a great deal of it will be destructive to all parties concerned, same as railroad competition was until the traffic grew up to sizable proportions. We expect this, and so far as I am concerned it will be taken as it comes and it will not in itself engender any ill will."[169] Although Keys wanted to do all he could to avoid direct competition, he was fully prepared to meet it head on.

Keys decided to make a trip to the West Coast to seek cooperation with Boeing Air Transport (BAT), which had won the San Francisco to Chicago air mail route, and Western Air Express, which had a route from Los Angeles to Salt Lake City. He was also still hopeful that BAT and Western Air Express could be persuaded to participate with TAT in some fashion, and so he wanted to speak with them directly. Despite Keys's efforts, the three airlines each decided to go their own separate ways, though BAT said it wanted to arrange close connections with NAT in Chicago. Keys reported to the TAT Executive Committee that TAT would likely face competition from Western Air Express from Los Angeles to Chicago but would proceed with its own plans regardless. BAT and Western's offer to cooperate on terminal facilities at large airports and on other infrastructure would be a direct benefit to TAT. "If capital expenditures are required in these places," Keys told the Executive Committee, "we shall probably be able to share those capital expenditures and their benefits with the Western lines. The net result of these negotiations will probably well repay the six weeks or so of time during which they occupied a large part of my time and a considerable part of Colonel Lindbergh's efforts."[170]

For someone who had studied different approaches to managing railroad systems, after TAT's formation, Keys appeared to be reluctant to address similar issues at his nascent airline. That summer Chester Cuthell, his lawyer, chided Keys for being away for so long and for failing to deal with a growing problem: the lack of organization at TAT. Cuthell wrote to Keys in early August, saying that "there is one point upon which everybody here is agreed and that is that they would like to see you home at the earliest possible date, because there are so many things that need your particular touch in order to secure the prompt action which seems to be of the greatest importance."[171] Since its formation in May, Keys had done little to set up a formal structure at TAT beyond appointing key officers. Cuthell expressed a growing concern about this situation. "On T.A.T. matters," he wrote Keys, "all of the parties I have met seem to feel that there is immediate need for the development of a functioning organization. There is need for adequate offices, a division of authority, and the appointment of sufficient men so as to develop all of the many different aspects of the enterprise."[172] By implication, Keys was taking on too much responsibility himself and was not coordinating the work of the various people he had assigned to different tasks on an informal basis.

For whatever reason, Keys took some months to address this issue. Contrary to Cuthell's recommendations, he did not add staff to TAT. He preferred to keep the number of employees to a minimum during this preliminary period, and he employed only those involved with surveying routes or improving the chosen airfields. Despite all of his other commitments, he continued to involve himself directly in all aspects of TAT's preparations, working with Lindbergh and Paul Henderson, and calling on Cuthell for all legal matters. Keys knew this arrangement wasn't ideal. Part of the problem in communication was the fact that he and Henderson were working in different offices, with Keys in New York and Henderson in Chicago still managing NAT. In October Keys wrote to Henderson that he and Lindbergh "find ourselves hampered by the fact that the only daily records of what is being done on the Transcontinental are scattered between this office and your office."[173] He asked Henderson to establish a routine whereby Henderson would send copies of memos and correspondence to Keys in New York at the end of each week, and Keys would do the same for him. Keys also instructed Henderson to prepare a weekly memorandum bringing Keys up to date on developments. That way, he said, "Lindbergh and myself can function intelligently in these matters."[174] From time to time, Keys wrote reports to TAT's Executive Committee supplementing information provided at committee meetings, using the updates Henderson provided. The formal structure and broader communication that Cuthell wanted to see came only later.

Over the summer there was steady, if slow, progress, though much remained to be done. When pressed, Keys continued to insist that TAT would begin operations only when everything was ready, and not a day before. The survey team returned from Europe with a wealth of information and data on European airline operations, having visited with Imperial Airways, KLM, Deutsche Lufthansa, and the French Air Union line. An extensive report included recommendations on fare structure, ticketing and traffic management, and provisions for passenger comfort in the airplanes and at terminal facilities, noting that "safety belts, cotton for ears, and paper bags for airsickness are all necessary and should be supplied in one form or another."[175]

The airline had chosen its first equipment. After a thorough study of the available alternatives and many test flights, Lindbergh recommended the Ford Trimotor as the airplane best suited to TAT's requirements. Based on Lindbergh's recommendation, Keys placed an order with the Ford Motor Company for ten Ford Trimotor airplanes to be delivered on or before February 1, 1929. Explaining his decision to the TAT Executive Committee, Keys said: "I think that everything possible has been done by the company to insure that the field has been thoroughly looked over and that the best available ship has been selected."[176]

Keys and Lindbergh spent the fall making a final determination of the transcontinental route and the timing of the schedule. Lindbergh wrote to Keys at the end of August to say that the route of what would be the Eastern Division of TAT had been decided, with the route starting at Columbus, Ohio, and ending in Waynoka, Oklahoma, with intermediate stops at Indianapolis, St. Louis, Kansas City, and Wichita. At Waynoka, passengers would board the Santa Fe Railroad for the overnight run to Clovis, New Mexico, where the Western Division of TAT began, transferring to the plane for the flight to Los Angeles via Albuquerque, New Mexico, Winslow, and Kingman in Arizona.[177] The schedule had to be carefully coordinated with the Pennsylvania and Santa Fe Railroads to ensure that sufficient time was allocated to transport passengers and their baggage to and from the airfields, while the daytime flight schedules had to incorporate stops at intermediate destinations, time on the ground for lunch or dinner, and arrival in time to connect with the railroads. The selection of airfields at some of the stops along the route was still an open question, and until this was done, no progress could be made on setting up hangars and terminals. These and a hundred other problems had to be dealt with.

Keys still had reservations concerning TAT's likely success, at least in the intermediate term. He referred to the air–rail venture as an experiment and insisted that it could not be viewed as anything else. Writing to Lewis W. Baldwin, president of the Missouri Pacific Railway, Keys said, "I regard this rail and air or all air passenger transport as a gigantic experiment, of very great promise but based on premises not yet entirely secure or sound. . . . The only doubt about the matter is, we believe, not the question of our ability but of the public to pay rates based upon the cost of the most expensive known means of transport."[178]

But Keys did have other real concerns, principally about the inadequacy of the airplanes then available. Keys viewed the Ford Trimotor and the Boeing

Model 80 as little more than stopgaps. Keys wrote to the directors that he, Lindbergh, and the Technical Committee "are disappointed only in the true carrying capacity of these ships. On long transcontinental journeys neither plane is capable of accommodating more than 12 passengers with the proper degree of luxury, comfort, and safety."[179] The degree of "luxury and comfort" in the Fords left much to be desired. As a contemporary account described air travel at the time, "the one remaining disagreeable feature about flying is the noise, particularly in the tri-motored airplane. With one unmuffled nine-cylinder engine at arm's length on the right, and another on the left, a boiler shop is quiet by comparison."[180]

Keys gave a detailed discussion of TAT's progress in his first formal report to stockholders, issued on February 20, 1929. Keys outlined the selection of the route to the west and described the landing fields that Lindbergh had selected, noting some of the problems that had been encountered. At certain cities the municipal authorities had arranged to fund development of the municipal field, but in some of the smaller communities, like Waynoka, Clovis, Winslow, and Kingman, TAT had to purchase land and build its own landing fields, in addition to the necessary hangars and passenger terminals. The airline had been working closely with the Radio Corporation of America to develop a radio that could be installed in the Ford Trimotors to allow constant communications with the ground stations. The AT & T Company installed a private teletype service for TAT linking together all twelve of the landing fields along the route. A weather service had been built up at points both along and to the north and south of TAT's route across the country so that the landing fields and the pilots would have up-to-date weather information.[181] TAT's preparatory period was coming to an end and Keys would not hurry, regardless of what others were doing. In a few more months Keys's own bold experiment would be underway.

5

SUSTAINING AIRLINE SYSTEMS

A New Framework for Commercial Air Transport

IN A SPEECH ON THE FUTURE OF AVIATION WHICH HE GAVE TO THE REPUBLICAN Club of New York City in February 1929, Clement Keys was most enthusiastic about the progress of the air mail service.[1] As Keys noted, there was no longer any doubt about the efficacy of transporting high-priced commodities like mail by air.[2] He predicted that in future an ever larger proportion of the nation's mail would travel by air. "We can see the time," he said, "when on the great main lines of the country and over long distances all first-class mail will be transported by air without any additional cost to the consumer over the regular rate, and with a regularity of operation about equal to the regularity of the Postoffice service."[3] The faith he had expressed in the future of the air mail to the Republican Club in New York was based on the complete turnaround in the fortunes of National Air Transport (NAT).

In its first two years NAT had focused on getting its air mail and air express operations underway on its routes, beginning with the Chicago–Dallas route in 1926, adding the New York–Chicago route in 1927, as well as a nighttime service between Chicago and Dallas and a service from Chicago to Kansas City in 1928.[4] In January 1928, the Executive Committee authorized NAT to set up a separate subsidiary, National Air Transport Flying Service, to provide sightseeing flights and a limited passenger service.[5] During February, the NAT Flying Service was averaging twelve paying passengers a day and

A National Air Transport Liberty-powered Curtiss Falcon in flight.
NAT purchased fourteen Curtiss Falcons equipped with landing lights
under the right wing to carry the air mail at night from Chicago to
Dallas. (United Air Lines, Box 1, Entry 19, U.S. Senate. 74th Congress.
Special Committee. Air-Ocean Mail Contracts, RG 46, NARA)

providing valuable experience.[6] In June 1928, the airline announced that it planned to begin a passenger service between New York and Chicago in early 1929.[7] None of the air mail and air express services, however, were profitable; for 1927, NAT had a net loss of $189,824.[8] The airline's fortunes changed after Representative Clyde Kelly introduced the Second Amendment to the Air Mail Act of 1926. The Amendment called for lowering the price of an air mail stamp from 10 cents per ounce to 5 cents, in the hope of stimulating the use of air mail, and it advocated granting longer-term route certificates to the air mail carriers to encourage investment.[9] The Amendment passed Congress on May 17, 1928, and the new air mail rates went into effect on August 1, 1928.

The effect on NAT was dramatic. The volume of mail carried over NAT's routes during the month of August increased 83 percent; to cover the increased loads, NAT frequently had to send out two airplanes on the New York to Chicago night run and on the Chicago–Dallas route.[10] The impact on profitability was equally impressive. For 1928, NAT made a net profit of $274,180, compared to the loss for 1927.[11] By the end of 1929, NAT was one of the few bright stars in the air transport firmament. The company reported impressive earnings for 1929. Total revenues were $2,562,932, an increase of 57 percent over 1928, with a net profit of $669,295.[12] The total volume of mail carried on the Chicago–Dallas and Chicago–New York routes increased by 75 percent; both routes showed a healthy increase, though the Chicago–New York volume was now four times greater than that on the Chicago–Dallas route.[13]

Not every air mail contractor, however, benefited from the increase in mail volume. Those contractors who had a high volume of mail, like NAT or Boeing Air Transport (BAT) carrying the mail from Chicago to San Francisco, or Colonial Air Transport (CAT) carrying the mail from New York to Boston, were profitable, but where the volume of mail was limited, the contractors were not doing well.[14] The system of paying the air mail contractors by the weight of mail carried regardless of distance had created disparities in compensation and had produced a structure that forced the contractor to maintain the high costs of service regardless of the volume of mail actually carried.[15] The circumstances for the air passenger lines were even more dire. Until the end of 1929, these lines were operating with payloads ranging from 16 percent to 40 percent of capacity; in the last two months of the year, most passenger lines drastically cut their fares to entice more passengers, but the increase in passenger traffic that followed could not compensate for the fact that air fares were substantially below the cost of operations.[16] For much of the year, the average passenger fare amounted to around ten cents per seat-mile; fare reductions at the end of the year cut this in half.[17] Total direct and indirect operating costs for a Ford Trimotor which many of the airlines, including Transcontinental Air Transport, used, amounted to sixty-nine cents per mile. At these low seat-mile fares, a Ford Trimotor could not cover its operating costs even if it was carrying a full load of ten passengers. As a result, by the end of 1929 all the passenger lines had operating deficits.[18]

In March 1929, Walter Folger Brown, a prominent Republican lawyer from Ohio and former assistant secretary of Commerce under then Secretary of Commerce Herbert Hoover, became postmaster general in the administration of newly elected President Hoover.[19] In his new position, Brown became "the chief architect of aviation policy during the Hoover presidency."[20] When Brown surveyed the nascent air transport industry, he found "a haphazard and problematic system calling out for a comprehensive overhaul."[21] Looking at the air transport companies at the end of 1929, Brown thought that "the very life of the passenger transport industry today is in the balance."[22]

Brown believed that commercial air transport was vital to the nation and to the future of the American aviation industry, necessary for commerce and national defense. Since the success of the aviation industry "must depend upon the will and habit of the people of our country to travel by air," he argued that the ultimate objective of the Post Office should be "the development of air passenger lines upon which the carrying of the mails should perhaps be only an incidental operation."[23] The problem with commercial air transport, Brown said, was that the enthusiasm of the promoters had exceeded the public's desire to fly. In comments that might well have been directly intended for Keys and his associates, Brown said that "the men who were most ambitious for the industry and who with infinite pains planned the longer passenger routes perhaps forgot their own experiences in the air, perhaps forgot that children creep before they walk, that they toddle from chair to chair before they engage in marathon races."[24] Brown wanted the government, through air mail payments and route certificates, to promote the stable and rational growth of passenger transportation.[25] He wanted "the airlines to acquire better, more modern equipment and to promote passenger travel that, when operating in the public interest, provided citizens with a new and faster method of transportation."[26]

Brown wanted to revolutionize the method of compensating air mail contractors and to rationalize the existing airway system in ways that would promote passenger services.[27] In effect, he wanted the government to use air mail payments to subsidize passenger services until the airlines could become self-sustaining through passenger revenues. Brown wanted to support the growth of passenger services by putting mail on passenger planes. This would enable the Post Office "to give immediate assistance to air passenger carriers on such routes that were deemed essential, by paying for carrying the mails a substantial sum, based upon a definite weight space preempted."[28] He proposed abandoning the system of compensating the air mail contractors by the weight of mail they carried and instead adopt compensation based on the space available in an airplane that could be used for carrying the mail.[29] A system based on space rather than the weight of mail would allow the Post Office to make payments to the struggling passenger airlines. The Post Office could, he argued, "give immediate assistance to air passenger carriers on such routes as were deemed essential, by paying for carrying the mails a substantial sum, based upon a definite weight space preempted."[30]

The government, Brown believed, should concentrate its support on air transportation routes that were deemed essential, rather than cover the entire country.[31] His primary concern was to establish several transcontinental routes across the country that followed the natural flow of people and mail. Moreover, preference in awarding air mail contracts "should be given to the pioneers in the air transport industry of good character and financial responsibility."[32] Brown saw the emerging consolidation of the aviation industry as a potentially positive force for promoting a strong system of commercial aviation. An oligopoly, under the control of the Post Office through its ability to award air mail contracts, could be made to serve the public interest, and for Brown oligopolies or monopolies "acting in the public interest were seen as good and deserved encouragement."[33]

Brown's inability to get the air mail contractors and airline operators to agree to a new system of compensation and route extensions persuaded him to seek new legislation that would give him the powers he believed he needed to rationalize and renew commercial aviation.[34] The result of his efforts was the Watres Act, signed into law on April 29, 1930.[35] This law, considered a "turning point in the history of the air mail service," gave Brown the authority to redraw the air mail map and directly regulate the commercial aviation industry.[36] The law provided for introducing the new system of compensation based on space and not weight, with bonuses for airlines that flew multiengine aircraft, carried radio, and flew at night—all steps to encourage safer passenger travel.[37] The legislation also allowed the air mail contractors to exchange their contracts for longer term route certificates, and it gave the postmaster general discretion to determine the condition of bidding on air mail contracts, as well as to consolidate and expand routes, the amount of mail to be carried over a route, and the type of equipment that would be used.[38] Brown's actions under the Watres Act would have a direct impact on all of Keys's air transport ventures.

The Struggle for National Air Transport

The aviation industry clearly understood that the purpose of the Watres Act was to "stimulate the development of air transport and particularly the carrying of passengers by air."[39] As the bill was under debate in Congress, the proposal to allow the postmaster general to give preference in the awarding of air mail contracts to passenger airlines prompted the directors of NAT to consider once again starting a passenger service along NAT's air mail routes.[40] The day after the House Committee on the Post Office and Post Roads reported the Watres Bill and sent it to the House, Colonel Paul Henderson, in his capacity as vice president of NAT, announced that NAT planned to introduce passenger services between New York and Chicago, as well as between

Chicago and Dallas.[41] While NAT did not indicate when the new passenger service would begin, the proposed line would fill a gap in the airline network providing passenger services connecting the two largest industrial and financial centers in America, Chicago and New York, and linking them with the largest cities in the Midwest and Southwest. Reporting on Henderson's announcement, the *New York Times* pointed out that the scheduled opening of Boeing Air Transport's twenty-hour service from San Francisco to Chicago later in the spring would mean the possibility of air travel across the country in thirty hours.[42] Within a few days, rumors began to circulate in the markets that the United Aircraft & Transport Corporation (UATC) was negotiating to acquire NAT to link the NAT line with UATC's Boeing Air Transport (BAT) subsidiary, which carried passengers from San Francisco to Chicago.[43]

The logic of linking BAT with NAT was not lost on Frederick Rentschler, president of UATC. Rentschler knew that Postmaster General Brown wanted to promote passenger airline development and to build sound transcontinental airline systems across the country.[44] With the air mail contract for the San Francisco–Chicago route, Boeing Air Transport was more than halfway there. In April 1929 UATC had purchased Stout Aerial Service, which operated a passenger line from Detroit to Chicago via Cleveland, giving BAT a stepping stone for moving further east, but it made more sense for BAT to acquire NAT rather than push its own route to New York, even if it could do so on its own. A BAT–NAT combination would be the first all-air transcontinental service. During February, Rentschler authorized United Aircraft to begin to buy NAT stock. In late March, shortly after Henderson's announcement that NAT was to initiate passenger service, Rentschler approached Earle Reynolds, NAT's president, with a proposal for an exchange of shares with UATC that would give UATC a controlling interest. Rentschler proposed an exchange on the basis of one share of United Aircraft & Transport Corporation for four shares of National Air Transport.[45]

NAT's Executive Committee met on March 26 to consider Rentschler's proposal. The members asked Keys to provide a brief analysis of the offer with figures that would help the directors form a judgment on its merits.[46] Keys presented a memo to the NAT Board of Directors on April 2 in which he pointed out that a comparison of the net earnings of the two companies showed that a four share exchange offer would mean NAT shareholders would be trading $4.12 in earnings from their four shares of NAT stock for $4.52 in earnings from one share of United Aircraft. This would be a clear benefit to NAT's shareholders.[47] Keys's view of the strengths and weaknesses of the two companies was balanced. He noted that the contract air mail rates that were then under review had not been fixed and that could affect NAT as well as Boeing Air Transport.

On balance, Keys was not opposed to the proposal. Beyond the business logic of a merger and the potential benefit to NAT shareholders, Keys had other reasons for not opposing United Aircraft's offer. William MacCracken, formerly assistant secretary for Aeronautics at the Department of Commerce, wrote to Harris Hansue of Western Air Express. "I do not know," McCracken said, "what C. M. Keys's attitude is on this particular proposal, but I do know that he is inclined to get his affairs in such a shape that in the future they will not require as much of his personal attention as they have in the past, and I also happen to know that he thinks very well of the United organization's management."[48] Even though Rentschler improved his initial offer by proposing an exchange of one share of United Aircraft for three and a half shares of NAT, the other NAT directors were unwilling to accept the change in control. Howard Coffin sent a letter to Rentschler on April 3 advising him that the NAT Board had voted unanimously to decline his offer, and that evening Coffin and Keys released a joint statement to the press to that effect.[49]

Rentschler then announced that he would submit his proposal directly to NAT's shareholders for their approval, bypassing the directors. In his own statement to the press, Rentschler said: "We are seeking control of NAT because we believe that great economies may be effected through the formation of a single transcontinental line, carrying mail, passengers and express. If the NAT line were to be paralleled by a United line, one line or the other might lose money. From an economical point of view, the air between the coasts is not big enough to be divided."[50] Rentschler then had United Aircraft send out a letter to all NAT stockholders offering to exchange shares of NAT stock and United Aircraft stock at 3½ to 1, and asking that they send in their shares and endorse a proxy for the annual meeting of stockholders, to be held on April 10.[51] To the delight of Wall Street, the struggle between Rentschler and the NAT directors for control of National Air Transport, which was how the contest was seen, boosted the share prices of both NAT and United Aircraft. United Aircraft climbed to $97 a share, a new high for the year, while NAT also reached a new high, advancing to $29½ a share. These movements in share price brought the stock of United Aircraft more in line with Rentschler's proposal of a 3½ for 1 exchange.[52]

NAT's directors responded to Rentschler's direct appeal on April 7 by rewriting the corporate bylaws to provide that board members could only be removed for cause and by a two-thirds majority of the stockholders. They further stipulated that future changes in the bylaws would also require a two-thirds majority vote.[53] Then, on April 9, NAT agreed to issue 300,000 new shares of common stock to North American Aviation, an investment company that Keys had established to invest in the aviation industry, in exchange for North American's holdings of Eastern Air Transport, another Keys company (see below).[54] Keys launched a public attack on United Aircraft's offer, bringing up the specter of monopoly and Wall Street's potential control of air transport, an ironic concern for one who had by then become one of the leading financiers of aviation. Keys also contrasted the efforts of the pioneers who founded National Air Transport with the aggressive "raid" by United Aircraft:

> Our fight has been altogether a defensive one. . . . We do not believe that it is to the best interests of the public, of our stockholders, or of the nation as a whole, that the main line from New York to Chicago should be controlled by United Aircraft & Transport Corporation, which also controls the main line from San Francisco to Chicago. No economic purpose can be served by such a consolidation and the very diversity of interest between the two has lead, and will lead to a competition in quality of service to the public that would be eliminated by such a consolidation.
>
> The theory upon which this raid was based was, in effect, that it was time to eliminate from aviation the pioneer group who have created most of the things that have been created so far in the United States in this art, and to put them under a financial control centering in one of the largest banking groups in Wall Street. Aviation has not reached that point and I think it will be a long time before it does reach it. A healthy, wholesome growth of this business seems to me to require the whole-hearted and friendly interest of a large number of stockholders, controlled not by ownership, but by good faith, integrity and honesty of purpose.[55]

Initially at least, it seemed that Coffin, Keys, and NAT's other directors had won the battle. At the annual meeting of stockholders, United Aircraft did not have enough proxy votes to oust the board. The two directors representing United Aircraft did not submit the proxies they had accumulated, but they sought to challenge the legality of the meeting by saying that a quorum was not present and that the recent amendment of the bylaws was illegal. But the board overruled the protest and declared that a sufficient number of proxies was present to constitute a quorum, which promptly voted to retain the existing board of directors. Howard Coffin issued a press release declaring victory in the battle for control.[56] But Rentschler had not given up. United Aircraft's lawyers filed an injunction in the Chancery Court in Wilmington, Delaware, seeking to prevent the exchange of shares between NAT and North American Aviation pending the court's determination of the legality of NAT's annual meeting.[57] The Chancery Court judge agreed to the injunction and issued an order restraining the exchange of stock and requiring the two parties to show cause on May 16 why the injunction should not be issued. In response, the NAT board elected Keys as president of the company, removing Earle Reynolds, and it adopted a resolution to defend the board's actions in the Chancery Court. United Aircraft then filed a suit in the same court, asking the court to void the annual meeting on the grounds that the meeting was illegally conducted. Rentschler then issued a call for a special meeting of NAT stockholders for May 7 to propose changes in the bylaws of the company. It appeared that a long, complicated, and expensive legal battle would ensue.[58]

Keys, however, had had a change of heart. He had no interest in an extended fight to retain control of NAT. Chester Cuthell had advised Keys that a prolonged struggle over NAT might interfere with many of Keys's other plans. Moreover, the idea went against Keys's view of what was good for the industry. On April 16 United Aircraft announced that it now controlled 57 percent of NAT's stock; there seemed no point to continuing the battle.[59] Three days later, Keys sent a telegram to Cuthell from St. Catherine's Island, off the Georgia coast where he was trying to get some rest from the increasing strain of the many demands on his time, telling Cuthell that the more he thought about the plan to retain control of NAT, the less he liked it. He told Cuthell the following:

> If United has bought control we should let them have it without trying to make it hard for them. The other policy is not in line with our usual corporate policies because it will do no good to aviation as a whole. We have plenty to do protecting our other interests and even if we won it would only be for a little while. We could not hold NAT.[60]

Keys managed to persuade Howard Coffin and a majority of the other directors to accept his position. Cuthell and Charles Lawrance, representing National Air Transport, met with Rentschler and other United Aircraft directors at Keys's offices on 39 Broadway on April 23, 1929. In a compromise, Rentschler agreed to drop the litigation in the Chancery Court and also agreed to an exchange of stock on the new basis of three shares of NAT stock for one share of United Aircraft. This was a better deal for NAT's stockholders, and in return the NAT board agreed to recommend the offer.[61] On April 26, Rentschler was elected chairman and president of National Air Transport; Keys and Howard Coffin resigned their positions as president and chairman, respectively.[62]

Losing One Airline, Gaining Another

Keys's original plan for a transcontinental passenger air transport system had been to develop feeder lines that would connect with the transcontinental airline he hoped to build. NAT, with its north–south route system now covering a good section of the Midwest, had been part of the plan, but NAT and its potential passenger flow was now lost to him. He had, however, concurrently been building another north–south airline system on the East Coast. Harold Pitcairn, whom Keys knew well as one of the original investors in National Air Transport and a fellow member of NAT's executive committee, had been the sole bidder on Contract Air Mail Route 19 between New York and Atlanta via Richmond through his company Pitcairn Aviation, Inc. and had been awarded the contract for CAM 19 on February 28, 1927.[63] Later that same year, on November 23, Postmaster General Harry S. New announced that Pitcairn Aviation had been awarded the contract to fly the mail between Atlanta and Miami via Jacksonville, which became CAM 25.[64] These two awards gave Pitcairn a route system of 1350 miles through the heart of the South, connecting New York with the popular tourist destinations in Florida. Pitcairn Aviation launched the New York to Atlanta service on May 1, 1928.[65] The new air mail rates introduced on August 1 brought about a doubling of mail volume on the service.[66] Then on December 1, Pitcairn began the Atlanta to Miami service under CAM 25. More dramatic news had come out on October 17, when Pitcairn Aviation announced that it would inaugurate a passenger service in the spring of 1929, beginning with service from Greensboro, North Carolina, to Atlanta, but ultimately extending from New York to Miami. Pitcairn Aviation placed an order for three Ford Trimotors to begin the service.[67]

In the fall of 1928, Keys and Pitcairn apparently began discussing the possibilities for a New York–Miami passenger line, which Pitcairn announced at the end of October. At some point in these discussions, Keys apparently expressed an interest in purchasing Pitcairn's growing airline.[68] Keys was clearly interested in a passenger line that would have financial support from profitable air mail contracts. Pitcairn was beginning to reconsider his own commitment to his airline venture. His real interest was in airplane design and manufacturing. He had also begun to realize that even though he had increased the authorized capital of Pitcairn Aviation to $1 million, the investment required for a full passenger service might well be more than he could provide, particularly in the face of competition from the larger and better financed aviation groups that were forming around him.[69]

Keys renewed his efforts to persuade Pitcairn to part with his airline. In early June, Pitcairn accepted an offer from Keys to purchase Pitcairn Aviation, Inc., and four flying service companies Pitcairn had established at Richmond, Greensboro, Spartanburg, and Atlanta for $2.5 million, leaving out Pitcairn's airplane manufacturing company.[70] In announcing the purchase, Keys made the following comment:

> We are going to carry on with the old staff the work which Mr. Pitcairn has begun and we hope to carry it on as well as he has. His studies indicate a good possibility for the development of rail and air passenger connections, perhaps seasonal in character, between the North and the South, and we intend to carry on these possibilities as far as seems reasonable. The lines connect at New York with the Colonial to Boston and Montreal and the National Air Transport to Chicago. At Atlanta they connect with the lines of the Southern Air Transport and at Miami with the lines of the Pan-American. We believe that ultimately there will be great value in these connections from the standpoint of mail, passenger and express, and we hope to be of good service to all of them.[71]

Keys moved quickly to bring Pitcairn Aviation under his control. Two weeks after he purchased the company, Keys sold Pitcairn Aviation to North American Aviation, the investment firm he had established, and appointed his own slate of directors.[72] He appointed Thomas B. Doe, a graduate of West Point and a capable businessman, to be vice president and general manager of the company.[73] In January 1930, Keys and the other directors agreed to change the name of Pitcairn Aviation, Inc. to Eastern Air Transport, Inc.[74] The following month Keys was elected chairman of the Board and Doe became president.[75]

In 1930, Eastern Air Transport placed an order for twelve Curtiss Kingbird
D-2 aircraft. EAT later acquired two additional Kingbirds. The D-2 model
could carry a pilot and seven passengers or six passengers and 200 pounds
of mail. EAT operated the Kingbirds along with the larger Curtiss Condor.
(2008-3-31_image_654_01, Peter M. Bowers Collection, Museum of Flight)

When Keys purchased Pitcairn Aviation, it had been his clear intention to
develop the airline into a passenger airline as well as an air mail carrier. With
passage of the Watres Act and Postmaster General Brown's clear intention to
develop passenger transport, Keys believed that Eastern now had a moral as
well as a legal obligation to go forward with developing passenger services.
More pointedly, as he wrote to Daniel Sheaffer at the Pennsylvania Railroad
Company, he believed that Eastern Air Transport had no choice but to go
ahead with developing a passenger service, "which the Postmaster General
has practically ordered us to establish."[76]

The first step in Eastern Air Transport's expansion was to increase the
number of daily flights between New York and the South. In March 1930,
Postmaster General Walter Folger Brown announced that Eastern Air Transport would be granted permission for an additional daily flight between

New York and Atlanta.[77] In April 1930, Eastern Air Transport announced that it was considering the inauguration of passenger services along three lines: Atlanta to New York, Atlanta to Miami, and Atlanta to Dallas, with services to be added incrementally. After several months of preparation, Thomas Doe announced on August 7, 1930, that Eastern Air Transport would begin a passenger service every weekday between New York and Richmond using its Ford Trimotors, with stops at Philadelphia, Camden, New Jersey, Baltimore, and Washington, DC.[78]

Postmaster General Brown was anxious that Eastern Air Transport begin service between New York and Washington with larger airplanes, and extend its service to the South to Atlanta and Miami. At Keys's request, Doe sent a letter to Brown stating that Eastern Air Transport intended to operate Curtiss Condors on the New York–Washington run, and Curtiss Kingbirds from Washington to Atlanta.[79] When Postmaster Brown awarded the airline renewed certificates for CAM 19 and CAM 25 in early November, Keys and the other members of the Executive Committee approved an additional capital investment of $650,300 to purchase three large Curtiss Condor twin-engine transport airplanes and twelve smaller Curtiss Kingbirds from the Curtiss-Wright Corporation, as well as additional mail planes and equipment to serve the new passenger services.[80]

Eastern announced that an extension of its passenger service from Richmond to Atlanta would begin on December 10, followed by a further extension to Miami beginning on January 1, 1931. Through connections with other airlines at New York, Atlanta, and Miami, passengers would be able to fly the length of the East Coast, link up with transcontinental services, and fly on to South America. Eastern Air Transport's two mail contracts were generating over $1.6 million in revenues and were growing, which provided needed support for the passenger services; the 35 percent growth in mail revenues during 1930 more than offset the 25 percent growth in operating expenses. This was what Keys had intended for Eastern Air Transport, having the air mail contracts support passenger services.[81]

Transcontinental Air Transport Takes Off

While Keys was working on Eastern Air Transport, TAT itself was moving closer to the start of the transcontinental air–rail service. Since its inception in May 1928, there had been, as Keys described it, "a long year of hard driving, of thousands of disappointments and delays, of little failures and big triumphs."[82] He had insisted that operations would only begin "the day we are properly ready." That day was approaching at last. Charles Lindbergh reported that TAT's Ford Trimotors had been equipped and delivered, the pilots and ground personnel hired, the coast-to-coast schedule set, and the meteorological stations installed along the line; hangars and passenger facilities would be ready by July 1, though work on the airfields was still continuing.[83] St. Louis had been chosen as TAT's operating headquarters. Paul Henderson remained in Chicago, where he continued to divide his time between running National Air Transport and overseeing TAT. Operations had been divided into two divisions, with the Eastern Division covering the route from St. Louis to Columbus, Ohio, and the Western Division covering the route from St. Louis to Waynoka, Oklahoma, and then from Clovis, New Mexico to Los Angeles. Lindbergh reported to Keys at the end of May that TAT would be ready to inaugurate the transcontinental service on July 8, 1929.[84]

While Lindbergh and others worked on the technical and administrative details of preparing for the start of TAT's passenger services, Keys worked to improve the airline's competitive position in California through a merger with Maddux Airlines. Formed in 1927 by car dealer Jack L. Maddux to carry passengers from Los Angeles to San Diego, by early 1929 Maddux was operating a fleet of thirteen Ford Trimotors on regular scheduled services between San Francisco, Los Angeles and San Diego, with extensions to Ensenada in Mexico, and he had won a concession to extend its route all the way to Mexico City.[85] A service from Los Angeles to Phoenix was also planned. As a financial reporter for the *Los Angeles Times* noted in December 1928, "there are logical reasons to suggest an ultimate expansion eastward by Maddux through an affiliation with the recently organized Transcontinental Air Transport, Inc."[86] From TAT's perspective, controlling Maddux Air Lines would feed passengers from north and south of Los Angeles onto the transcontinental line and would allow TAT to offer a through service from New York to San Francisco. Despite the steadily growing volume of passengers, Maddux was operating at a loss due to the aggressive competition on the Los Angeles to San Francisco route, which limited the company's ability increase its fares.[87] Linking-up with a well-connected and well-financed partner had obvious advantages.

From his own research and discussions with Jack Maddox, Keys knew that Maddux Air Lines had built up a strong traffic organization along the West Coast. While the airline was losing money on operations, Keys did not believe these losses would continue for an extended period. Acquiring the airline would be of real benefit to TAT. At the TAT Executive Committee meeting on May 14, 1929, Keys presented a proposal to acquire 51 percent of the stock of Maddux Air Lines Company, the holding company that owned Maddux Air Lines. The Executive Committee recommended the acquisition to the TAT Board of Directors, and on June 1, 1929, Jack Maddux announced that the directors of Maddux Air Lines had approved the plan for consolidation with TAT.[88]

Keys had promised that Transcontinental Air Transport would not begin service until "the day we are properly ready." That day finally arrived on July 7, 1929. Practice flights along the TAT route had begun on June 24 in a final test of the system. Over the next eight days, the airline flew 40,000 miles and carried 200 guest passengers. One reporter, who traveled the route by plane and rail, commented that "no one who has made the trip, be he an experienced or inexperienced air traveler, has failed to be impressed with the results of the efforts of the company to create this new form of transcontinental travel. . . . It has been said that two consecutive days of air travel with its confinement and the continual roar of the engines would be more than the average traveler could enjoy. This did not prove to be the case in the writer's experience and that of his nine companions."[89] Elaborate plans had been made for the inauguration of the air–rail service, which would start simultaneously from the East and West Coasts.

Keys chose not to be present at the culmination of his greatest achievement, the inauguration of coast-to-coast services on Transcontinental Air Transport. He had agreed once again to be the American representative at the air transport meetings at the Fifth Congress of the International Chamber of Commerce in Amsterdam opening on July 8. On June 29 he and his wife sailed on the Dutch liner *Statendam* as part of a group of 150 American businessmen with Thomas W. Lamont, a senior banker at J. P. Morgan & Company, leading the delegation.[90] With characteristic reticence, Keys sought to avoid the limelight. He had left Transcontinental Air Transport in capable hands.

At 4:05 p.m. on July 7, 1929, Colonel Charles Lindbergh, sitting in the Los Angeles offices of the governor of California, pressed a button that sent a signal to Pennsylvania Station in New York, turning on a white globe at 7:05 p.m. above the Pennsylvania Railroad train "The Airway Limited." More than a thousand people witnessed the train's departure. Congratulatory telegrams poured in from air-minded officials across the country. General William Atterbury sent congratulations to Keys, saying "My heartiest congratulations in the magnificent work you have done in promoting commercial aviation and bringing this into harmony with established railroad service. . . . May it achieve the full measure of success that we have visualized for it." Keys sent a reply to Atterbury from Amsterdam:

> Never has an older form of transportation extended to a new form such generous and unstinting co-operation as you and your fellow officers of the Pennsylvania system have given to Transcontinental Air Transport in its task of forming the "Lindbergh Line" across the continent. Forty-eight hours from coast to coast is an achievement to fire the imagination. Few realize the enormous care required and the vast groundwork organization to be perfected before we could with confidence give the signal to start. "The Airway Limited," starting tonight, carries our hopes and the best we have been able to give to the task.[91]

In his November 1929 speech to the Republican Club of New York City, Keys said that the passenger business was still in the pioneering stage; it required "a lot of courage, hard work, and patience."[92] Courage, hard work, and patience were exactly what were needed at Transcontinental Air Transport because the airline was struggling badly for the very reasons Keys had outlined. The traveling public had not taken to air transport in the volume that had been expected, and it was well-below what was required.[93] A month into the service, TAT reported that it had carried 433 passengers along the line, averaging a mere 38 percent of capacity.[94] Although passengers were generally enthusiastic about the service, traffic grew slowly in August, pushing up the load factor to just 52.5 percent in the last week of the month.[95] A month later the worst happened. On the morning of September 3, 1929, the TAT Ford Trimotor "City of San Francisco" crashed in a storm on its way from Albuquerque, New Mexico, to Winslow, Arizona, killing five passengers and the crew of three. Full service resumed on September 16. Traffic was slow to return, in part due to the end of the vacation season, but also due to public alarm over the TAT crash.[96]

In October, TAT's Executive Committee decided to merge the operations of Maddux Air Lines with TAT's Western Division to reduce expenses by eliminating duplicate repair shops, overhead, personnel, and traffic departments at Los Angeles and San Francisco. The integration of Maddux more fully into TAT's operations would also improve the ability to compete with Western Air Express, the main competition. At the same time, the Executive

Committee launched a serious debate on the strategy TAT should pursue going forward. The Executive Committee believed that with the growing competition from Chicago to the West Coast from Boeing Air Transport and Western Air Express, TAT needed an intensive development of traffic from outlying points to link up with the main transcontinental route. These steps, it was thought, would provide TAT with a reasonable prospect of operating with a profit. The Executive Committee appointed a special subcommittee to study ways of expanding traffic on the Eastern Division to feed into TAT's main transcontinental line.[97]

Compounding the lack of traffic growth was the inadequacy of the current equipment in service. TAT–Maddux now had a fleet of twenty-three Ford Trimotors. Responding to Lindbergh's report on possible alternative transport planes for TAT, Keys said in no uncertain terms that "we have operated enough to know that Trimotor ships in this service are almost impossible from the standpoint of economy under present conditions and would probably add to rather than decrease the loss of the system in operation."[98] Keys told Lindbergh that TAT was unlikely to order new aircraft any time soon.

In a move designed to boost traffic during the slow winter season in the middle of November, TAT announced a 25 percent reduction in fares to $267.43, thereby reducing the price gap between the air–rail service and the regular all-rail fare. Commenting on the fare reduction as a positive step, the magazine *Air Transportation* reported that lower fares would lead to greater traffic volume, which would mean lower operating expenses per passenger. It noted hopefully but prophetically that while profits might well follow as volume increased, "at first these lowered fares can only show greater losses than the higher fares did."[99] Other airlines soon followed TAT. On January 10, 1930, Keys announced that TAT–Maddux had slashed its New York to Los Angeles fare to $159.92.[100] This new fare was lower than the equivalent fare for a Pullman train car across country, but less than half of what TAT had charged when it began service the previous July. Putting a brave face on it, Keys said that "reducing fares to the equivalent of railroad tariffs will certainly be in response to a demand from the general public. And with this reduction air transportation will have been made available to the entire traveling public."[101] Before these new fares could have an impact on traffic volume, however, TAT suffered its second fatal accident in less than five months. A TAT–Maddux Ford Trimotor returning to Los Angeles from Agua Caliente in Mexico in the early evening of January 19 crashed, killing the fourteen passengers and crew aboard. Once again a plane crash made headlines across the country, creating publicity that TAT–Maddux did not need.[102]

This second accident, together with issues centering on flying discipline and TAT–Maddux's traffic problems, triggered a change in management at the top of the company. It was now clear to many of the directors that TAT–Maddux needed an active president to direct operations on a daily basis, something that Keys simply could not do. The arrangement of having Keys as president based in New York and Henderson as vice president based in Chicago running TAT–Maddux and National Air Transport may have been appropriate during TAT's formative stage, but no longer. Apparently, Charles Lindbergh, for one, had suggested the need for a more active president while TAT was being organized and had continued to argue for this afterward.[103] J. Cheever Cowdin, Keys's financial associate and a TAT–Maddux director, recommended that Jack Maddux become president and that Daniel M. Sheaffer, vice president in charge of traffic at the Pennsylvania Railroad and also a TAT–Maddux director, become chairman of the Executive Committee. Cowdin wanted Lindbergh to become chief of operations, but Lindbergh declined as he wanted to remain a consulting aeronautical engineer.[104]

On January 26, 1930, it was announced that C. M. Keys was retiring as president of TAT—Maddux Air Lines to become the chairman of the board of directors and that Jack Maddux would become president.[105] As Cowdin had recommended, Daniel Sheaffer was appointed chairman of the Executive Committee. Lindbergh was reelected as chairman of the Technical Committee, and Paul Henderson continued as executive vice president. Keys took the move with equanimity, or at least he never complained about being in effect booted upstairs. With all his other commitments, Keys had too little time to devote to TAT–Maddux, nor, despite all his expertise in transportation and his years of study of the railroads, did he have actual experience in running a transportation company. Given his character, he was likely practical enough to recognize that the new arrangement was not only a recognition of the reality of the situation, but also ultimately in the best interests of the airline.

The reduction in air fares put through in January had been successful in boosting traffic over the line, but as Keys reported in the Transcontinental Air Transport annual report on March 13, 1930, the overall financial results were still disappointing. For the full year TAT–Maddux posted a loss of $986,591.[106] Traffic volume had proved to be less than expected, while passenger fares were well-below operating costs. The lesson of the first six months seemed to be that most passengers were unwilling to pay nearly double the price of a cross-country rail ticket in order to save a day's travel. Reducing fares had only compounded the problem. More people were flying along the line, with many flights operating at full capacity, but TAT still wasn't making money. During 1929, TAT's revenues were a mere 27 percent of operating expenses, a figure that must have been not just shocking to Keys and the other TAT directors, but positively frightening.[107] On an annualized basis TAT–Maddux was losing almost $2.6 million a year. If that rate of operating loss continued, in less than three years the airline would burn through its capital and reserves and be wiped out.

When he organized TAT, Keys had little doubt that there would be operating losses during the initial period of the airline's operations. TAT was, after all, a bold experiment, the first airline to attempt to operate profitably on passenger revenues alone. But the losses posted for the first six months of operation had been greater than expected. The experiment did not appear to be working. Talking with Thomas Eastland, the San Francisco banker now working with Goldman, Sachs & Co. in New York, Keys confided that in the absence of some arrangement with TAT–Maddux's competitors which would eliminate the duplication of service, Keys saw "very little hope for the air transport companies."[108] There were, however, two solutions to the problem: drastically lower operating costs, or get an indirect subsidy from the government.[109] Keys and his fellow directors would pursue this latter course, though it would take months of negotiations to achieve the result.

Searching for Solid Ground

TAT–Maddux was clearly struggling. A solution would emerge from Postmaster General Brown's intent to reshape the airways through his newly acquired powers under the Watres Act. Brown wanted to establish a group of airlines that could each provide mail and passenger services across a broad area, avoiding wasteful duplication by linking the main centers of the country through alternative routes.[110] Brown called representatives of the contract air mail lines and the passenger airlines to a meeting in Washington on May 19, 1930, to discuss the implications of the Watres Act and to work out how to save the passenger airlines. Representatives of all the main air transport companies attended, including United Aircraft & Transport Corporation, fresh from its takeover of National Air Transport, Western Air Express, and TAT–Maddux, among others. Brown told the assembled group that he wanted them to cooperate and to agree to an equitable division of air mail and passenger services across the country, based on their existing routes and operations. With UATC having effectively established a northern transcontinental route from San Francisco through Chicago to New York, Brown let it be known that he wanted a southern transcontinental route from Los Angeles to the south, and critically for TAT–Maddux, a central route from Los Angeles to New York via St. Louis. But as this latter route appealed to several airlines, the group could not reach agreement on how to allocate it, although it was generally recognized that both TAT–Maddux and Western Air Express had a claim to the route.[111] Western Air Express was then operating a passenger service from Los Angeles to Kansas City, whereas TAT–Maddux was operating essentially the same route and connecting to New York via its air–rail service.

Brown's solution to the problem was simple: he wanted one company to operate the central transcontinental route, and so he accordingly suggested that TAT–Maddux and Western Air Express either consolidate their operations or work out an operating arrangement that would achieve the same end.[112] Early in the year, TAT executives had once again approached Western, but the effort came to nothing after several months of talks.[113] Brown, however, was not to be denied; he wanted to preserve TAT–Maddux by granting the airline a mail contract, and he indicated to the two parties that "he could and would arrange so that an air mail contract award would be properly made to the central transcontinental providing the two companies organized for the operation of the service."[114]

Keys did not participate in the difficult negotiations that followed. Daniel Sheaffer, in his capacity as chairman of the Executive Committee, held a series of meetings with Harris Hanshue, president of Western Air Express, and Postmaster General Brown to work out an agreement. On July 15, 1930, Sheaffer, in a memorandum to TAT's Executive Committee, outlined the terms of the agreement he had worked out with Hanshue. TAT–Maddux's and Western Air Express' assets were to be transferred to a new company, Transcontinental and Western Air (TWA), and in return both would receive an equal amount of shares in TWA. At the last minute a third party, the Pittsburgh Aviation Industries Corporation (PAIC), which had been seeking to develop the air route between New York and Columbus, Ohio, inserted itself in the negotiations and was reluctantly given a minority share in TWA. Because it had relied on the Pennsylvania rail service to carry passengers from New York to Columbus, TAT could not claim any right to an air route

The Pennsylvania Railroad's train "The Airway Limited" about to inaugurate Transcontinental Air Transport's air–rail service on July 7, 1929, from Pennsylvania Station in New York City. Paul Henderson, Vice President and General Manager of TAT, stands second from right. (Author)

between the two cities, which PAIC could.[115] The new TWA was formed on July 24, 1930.[116]

On August 25, 1930, Assistant Postmaster General W. Irving Glover announced the opening of bids for the new central transcontinental route, which would operate from New York to Los Angeles via Philadelphia, Pittsburgh, Columbus, Indianapolis, St. Louis, Kansas City, Amarillo, and Albuquerque. TAT–Maddux and Western Air Express submitted a joint bid, which was awarded to the two companies on October 1 as Contract Air Mail Route 34. TAT–Maddux and Western Air Express then transferred their awards to Transcontinental and Western Air, Inc. Keys was appointed chairman of the board of the new company, with Harris Hanshue as president, Jack Maddux as vice-president, and Daniel Sheaffer as chairman of the Executive Committee. In a letter to TAT's stockholders, Keys explained the new arrangement and the rationale behind it. While noting the public's increasing use of air transportation, Keys had to acknowledge that "the inevitable cost of an enterprise of this sort has made it most important that your company seek additional sources of revenue. With the passage of the [Watres] Bill referred to, the opportunity for obtaining an air mail contract was one that your company could not afford to overlook."[117]

On October 25, 1930, Keys's bold experiment in a pure passenger airline came to an end when TWA inaugurated its combined mail and passenger transcontinental service, the first all-air transcontinental service in America. Complementing, but bettering the forty-eight-hour air–rail service that continued in operation, the new service offered a thirty-six-hour crossing of the continent, with flights leaving New York and Los Angeles in the morning and stopping overnight at Kansas City before continuing on to the east and west the next morning. As Keys had suspected when he began organizing Transcontinental Air Transport, the effort was a few years premature; the public was not as yet sufficiently air minded, and the available equipment was uneconomic at the rates needed to entice the public to travel by air. To make such a start in the face of these obstacles had taken courage, hard work, and perseverance of a high order. While there were still many hurdles to overcome, not the least of which were the lack of equipment, Keys's transcontinental passenger venture appeared to be on the way to a more solid footing. In its first two weeks of operation, the new air service carried 2041 passengers, with many flights operating at capacity.[118] But far more important, as Daniel Sheaffer pointed out to one of his associates, the new air mail contract "would produce not less than $1,000,000 additional gross revenue per annum and with prospects of this figure being considerably increased as the mail traffic develops."[119]

Pressing On

As 1931 dawned, TWA faced challenges with management of the new venture and the seemingly intractable problem of the imbalance between revenues and operating expenses. The issue of management at TWA was thrust on Keys early in the new year. On January 20, J. Cheever Cowdin wrote a long, blistering letter to Keys, Daniel Sheaffer, and Harris Hanshue, with copies to all other directors, that was sharply critical of the airline's operations and its senior management. "A close study of our operating expenses will," Cowdin said, "prove to each member of this Board that the existing operating costs are out of all proportion or necessity to our probable revenue and that only under most favorable circumstances would it be possible to receive revenue sufficient to cover such costs, which means that if we continue to operate as at present we will very likely be faced with continuing deficits"[120] Cowdin was critical of the fact that, as he claimed, senior officers of the company were working at the same time for other, competing companies (several senior managers held concurrent positions with the Pennsylvania Railroad and, like Harris Hanshue, at Western Air Express). Cowdin insisted that these officers "devote their entire time, thoughts and energies to the welfare of this company.... We know the operation of this line is sufficiently large and important to command the entire time of those in charge of its operations."[121]

Cowdin further recommended that the airline hire an individual with extensive business experience in a senior capacity. As he went on to explain:

> From my association with the industry, the general theory that, as this is aviation we must have an executive brought up in aviation, is a complete myth. Besides, in looking over the executives in aviation how many are there who would be considered real successes as business executives and of great ability and general experience or who, prior to their association with aviation have had the responsibility of the successful operation of a corporation comparable in size and money invested with this corporation, or with the keen competition encountered by this company with its activities from one coast to the other.[122]

This was harsh criticism. Keys did not reply directly, but Harris Hanshue wrote to Cowdin that, while he agreed with many of Cowdin's observations, "as to the others, it seems clear that you are laboring under misinformation as to the facts."[123]

Cowdin had urged that the company obtain the services of men with broader business experience. The answer to his request came from an unexpected source: General Motors. To become involved in aviation, the automotive giant had purchased a 40 percent interest in the Fokker Aircraft Company in 1929.[124] In May 1930, Harris Hanshue, who was president of the Fokker Aircraft Company as well as Western Air Express, announced that Fokker would come under a new General Motors-controlled holding company called the General Aviation Corporation.[125] On March 4, 1931, General Aviation purchased a 24 percent interest in Western Air Express, which amounted to 24 percent of Western's outstanding shares.[126] This purchase gave General Aviation the opportunity to put its own directors on the board of Transcontinental and Western Air, replacing several directors from Western Air Express.

On March 12, Keys announced that Frederick Fisher and C. E. Wilson, both vice presidents of General Motors and directors of General Aviation Corporation, and J. M. Schoonmaker Jr., president of General Aviation Corporation, had been appointed directors of TWA, and that General Aviation appointed six other men as alternate directors, including the general counsel of General Motors, all replacing men from Western Air Express.[127] Keys wrote to Charles Lawrance a few weeks later saying that

> Schoonmaker and Wilson and Mr. Hogan, Counsel of General Motors, are working very closely with our group. They are starting on Tuesday to study the administration of Western Air itself. We hope that we shall be able with them to get management really moving. . . . This has been a difficult job, somewhat similar in character, but much bigger in scale than the reorganization of NAT in the summer of 1928, and the shakeup of TAT in February 1930. It has been terribly complicated by the diversity of interests between Western Air and T.A.T., which is now rapidly disappearing.[128]

Despite the positive tone in his letter to Charles Lawrance on the changes at TWA, Keys apparently continued to have concerns about TWA's management. As he wrote to one correspondent: "[W]e are engaged at the present time in shaking down a combination management of Western Air and TAT into a working organization. What we need more than anything else is close, economical old-fashioned business management."[129] After the announcement of the changes in the TWA board, Keys met with General William Atterbury, president of the Pennsylvania Railroad, to discuss the possibility of a joint effort by North American Aviation, the investment trust Keys had established in 1928, and the Pennsylvania Railroad to increase their control over TAT through additional share purchases. This would provide TAT with sufficient cash to meet its capital contributions to TWA, to increase its ownership in TWA, and, as Keys pointed out to Atterbury, "through that control to cause Transcontinental & Western Air, Inc., to install an efficient management."[130] Within the Pennsylvania Railroad, however, there were questions as to what benefit the railroad would reap from such a move. While new management might improve the airline's performance even if the airline were successful, it would most likely mean taking business away from the railroad.[131] The Pennsylvania Railroad decided not to pursue Keys's proposal to increase their ownership of TAT.

A change in management came through another vector. The extent to which Keys was involved is uncertain. As will be seen, his financial problems mounted as the Depression worsened, as did the problems at many of the companies he was involved with. These companies proved to be a time- and energy-consuming distraction. In June 1930, General Motors replaced C. E. Wilson as a director of TWA with Ernest Breech.[132] As he became more involved with TWA, Breech was astonished at the lack of direction among top management, and he was critical of Hanshue's lack of attention to his executive responsibilities.[133] Based on Breech's observations, General Motors recommended reducing the size of the board, appointing a more capable Executive Committee, and removing Hanshue as president.[134] TWA's directors accepted General Motors' proposals without dissent. On July 8, 1931, Keys announced that Harris Hanshue had resigned as president of Transcontinental & Western Air to devote all his time to Western Air Express and that Richard Robbins, a director who had come to TWA from PAIC, would be the airline's acting president.[135]

Robbins immediately tried to address the continuing imbalance between TWA's revenues and expenses. During its first three months of operations, TWA had revenues of $516,029, of which 29 percent came from carrying the air mail, but an operating loss of $664,936.[136] TWA needed more mail revenue and lower costs. Robbins sought to address the imbalance of revenues and expenses with two approaches: pushing the Post Office Department to increase TWA's allocation of the air mail budget, and cutting expenses at TWA. Over the summer, Robbins met with Postmaster General Brown and Second Assistant Postmaster General Irving Glover, but to no avail. In July, Brown told Robbins that the Post Office had no money available to increase its payments to TWA, even though TWA was about to start its faster night mail service between New York and Kansas City.[137]

Robbins enlisted Keys's help with the problem of cutting expenses at TWA. Robbins had apparently concluded that the best that could be done in the short term was to cut salaries at TWA across the board; senior managers had already taken a one-third cut in salary in March, but clearly, more needed to be done. To prepare the way for these cuts, Robbins asked Keys, Daniel Sheaffer, and J. M. Schoonmaker to write letters to Robbins expressing their concern about the situation, which Robbins then sent out to all staff. In response, Keys wrote the following:

> My opinion is that we have done about all we can do at this time to
> set up the T.& W.A. and that the salvation of that system depends

entirely on its administration. The most fundamental thing is the reduction of expenses. Our costs are very high as compared with costs for the same services of several other lines. These costs must be reduced. . . . We must handle the T.& W.A. not as though it were a pioneering adventure but as though it were established business passing through extreme depression. Amounts of money paid for services in this trade as a whole are greater than are paid in any standard from of transportation with which we compete. And it is essential in our judgment, both for the welfare of the capital invested and for the permanence in the work of the men themselves that readjustments be made in these expenditures as quickly as possible.

We still have too much expense for the work we are doing and the revenues we are receiving. The drain upon capital is very severe and the resources of capital are close to exhaustion. We do not know from what source new capital could be derived at this time, especially for the purpose of paying losses.[138]

On September 22, Robbins sent out a letter to all TWA employees announcing a reduction in salaries. Robbins enclosed the letters that Keys, Sheaffer, and Schoonmaker had written on TWA's situation. The salary cuts at TWA, which ranged from 2½ to 20 percent, were scheduled to go into effect on October 1.[139] Shortly after Robbins sent out this letter, Keys made a formal announcement that Robbins had been appointed president of Transcontinental & Western Air. Upon assuming the office after three months as acting president, Robbins wrote to Lindbergh that "I am more and more impressed with the amount of work ahead of us before we have what I consider a sound business."[140]

The operating costs of TWA's airplanes was an intractable problem. At a time when the airline's Ford Trimotors were earning $0.44 a mile, the Ford airplane's operating costs were more than $1.00 per mile.[141] The difficulty was that, at the time, there were simply no suitable replacements for the airline's Ford Trimotors and Fokker F-10 aircraft. Several directors came to believe that fast, single-engine airplanes could be the answer to TWA's cost problems.[142] In March 1931, the TWA Executive Committee, apparently following discussions in Washington with the Post Office, decided to begin a fast night air mail service from Kansas City to Los Angeles with single-engine airplanes, with a further plan to extend the service from Kansas City to New York. The airline purchased five single-engine, low-wing monoplane Northrop Alpha airplanes from the Northrop Corporation that were 35 miles an hour faster than

Passengers waiting to board a Transcontinental & Western Air (TWA) Ford Trimotor. Maddux Air Lines acquired this airplane, NC 8413, in June 1929 before the airline merged with TAT. TAT-Maddux transferred the airplane to TWA in April 1931. (2008-3-31_image_659_01, Peter M. Bowers Collection, Museum of Flight)

the Ford Trimotor.[143] With one engine instead of three, the Alphas had lower operating costs than the Ford, and with faster speeds, higher utilization.

Keys was against this push for more speed, and he also opposed the use of single-engine aircraft for passenger transportation. He believed that, with a traveling public still nowhere near sold on the idea of air travel, safety would likely be the predominant concern of most air travelers and would outweigh the benefits of greater speed. Single-engine airplanes, while attractive from the standpoint of operating costs, were a dead end; Keys believed that speed and safety would come in time. He told Robbins that Postmaster General Brown had stated unequivocally that safety was his major concern. Brown, he said, "is going to insist upon safety as the first consideration and that he is not interested at the present time in greater speed than now prevails."[144] Keys was concerned about controlling an airplane at the low speeds required for landing. He assumed, unnecessarily as it transpired, that adding retractable landing gear to reduce drag and increase speed, and flaps to slow a plane down for landing would complicate the pilot's tasks on landing, thus increasing the danger.[145] While concerned that at the present time single-engine airplanes did not have the necessary combination of speed with safety, he believed that airplanes that met the dual need would be developed. "This does not mean," he said, "that these ships will not come, probably within a year or two and we shall be carrying 700 hp to 800 hp in a motor with no more weight than the present 500 hp and it is quite possible that under these conditions a sufficiently high flying speed may be maintained with big enough wing and control areas, to afford a safe control at low speed."[146]

The airplane that Keys envisioned materialized two years later in the form of the Douglas DC-1. The improved Douglas DC-2 featured better aerodynamics and more powerful engines than the Ford Trimotor, retractable wheels, wing flaps that allowed slower landing speeds, and a two-man crew; the DC-2 could cruise at 170 mph, more than 70 mph faster than the Ford, but it had approximately the same landing speed.[147] The DC-2 entered service with TWA in 1934, but Keys would not be there to welcome it.

Foreign Ventures

Keys's vision of aviation did not stop at the borders of the United States. As a man of his times, he was a believer in "progress." In an article written for *The World's Work* in 1907, Keys had stated that "there is nothing in the world more astonishing than the rapidity with which the outlying and until lately unprogressive corners of the earth are yielding to the uplifting influences of Western civilization and commerce."[148] When he wrote these words, he saw the railroad as the emblem of the Age of Progress, the Advance Agent of Prosperity as he called it in another article.[149] Now, in the 1920s, the airplane was replacing the railroad as the conqueror of distance and time. The airplane had just as much potential in other parts of the world as it did in America, particularly in areas that were poorly served by other means of transportation, and that meant the potential for many profitable ventures.

Keys had a longstanding interest in the larger world around him, going back to his years as a journalist with *The World's Work* where he wrote several articles on foreign trade and the work of American companies abroad. The Curtiss Company had been selling aircraft in Asia and Latin America and had, at different times, operated flying training schools in Argentina, Brazil, and the Philippines. Keys was keen to build sales of Curtiss airplanes and engines abroad, but he also wanted to pursue opportunities for ventures in air trans-portation. With several of his associates, he began studying aviation activity in Latin America, looking at the fledgling airlines in Argentina, Brazil, Chile, Peru, Colombia, and Mexico.[150] Among the many potential opportunities that Keys was considering, China seemed to hold particular promise. During 1928, in order to expand his knowledge of Asia, Keys had joined the China Society of America and the Japan Society in New York. No doubt at these forums he would have learned that Chiang Kai-Shek and the Kuomintang Party had that year established a new national government in Nanking exercising nominal control over the country. The new government soon developed an interest in air transportation as a means of improving internal communications within China at a lower cost than the formidable expense of building railroads.[151]

The question that arose was how, from an organizational standpoint, to identify and exploit these potential opportunities. Operating on the principle that "different types of men are required for different branches of the business," in the fall of 1928 Keys began working to set up a new organization that could pursue new ventures in manufacturing or air transportation that would be sep-arate from Curtiss's own entity for foreign sales, the Curtiss Aeroplane Export Corporation, and its efforts to sell Curtiss products in foreign markets. Keys envisioned two interrelated companies working in concert. The first company would be an exploration company charged with discovering and analyzing promising foreign ventures around the world. The exploration company would carry out the preliminary investigation of a project, determine the amount of capital required, the best ownership structure (direct ownership, joint venture, or minority interest), and work to get a completed contract with the proposed partners. The second company, with substantially more capital, would provide the financing for the projects that the exploration company recommended. The financing company, in addition to providing financing, would become a holding company owning the stocks of the foreign projects. In October Keys began discussing the idea for the two companies with Clarence Dillon, an influential investment banker and founder of Dillon, Read & Company, who supported the idea and had access to interested investors.[152]

In early 1929, Keys and Dillon organized Aviation Exploration, Inc., with Keys and several of his associates providing half the paid-in capital of $500,000 and Dillon and his associates the rest. Keys once again took on the role of chairman and appointed William B. Robertson of the Curtiss-Robertson Airplane Manufacturing Company as president. In March 1929, Keys secured the board's approval of his aviation investment trust, North

American Aviation, to invest in a new company, Intercontinent Aviation, Inc. This company would be the financing vehicle for the foreign ventures identified by Aviation Exploration.[153]

Keys had written to Clarence Dillon of "the tremendous number of small local very profitable projects that are likely to be available in South America and in Asia."[154] Among the many potential opportunities that Keys was considering, China seemed to hold particular promise. Keys asked Robertson, as president of Aviation Exploration, Inc., to head up an expedition to China to examine the potential for an airline.[155] "In China," Keys wrote, "you will look very carefully into the general subject of air transportation and decide whether or not conditions justify the expectation that air transportation will be profitable in that country."[156] Keys wanted Robertson to investigate the potential for air mail contracts, passengers, and, interestingly, the possibilities for the trade and transport of high-value commodities where there were difficulties transporting these commodities to markets because of distance or lack of other means of transportation. Keys also showed no hesitation to linking up with local Chinese partners. He explained to Robertson that his policy in dealing with foreign enterprises was straightforward. Keys said he had "no desire to dominate with American capital the transportation enterprises of any country. If in any country there are reasonable business groups able and willing to participate in aviation enterprises in that country or to supply the majority of the capital required, my interests will be glad to cooperate with these gentlemen for the carrying on of enterprise that seems to us to be promising. This may be particularly true in China."[157] Keys also wanted to establish a policy of using Chinese personnel in any venture to the extent possible. Keys was optimistic; as he told Robertson, "there is no limit to what we shall do in China if the Chinese Government really wants us to cooperate and the Chinese commercial interests welcome us to China."[158]

Robertson spent two and a half months in China conducting survey flights and negotiating with the Chinese government. He returned to New York in May with a contract between Aviation Exploration and the China National Aviation Corporation, which the Chinese government had created to deal with aviation in China. The contract gave Aviation Exploration the exclusive right to operate three air mail routes in China for a period of ten years. Under the terms of the contract, the new service was to begin in October 1929. The Chinese government agreed to turn over all cash receipts from the air mail to the operating company, and it would make up any deficit with bonds the government would issue bearing interest at 8 percent. On

June 27, 1929, Keys presented a memorandum to North American Aviation's Executive Committee on the China project, saying that he and others had studied the contract thoroughly and that airplanes had been ordered for the planned service. The contract would probably require an additional investment of $1 million to develop airfields along the three routes, but Keys had carefully investigated the risks with the Standard Oil Company of New York, which had extensive experience dealing in China and thought these risks were acceptable. There was some risk in taking Chinese government bonds in payment, but Keys believed that the rates Robertson had set in the contract were more than sufficient, as he estimated that a rate of 75 cents per mile would ensure a profit. Keys was not unmindful of the risks involved, but he believed the opportunity outweighed the danger. "The basis upon which these contracts have been accepted," he argued, "is that, if successful and if the Government stands up, they constitute the most desirable possible entrance into what will undoubtedly become one of the greatest of all commercial aviation fields."[159]

At Keys's suggestion, Aviation Exploration also sent a mission to Peru and Chile under C. W. Webster, an officer of the Curtiss Aircraft Export Corporation, to investigate the best means of getting involved in those countries. Through this mission, Keys had entered into a contract to buy 2500 shares of Cia. De Aviacon Faucett S.A., a company founded by a former Curtiss employee, Elmer Faucett, for $100,000. Faucett had received a concession to operate an airline over the length of Peru with the backing of some of the leading businessmen and sought financing from Keys for the purchase of an airplane for the new service.[160] Keys offered the stock to North American, and the Executive Committee agreed to purchase it at cost. In Chile Webster proposed a commercial venture to the Chilean government that would involve setting up a commercial airline and an aircraft factory. The government was receptive to the idea, particularly to setting up an aircraft factory. Both ventures would be through a Chilean company that would be owned by Intercontinent Aviation.[161]

Based on these pending opportunities in China, Chile, and Peru, Keys decided to formally establish Intercontinent Aviation and arrange its financing. At the meeting of the North American Aviation Executive Committee on June 27, 1929, he recommended that North American Aviation subscribe up to $2.5 million in Intercontinent's preferred shares. The Executive Committee agreed with the recommendation and appointed Keys and all the other members of the Committee as directors of Intercontinent. Intercontinent Aviation was incorporated in August 1929 with capital of $2.5 million. On August 6,

1929, 250,000 shares were issued at $10 a share, with North American Aviation purchasing 125,000 shares and the remaining shares sold to other investment trusts and banks. Keys received 25,000 shares of Intercontinent in exchange for his interest in Aviation Exploration, Inc.[162]

Keys moved quickly to put Intercontinent's capital to work. On August 26, 1929, Intercontinent Aviation established a wholly owned subsidiary, China Airways Federal, to operate the airline in China under the previous agreement between Aviation Exploration and the Chinese government.[163] China Airways signed a contract with the Ministry of Communications to jointly organize a company to be called China National Aviation Corporation (replacing the initial company so named). The company would fly three routes from Shanghai to Chengtu in the west via Hankow, from Nanking to Beijing in the north, and from Shanghai south to Canton, beginning with the route from Shanghai to Hankow.[164] A month before, a party consisting of an operations manager, three pilots, five mechanics, five radio operators, and five Loening amphibian aircraft in crates had left Seattle, Washington, for Shanghai. In September Keys organized Compania Nacional Cubana de Aviacion Curtiss S.A. as a subsidiary of Intercontinent Aviation. Modeled after the Curtiss Flying Service, the company was intended to act as a sales agency and to provide flying schools, an air taxi service, and air mail and passenger service within Cuba and between Cuba, the United States, and other countries in the Caribbean.[165] The directors consisted of several prominent Cuban businessmen, along with Keys and other members of the North American Aviation Executive Committee.[166] By the end of September 1929, Intercontinent had investments of $500,000 in China Airways Federal, $500,000 in Compania Nacional de Cubana de Aviacion Curtiss, and $100,000 in Compania de Aviacion Faucett.[167] Keys had yet other initiatives underway. Negotiations with the government of Chile were continuing, while his representatives were discussing with the government of Turkey the possibility of setting up a semigovernmental air service company in that country.[168]

In less than a year, Keys had expanded his involvement in aviation around the globe, but even then he was looking beyond these projects to more ambitious participation in international air transport. Returning from an extended trip to Europe that summer, during which he had talks with airline and civil aviation officials in Belgium, England, France, and the Netherlands, Keys recommended to North American Aviation's Executive Committee that North American "lay down a program of study looking to the ultimate development of purely international routes without subsidy on the main lines of international transport throughout the world."[169] Keys viewed all aviation in Europe,

Africa, Asia, and Latin America as international aviation. None of the countries in these regions, however, with the possible exception of China, had a market big enough to support a viable aviation industry on a solely national scale; only the United States had a large-scale national aviation market. The British, French, and Dutch governments were preparing subsidized airlines from Europe to Asia, Africa, and Latin America. These airlines, Keys believed, "would lay a very heavy burden for years on the taxpayers of all the countries involved and will be unprofitable and probably second-rate lines."[170]

Keys had expressed the view to his European counterparts that if Europe ever decided to internationalize these trunk lines and put them on a commercial basis, then he and his colleagues might be interested in cooperating in the venture. Such a development, he told the Executive Committee, "will come within three or four years and when it comes we should certainly play a part in it rather than let someone else in the United States do so."[171] Characteristically, Keys said that "it seems to me that an attitude of very watchful waiting and of very careful attention to details is essential."[172] He doubted that either European law or public opinion would ever allow American control of aviation in Europe, but he suggested that "a process of slow infiltration . . . will pick up a good many opportunities that could be profitable and might result in the long run in the Americanization of civil aviation in Europe."[173]

The next two years were years of mixed fortunes for Keys's international airline ventures. In Cuba, Compañía Nacional Cubana de Aviación S.A., which Keys had organized in October 1929, was now flying a regular scheduled service providing daily flights to fourteen cities along its 740-mile route stretching from Havana in the western part of the island to Baracoa on the east end of Cuba, using eleven Curtiss Robin aircraft.[174] Keys's venture in China had a frustrating year. The original contract signed between Aviation Exploration, Inc. and the Chinese government, which created the China National Aviation Corporation to represent the government and China Airways Federal, had been negotiated through the Minster of Railroads. This created opposition from the Ministry of War and the Ministry of Communications, both of which wanted control over commercial aviation. The competition for control between different ministries within the Chinese government continued unabated, with deleterious effects on China Airways Federal.

The airline had begun service between Shanghai and Hankow in October 1929, just before the stock market crash in New York. A month later, when China Airways tried to collect its revenues due under its contract with the Chinese government for carrying the air mail, the minister of communications refused

Chinese and American personnel of the China National Aviation Corporation (C.AC) stand in front of one of the airline's Loening Air Yachts circa 1929–1930. The Loening could carry six passengers. CNAC used a Loening to inaugurate passenger service from Shanghai to Hankow on October 21, 1929. (0214 AVG/ China Collection, San Diego Air & Space Museum, San Diego, CA)

to release payments from the Post Office and criticized the terms of the contract itself. The delay in payment continued into December. With the collapse in the stock market and growing financial problems, Keys could give China Airways only limited financial assistance. He reported that it was "utterly impossible" to raise additional funds and advised the airline of the "need for extreme conservatism and extreme economy" in the airline's operations.[175] In January 1930, the Chinese State Council decided to unilaterally cancel the contract. This action produced a storm of protest from foreign business and diplomatic interests in China, which then forced the State Council to cancel the cancellation.[176] The following two years were times of "alternating hope and despair over the prospects of developing commercial aviation in China."[177] On the positive side, the Chinese government agreed to a new contract creating a new entity. It was also named the China National Aviation Corporation, incorporating China Airways Federal and the former China National Aviation Corporation, but with joint Chinese and American ownership.[178] But operational and management problems continued, and the airline's fortunes did not improve. Keys sent his own representatives to report on the situation in China and attempt to address the many problems, but, as will be seen, beset with mounting crises across his many business ventures following the Crash of 1929, he was too far removed and too preoccupied to devote his full attention to the airline in China.

6

BUILDING AN AVIATION GROUP

———————■———————

A Vision for the Business of Aviation

"INDEED IT HAS BEEN A BUSY WORLD! IN THE PAST TWELVE MONTHS I HAVE been busily engaged in acquiring the ways and means to carry out half a dozen plans that have been in my mind for years as things that ought to be done when the right time came."[1] In November 1928, Keys wrote these words to Charles Gray, editor of *The Aeroplane,* one of the leading English aviation magazines. Since the passage of the Army and Navy acts in 1926, Keys had overseen the rapid expansion of the Curtiss Aeroplane & Motor Company, with the establishment of new subsidiary and associated companies across a range of aviation activities and the start of National Air Transport, and he was working on setting up Transcontinental Air Transport. Keys had also been actively involved in raising the financing for all these ventures and underwriting their securities through C. M. Keys & Company.

By the middle of 1929, there were more than twenty companies associated with Keys and the Curtiss Aeroplane & Motor Company. They were engaged in research and engineering, manufacturing a broad range of military and commercial aircraft, selling aircraft and aircraft engines in the domestic market and abroad, air transportation, consumer credit, training and commercial flying activities, and finance and investment in all aspects of aviation.[2] The market value of the major companies that Keys was involved with as chairman, president, or managing director amounted to some $175 million.[3] As a contemporary securities report said of Keys, "In 1920, when he acquired control of the Curtiss Aeroplane & Motor Corporation, it required high imaginative powers to conceive of aviation as an organized and profitable business. Mr. Keys not only thus conceived it but possessed the business insight, the driving force, and the ability to associate all of the technical and financial assistance to bring his vision to reality."[4]

Keys had a broad vision for aviation as a business, broader than that of most of his contemporaries. He saw business opportunities in all aspects of aviation. Once the American government adopted a sound policy for military aviation, the Army and the Navy would need a variety of aircraft ranging from fast pursuits to observation and attack aircraft to large bombers and patrol planes, as well as the engines to power them. Commercial air transport would need fast mail planes and multimotored airliners to carry passengers in safety and comfort. Airplanes could provide many other services, such as air taxi and charter services, aerial photography, air express, and a host of other uses. In addition, individuals who wanted to take advantage of the speed and savings in time the airplane could provide, or who were simply entranced by the romance of flight, would want small private airplanes that were relatively inexpensive to own and operate.

There was a need for organized services ranging from flight training to the sale and distribution of aircraft to the management of airlines and the ownership and management of airports. And there would be a need for finance at all levels, from the individual looking to finance the purchase of a private plane or a manufacturing company or airline venture looking for capital to expand or begin a new service. These opportunities would exist not only in the United States, but anywhere around the world where it was practical to operate airplanes. Keys wanted to expand horizontally to manufacture a range of airplanes for the military, commercial, and private markets, and vertically into the marketing, sales and service, and distribution of aircraft—and add to this mix, vehicles for finance and investment. Keys summarized his vision to a journalist in the spring of 1929:

> The principles on which we have built and are building are simple: Let the best engineering brains design the product. Develop different personnel units for different tasks. Develop a distributor system which is also educational-through schools and through service to the business world. Create an opportunity for the airplane to demonstrate itself. In building air lines, use multiple-engined transports,

2 pilots, provide constant radio communication, to include such a complete meteorological service as will keep the pilot informed of weather conditions over a 100-mile radius—in short, eliminate the necessity of forced landings. Build a transcontinental trunk line, and let the "feeders" grow according to traffic demands. And for every part of the industry, provide for constant research, and have ready at hand means of financing a pioneering effort.[5]

The creation of this group of companies was, however, less the unfolding of a grand design and more an evolutionary process taking advantage of opportunities as they emerged, as one venture created the need or opportunity for a second. The vision was his and the logic was his, but the actual process of creation depended on many other factors—the demand for aircraft, for one thing, and the availability of men willing to invest capital in this fledgling industry, for another. Keys saw nothing particularly innovative in what he was planning. "We have been doing," he said, "step by step, what seemed necessary; meeting the problems of aviation's needs as they arise."[6] He had few illusions about the challenges he and his colleagues faced. "Too many glowing words are written and uttered about aviation, its growth, and its future," he claimed. "This romantic element had obscured somewhat the fact that *building* aviation is a difficult business task, to be done by businessmen, using the best business principles."[7]

Expanding Horizontally

Keys could not have attempted to realize his vision for the aviation business had the financial health and business prospects of the Curtiss Aeroplane & Motor Company remained uncertain. Fortunately for Keys, the Curtiss Company's financial condition was steadily improving. In the months following passage of the Army Air Corps Act in 1926, Curtiss received orders for pursuit planes and advanced training planes, with additional orders following in 1927 and 1928. Curtiss had equal success with its observation and attack airplanes for the Army Air Corps and an order for twelve large twin-engine bombers. After several years of acrimony, the Navy once again turned to Curtiss, ordering single and two-seat fighters in the two years following passage of the 1926 Navy Act. With the orders for all these airplanes came additional orders for large numbers of Curtiss D-12 engines.[8]

The financial results were gratifying. Sales had nearly doubled over the prior year, and since none of the company's government business had been awarded through the damaging competitive bidding process, there was a near-threefold increase in net profits over 1925. In the annual report for 1926, issued on March 1, 1927, Keys reported to the stockholders that the Curtiss Aeroplane & Motor Company had had its best year since the end of the war. As Keys noted in his report, the Army Air Corps and Navy Acts "changed the whole status of American companies engaged in the design and manufacture of military machines. The five-year program itself assures a fairly continuous flow of business in moderate volume for the next five years . . . and in general makes this branch of the aeroplane industry a business instead of an adventure."[9]

With a clean balance sheet, increasing profits, and a growing backlog of military business at Curtiss Aeroplane, Keys could now afford to turn his energies to other challenges. It was now time to reenter the growing commercial market. The demand for commercial aircraft had started to expand in 1926 with the transfer of the air mail service from the U.S. Post Office to private companies, and demand accelerated following Lindbergh's historic flight to Paris, which sparked an explosion of interest in private flying. During 1927, the resurgent aviation industry built twice as many commercial aircraft as it did military aircraft.[10] Curtiss had taken tentative steps toward the commercial market, but in the months following Lindbergh's flight, the Curtiss Company had little to offer the commercial market. Then an opportunity arose that provided an opening.

Toward the end of August 1927, Keys received a visit from Major William Robertson, president of the Robertson Aircraft Corporation in St. Louis. Robertson and his brother had founded Robertson Aircraft Corporation in 1919 and by 1927 had built up a steady business operating a flying service, a flying school, and the sale and distribution of aircraft, engines, and parts; Robertson Aircraft had been the Midwestern distributor for Curtiss for eight years. Robertson wanted to set up a company in St. Louis to manufacture commercial airplanes, which the Curtiss Aeroplane & Motor Company would design. The plan would be to start with production of a two- to three-passenger airplane, and then move into larger single- and multiengine airplanes for carrying mail, passengers, and express. Robertson Aircraft could offer the new venture its knowledge of commercial aviation and its sales organization in the Midwest. Robertson proposed to Keys that the new company be named the Curtiss-Robertson Airplane Manufacturing Company. The first airplane to be built in St. Louis would be the new three-seat cabin monoplane Curtiss was developing as the Curtiss Robin. Robertson believed that he and his partners could raise the capital needed from investors in St. Louis.[11]

Clement Keys stands next to one of the early prototypes of the Curtiss Robin at Garden City. The early Robin aircraft, with a Curtiss OX-5 engine, could carry a pilot and two passengers. Designed at Garden City, Curtiss transferred production to the Curtiss-Robertson Aircraft Manufacturing Company in St. Louis, ultimately selling 769 Robins. (Folder 13, Box 33, Curtiss-Wright Photos, Museum of Flight)

Keys was immediately attracted to the proposition. Robertson was offering the Curtiss Aeroplane & Motor Company a means of entering the commercial aircraft market without having to invest the company's own capital, though Curtiss would get a controlling interest in the new venture through its share of the common stock. The Curtiss Company would provide a factory superintendent, chief inspector, and an engineer to oversee production in St. Louis so that the airplane could be sold as a Curtiss product. Keys also approved of the people he was dealing with. Not surprisingly, whenever he entered a new venture Keys wanted to ensure the competence of the management. Robertson had a proven track record with Curtiss, having earned a reputation as one of the best distributors the Curtiss Company had ever had, while his partners, Harold Bixby and Harry Knight, were solid businessmen with good connections in their community.[12]

Keys wrote to Robertson on September 9, 1927, saying that the entire proposition was "quite attractive to all my people."[13] Three weeks later, having clarified the relationship between Curtiss and the new St. Louis venture and worked out the new firm's financial structure, Keys wired Harold Bixby that the Curtiss Aeroplane board of directors had approved the proposal in principle. William Robertson apparently arranged for Lindbergh, who had worked as a mail pilot for Robertson before the Paris flight, to meet with Keys in New York—apparently the first time the two would meet—and spend several days with the Curtiss Engineering staff at Garden City going over the plans for the new monoplane.[14] With plans for the airplane and the factory moving ahead, the formation of the Curtiss-Robertson Airplane Manufacturing Corporation was announced in November, and the new company was formally established early in 1928. William Robertson was appointed president of Curtiss-Robertson, with Keys becoming a vice president and director.[15]

Keys and Robertson had received inquiries about the availability of stock in the new venture from investors in Chicago and New York. It soon became apparent that there were benefits to broadening both the underwriting group and the investor base. Other investment banks in St. Louis wanted to be included. Keys thought it would be a good idea to have a large number of stockholders and had no objections to bringing in other investment firms to help underwrite the stock. Two new banking firms were brought into the syndicate, and Keys agreed to underwrite a portion of the preferred through C. M. Keys & Company. One of the new firms was the investment banking firm of James C. Willson of Lexington, Kentucky, who would become a partner with Keys in underwriting many aviation securities over the next two years. In December 1927, Willson proposed that a syndicate consisting of his firm, along with Stifel, Nicolaus & Company and Knight, Dysart & Gamble, both in St. Louis, and C. M. Keys & Company agree to underwrite $350,000 in preferred stock and 25 percent of the common stock immediately, and commit to underwriting an additional $150,000 in preferred stock and 25 percent of an additional 9000 shares of common stock by May 1, 1928. There was brisk demand for the stock of the Curtiss-Robertson Airplane Manufacturing Corporation; Keys managed to sell his share of the underwriting before the end of January 1928.[16]

Keys wanted to expand further into manufacturing commercial aircraft at both ends of the commercial aircraft spectrum to have a complete range of products for the commercial market. To acquire a light aircraft, that fall Keys had the Curtiss Aeroplane & Motor Company acquire 51 percent of the

Clement Keys at his office on Broadway sometime in the
mid-1920s. Keys regularly corresponded to his colleagues,
investors, and government officials. To keep up to date on
the industry, he maintained clipping services relating to his
aviation interests and companies. (Courtesy C. Roy Keys Jr.)

stock of Reid Aircraft Company in Montreal, Canada, which built a small two-
seat biplane named the Reid Rambler, which was similar to the de Havilland
Gypsy Moth. The new company became the Curtiss-Reid Aircraft Company,
Ltd. As a Canadian company, the Curtiss-Reid Aircraft Company had access
to other British Commonwealth countries that imposed high tariff barriers on
American-made aircraft.[17]

Keys's first venture into the area of large commercial aircraft was an investment in the Sikorsky Aircraft Corporation. Founded by the Russian émigré
Igor Sikorsky, the Sikorsky Aircraft Corporation had recently introduced the

S-38, a twin-engine seaplane that could carry up to ten passengers and was working on a design for a larger four-engine, twenty-four passenger seaplane. During October 1928, Sikorsky's company was restructured, and its finances were reorganized to accommodate a planned increase in production of the S-38, allowing Keys to acquire a substantial minority interest in the company. Then, at the end of the year, Keys formed a joint venture with the Italian airplane designer Gianni Caproni, founder of the Societa Aeroplani Caproni of Milan, to acquire an organization specializing in the manufacture of large aircraft. The Caproni Company had built large bombers for the Allies during World War I and in the postwar period had become known for its multiengine bomber and transport aircraft. Keys had met Caproni while attending the International Civil Aeronautics Conference held in Washington, DC between December 12 and 14, 1928. Conferences between Caproni, Keys, Curtiss engineers, and bankers soon followed, leading to an agreement on a joint venture to build large airplanes in the United States. Keys announced the formation of the Curtiss-Caproni Corporation on January 16, 1929.[18]

Expanding Vertically

It was evident to Keys that Curtiss Aeroplane's expansion into the commercial aircraft market and its growing success in selling military aircraft required a parallel expansion in the Curtiss sales and distribution organization, both at home and abroad. As he noted in the Curtiss Aeroplane & Motor Company Annual Report for 1926, "most nations have now reached a point where war surplus machines and motors are no longer good enough, and are becoming substantial buyers of new modern machines. It is, therefore, a logical time to push forward the development of these markets and plans are well underway for this development."[19] In addition, with a rapidly growing interest in private flying, there was a need to provide a range of services to the private flyer and the private airplane owner beyond the sale of airplanes. Fortunately, Keys had an existing framework to build on in the Curtiss Aeroplane Export Corporation and the Curtiss Flying Service.

Incorporated on July 25, 1921, the Curtiss Aeroplane Export Corporation was the first dedicated airplane export company in America, with a long-term contract to act as the exclusive sales agent in Latin America for all Curtiss products.[20] During 1927, Keys learned that Argentina and Brazil were in the market for new military planes and that several countries were arranging financing for new equipment. There were other potential opportunities outside of Latin America. During the fall of 1927, Keys learned that several European countries, including Poland, Belgium, and Switzerland, had decided to move away from France as their principal supplier of aircraft and to consider American products. Also possible were sales to the Dutch East Indies, Siam, and Japan, all of which had previously bought airplanes from France or England.[21]

Capitalizing on these geographically diverse opportunities would require a stronger organization that could mount an aggressive sales campaign. What Keys had in mind was an arrangement whereby the new export company would become the exclusive selling agent for Curtiss products not just in Latin America, but in Asia and Europe as well. During January 1928 Keys arranged a recapitalization of the company. The Corporation sold 5000 shares of 6% preferred stock and issued 100,000 shares of new common stock, with 50,000 shares going to the Curtiss Aeroplane & Motor Corporation and 15,000 shares sold to the market. The major portion of the new preferred issue, placed through the investment banking firm of James C. Willson & Co., went to six major banks and investment houses that were involved in financing export trade or underwriting the securities of foreign governments and that were supportive of the Curtiss Aeroplane & Motor Company. In a letter to the directors of the Curtiss Aeroplane & Motor Company, Keys said that he was glad to have both classes of investors in the Curtiss Aeroplane Export Corporation, believing "in the first instance that the international banking houses will assist us very greatly in the carrying on of the business of the Export Company, both with credit facilities and with contacts; and believing further that the widespread distribution of the stock of any company makes the soundest foundation of any business."[22]

Next Keys turned to reorganizing the Curtiss Flying Service, which had grown out of the older Curtiss Exhibition Company that Glenn Curtiss had organized in 1910. Based at Mineola Field on Long Island, the Curtiss Flying Service had a small but profitable business providing air charters, air taxi services, aerial photography, and flying instruction around the New York area, with Charles S. "Casey" Jones, a noted flyer, serving as president. Preoccupied as he had been ensuring the future of the Curtiss Aeroplane & Motor Company, Keys had paid little attention to the Curtiss Flying Service, which had operated quite happily on its own.[23] But now new opportunities were emerging, and Keys wanted "some division of our business devoted to *use* of planes, as distinct from manufacturing or airline operation."[24] He wanted to transform the Curtiss Flying Service from a local New York area company to a nationwide operation.

Keys wanted to establish complementary operations across the country that would greatly expand the commercial market for the Curtiss Aeroplane & Motor Company. The Curtiss Flying Service would set up flying schools in some twenty-five of the most important cities across the country to provide primary and advanced flying training, taking advantage of the phenomenal increase in the demand for flying training after Lindbergh's flight to Paris. At the same time, the Curtiss Flying Service would set up a nationwide network of exclusive sales agencies to sell the complete line of Curtiss Aeroplane & Motor Company products. Keys planned that the Curtiss Aeroplane & Motor Company's engineering development department would design new commercial airplanes, the Curtiss-Robertson Company would build them, and the Curtiss Flying Service would train individuals keen to fly at a Curtiss Flying Service school. Once trained, these individuals could buy a private plane through the same office and have their plane maintained and serviced at the same airfield facility. The primary and advanced training schools would also provide a pipeline of qualified pilots and mechanics for the airline ventures associated with the group.[25]

On September 6, 1928, Keys announced the formation of a new corporation, the Curtiss Flying Service, Inc., with Casey Jones as president and Keys as chairman, and a distinguished Board of Directors. Keys arranged to sell 675,000 shares of common stock in the new company through a private placement raising $7.5 million in capital. Two days after the sale of its stock the Curtiss Flying Service placed orders with the Curtiss Aeroplane & Motor Company for 150 new training aircraft based on the Curtiss N2C-1 Fledgling designed for the U.S. Navy. The selection of flying fields had already been under study for some time. At the end of November, Keys announced that the Curtiss Flying Service was scheduled to begin operations at eleven airfields in the East and the Midwest in the spring of 1929 and was negotiating to establish operations in seven more cities across the country by summer.[26]

In the structure he was creating for the commercial aviation market, Keys now had the capacity to manufacture airplanes, a sales and distribution system for these products, and a training organization to create demand. Lacking were modern airports for the private flyer and the newly emerging airlines with up-to-date facilities for maintenance and operation by day and by night. The 1926 Air Commerce Act assigned responsibility for constructing airports to local municipalities.[27] Many municipalities began planning new airports, but such was the demand for airport facilities from cities across the country and from the airlines that there was apparently more than enough room for commercial ventures to enter the field. During February 1928, Keys formed National Airways Terminals, Inc. to build a chain of airports up and down the East Coast, with the first to be built in Bridgeport, Connecticut and Portland, Maine, with a total planned investment of $5 million; Keys became one of the new company's directors.[28]

National Airways Terminals was only a precursor to a much more ambitious airport venture that Keys organized a year later. On May 6, 1929, Keys formed the Curtiss Airports Corporation with the stated purpose of acquiring land to build, own, and operate a group of airports strategically located along the main lines of air transportation across the country.[29] The Corporation sought to purchase or lease land or existing airports in the New York area, Philadelphia, Baltimore, Pittsburgh, Chicago, Cleveland, Louisville, St. Louis, San Francisco, and Los Angeles. Revenues were expected to come from rental of the airport facilities, charges for flying operations, agency concessions for restaurants and hotels, and miscellaneous charges for the accommodation of spectators and parking spaces. Keys was confident that these sources would provide substantial earnings for the new company. Walter S. Marvin, a partner in the investment banking firm of Hemphill, Noyes & Company, was appointed president of the company with Keys as chairman. On May 8 the Curtiss Airports Corporation issued 2.5 million shares of common stock, raising close to $30 million for investment in land and facilities.[30]

Providing Financing for Aviation Ventures

Keys was keen to ensure that capital would be available to finance not only his Curtiss ventures but other aviation endeavors as well. He differed from many of his contemporaries in the aviation industry in his pursuit of vehicles to finance his ventures in the industry. Keys understood clearly that commercial aviation at its present stage of development was a purely speculative venture, with substantial risk. As the *Wall Street Journal* commented in an editorial at the time, "it is in the nature of things that securities of all corporations connected with the aircraft industry and air transport service are still highly speculative. They could not be otherwise, with aircraft design still in its development stages and with commercial utilization of the air little more than begun."[31] Keys knew from experience that new and speculative industries had neither the sustained earnings power nor the seasoned performance that would attract capital from the usual sources. Describing these potential sources some years later, he noted that neither "savings banks, trusts, commercial banks, life insurance companies or even fire and casualty companies, can find anything in the trade that suits their portfolios and their policies. The private investment market is always unreliable and uncertain and is never there at all in times of crisis."[32]

What Keys wanted was a source of financing that "would be strong enough and courageous enough to take care of any critical needs for money that might arise."[33] His idea was to form investment trusts that would specialize in aviation securities. The investment trust was "an organization for the collective investment of the funds of numerous individuals in numerous securities."[34] An individual would purchase a share in the investment trust, which would then invest the collected funds of many individual shareholders in a basket of securities representing a broad range of industries or in a specialized sector. The investment trust would provide, theoretically, the analysis and selection of profitable investment opportunities for its investors. Unlike a holding company, the standard investment trust did not exercise management control over the companies the trust invested in. During the 1920s, the investment trust became a popular investment vehicle primarily because of the phenomenal increase in American investors, with the number of organized investment trusts growing from 40 in 1921 to 770 in 1929 with total assets exceeding $7 billion.[35]

Keys wanted to bring into his aviation group "sound and promising enterprises connected with every phase of the aviation industry."[36] The investment trust would be the financial vehicle to achieve this goal. As he conceived it, the role of an investment trust in the group would be to provide a "unifying element." The functions of this entity, however, would be broader than the standard investment trust, specifically "to investigate, organize, and finance aviation enterprises, underwrite and market their securities, and acquire and hold an interest in them."[37]

The National Aviation Corporation, established in June 1928, was the first specialized investment trust that Keys organized. He wanted to raise a substantial fund of money to invest in aviation, while at the same time gathering a group of men from the aviation and banking industries to serve as directors. Active management of the company was in the hands of the executive committee, which combined men with extensive experience in manufacturing and men with backgrounds in investment banking and finance. The executive committee consisted of Keys as chairman, James C. Willson, of the investment banking firm of James C. Willson & Co., as president, Sherman Fairchild, president of Fairchild Aviation Corporation, Richard Hoyt, a senior partner in the investment banking firm of Hayden, Stone & Co. and chairman of Wright Aeronautical Corporation, Leonard Kennedy, vice president of Curtiss Aeroplane & Motor Company, Chester Cuthell, Keys's lawyer, and Grayson M.-P. Murphy, a well-known director of companies, private banker, and senior partner in the banking firm G.M.-P. Murphy & Co.

The stated purpose of the company was straightforward: "to place part of its resources at all times in the securities of established companies in the various lines of aviation activity, both manufacturing and transportation" and "to [promote] the development of such new aviation projects as, in the opinion of the Board of Directors, supplemented by the report of the company's technical staff, merit such assistance."[38] As Keys explained in a letter to William E. Boeing, president of the Boeing Aircraft Company, "I have in mind that this company may be useful not only through stirring up and maintaining the interest of a large number of wealthy and sensible people, but also possibly as a future underwriter for the financing of the principal companies, yours included."[39] Keys added that if Boeing should choose to split its operations into different units, or need to increase its capital, "please do not hesitate to consider that this company might be very glad to be a small minority holder in your enterprise, if you wanted it to be. This sort of thing is really the main reason why I have gone into this enterprise and will help to guide its affairs."[40]

The National Aviation Corporation raised $3 million for investment. The company participated in the underwriting for the Curtiss Flying Service in September, and it provided financing for the Sikorsky Aviation Corporation and the Curtiss-Caproni Corporation soon after the latter's formation. The company also participated in the formation of three regional investment trust companies—the Aviation Securities Corporation in Chicago, the Aviation Corporation of California, and the Aviation Securities Corporation of New England—as part of a plan to establish financial units across the country. Collectively, the three regional investment trusts raised $8.5 million to invest in the securities of aviation companies, both manufacturing and transportation, in their respective regions. As with the National Aviation Corporation, the boards of directors of each of the three trusts consisted of prominent local businessmen and bankers who were identified with the aviation industry and representatives from local aviation companies. Keys was named a director of the Aviation Securities Corporation and the Aviation Corporation of California.[41]

North American Aviation: A Different Kind of Animal

Keys knew that while he was building an aviation group, others were doing the same thing. New companies and new airlines were springing up seemingly right and left, with new combinations forming or coalescing with strong financial support from various banking groups. In the space of a year, the aviation industry had undergone a transformation. As one commentator put it, "at the beginning of the year the aviation industry was owned and controlled by a comparatively small number of men who were vitally interested, and often the sole owners of the concerns which they managed. At the end of the year we see the industry taken over by the public and under the control of various banking groups. Broadly speaking there is an entirely new lineup in which high finance plays the leading role. . . . 1928 has seen the aeronautical business change from what might be called a private family affair into a business which is owned by the public and controlled by bankers."[42] The market expected an accelerated movement toward merger and combination, as had happened in other industries.

The implications of this transformation were not lost on Keys. As he explained a few years later, "we did not control any of these companies either individually or as a group. The new groups in aviation, backed by powerful banking interests, were obviously dangerous to us, not only in that, in their eager quest for companies to control, they could at any time buy the control of our companies, but also because in pushing the trade forward we were in constant competition with them in the formation of new companies and in the possible acquisition of existing companies."[43] Keys needed an ally, and a rich one, who could support the companies he was involved with through share ownership and financing in time of need. And what better way to find such an ally than to create one that would be under his complete control. What emerged from this concern was North American Aviation, Inc. , which Keys formed for two strategic reasons: "first, to protect, so far as possible, the things we had built up; and, second, to give us additional capital to keep pace with the development of aviation in the country."[44]

When Keys announced the formation of North American Aviation on December 6, 1928, he described it as a holding company that "plans to take part in furthering the expansion of aviation, especially in the commercial field."[45] The new company was authorized to buy, sell, or trade the securities of aviation companies both domestic and foreign, but its primary purpose was "to make more or less permanent investments in aviation companies."[46] The general policy for investments was "to own substantial blocks of stocks of the more important aviation companies, in both manufacture and transport, when these stocks could be bought at prices that would offer a reasonable chance for income return and for profits; to own outright, either in whole or in part, the stocks of key companies, whose products are extensively used in aviation and which promise ultimate development and expanding markets; to own small amounts of a diversified list of stocks representing aviation industries, preferably those manufacturing a diversified line of products; and to assist in financing or developing new companies, either by investing in their stocks or by assisting in underwriting them."[47]

Keys became president of the new company, but he wanted to create broad geographic support for North American Aviation in the business community as well as have the ability to tap into a wide network of financiers. The Board of Directors of the new company consisted of forty men from outside the Curtiss Aeroplane & Motor Company, more than half of whom were bankers or capitalists drawn from New York, Boston, Philadelphia, Chicago, Los Angeles, and San Francisco. Almost all of these men were involved with aviation in some capacity, and a number were already directors of other Curtiss companies or National Air Transport. A few weeks after the incorporation of North American, Keys added more prominent businessmen as directors including R. D. Chapin, chairman of the Hudson Motor Car Company, R. R. McCormick, president of the Chicago Tribune, William B. Mayo, chief engineer of the Ford Motor Company, Frank Phillips, president of the Phillips Petroleum Company, and James A. Talbot, president of the Richfield Oil Company, among others.[48] The executive committee of North American, however, was composed of some of Keys's closest associates: J. Cheever Cowdin of Blair & Company and James C. Willson, his principal bankers; Chester Cuthell, his lawyer; J. A. B. Smith, Curtiss Aeroplane secretary and treasurer; and Leonard Kennedy, Curtiss Aeroplane vice president.[49]

North American entered into an agreement with Blair & Company to sell two million shares of common stock to Blair and a syndicate of banks. The underwriting and sale of the stock in the Curb Market was completed in less than two weeks, raising $25 million for North American Aviation, the largest amount of money raised to that time for investment in the aviation industry.

Even before the sale of North American Aviation shares had been completed, Keys was actively investing according to the objectives he had set for the company, placing orders for shares of Curtiss Aeroplane & Motor Company, Curtiss Flying Service, Transcontinental Air Transport, and Douglas Aircraft Company for a total of $4,493,537.[50] Over the next several months North

American rapidly built up a portfolio of investments and core holdings. In late December, North American purchased all the outstanding capital stock of the Sperry Gyroscope Company, makers of the Sperry gyroscope and other aeronautical instruments, shares of the Travel Air Manufacturing Company, and shares in the Niles-Bement-Pond Company, one of the world's largest machine toolmakers.[51] In February came purchases of shares of the Curtiss-Caproni Corporation and the Curtiss-Reid Aircraft Company, and an initial investment in the Aviation Credit Corporation, a finance company venture set up to provide financing for the purchase of small private airplanes. In March the executive committee approved additional investments in National Air Transport, United Aircraft & Transport Corporation, the Irving Parachute Company, and its first foreign investment, debentures and common stock of the Fairey Aviation Company Ltd. By the middle of March, North American had investments of $7,271,891 with a market value of $8,303,316, in addition to Sperry Gyroscope, which was worth $3,766,955; the company's remaining funds were on call with Blair & Company and C. M. Keys & Company.[52]

The Stock Market Takes to Aviation

Keys could not have expanded his aviation group without financing, and he proved adept at raising capital in the stock market. Keys was consistently successful in obtaining strong financial backing for his ventures. He had the reputation, as well as the persuasiveness, to attract prominent businessmen and bankers as directors of the companies he formed, which enhanced the credibility of these companies in the stock market. Keys established strong working relationships with a small core of investment bankers from several of the more aggressive medium and smaller-size private banks and securities affiliates of national banks. These men served as his advisors, sat on the boards of his companies, and organized the syndicates selling the stock of the Curtiss companies and his other ventures. But most of all, Keys was exceptionally fortunate in his timing.

After 1925, the stock market experienced a sea change in its appetite for aviation stocks. Prior to the Air Mail Act in 1925, the Army and Navy Acts of 1926, and the Air Commerce Act, the nascent aviation industry evoked next to no interest on Wall Street. Lindbergh's flight to Paris in 1927, however, had an immediate positive impact on the public's perception of aviation and on aviation stock prices. As the *Aircraft Yearbook for 1928* said, "public recognition and confidence seemed to come in an instant."[53] Aviation stocks rose sharply in the months following Lindbergh's flight in expectation of an expansion in the American aviation industry. Curtiss Aeroplane's common stock climbed from $22 a share in May 1927 to $46 a share by early September, and Keys finally obtained a listing for the Curtiss Aeroplane & Motor Company on the New York Stock Exchange in October, a sign of the company's new standing.[54] Articles began appearing in the nation's leading business magazines on the promising outlook for commercial aviation.[55]

More critically for Keys and others in the industry, investment houses and banks across the country began to get involved in aviation projects. This had a ripple effect on both smaller financial institutions and individual investors. As announcements appeared concerning more stock issues and syndicates forming around them, more investment houses, brokerage firms, and banks joined in syndications and became purchasers of aviation stocks. This "fortified the hitherto rather timid interest of private investors."[56] The result was the so-called Lindbergh boom, coinciding with the bull markets of 1928 and 1929, which saw a seemingly insatiable demand for aviation stocks, "sudden, entirely unexpected and more or less sentimental" among a broad range of investors.[57] During 1928, the number of new securities issued for aviation companies rose from one new issue in March to seven in October and twelve in December, while the amounts raised through these flotations rose from $11.4 million between January and August, to $76.9 million in the last four months of the year.[58] From January 1928 to March 1929, financing for the aviation industry totaled $205,384,533, a remarkable amount of money for an industry that ranked 144th out of all other manufacturing industries in the United States.[59]

Keys took full advantage of this turnaround in market sentiment to raise financing for his aviation ventures. However, he wanted to finance his ventures in aviation using only common stock. From his many years of analyzing the railroad industry, he knew at first hand that excessive debt had broken the backs of far too many railroads as the interest and principal payments on their bonds and mortgages overwhelmed operating profits during the inevitable lean years. The history of railroads, railroad bankruptcies, and defaults on railroad bond issues was common knowledge among many of the businessmen of Keys's generation. Weak capitalization with too much debt and too little equity had been among the more flagrant abuses of the early railroads. It was far better, and more prudent, for a company to avoid debt entirely as a means of financing and to be free of the burden of fixed charges, particularly a company in a new industry with an unpredictable and uncertain future. Keys and other pioneers in air transport received numerous warnings on this score and were urged to finance their companies with equity, not debt, to avoid the errors of the early railroads.[60]

Keys also knew that defaults on railroad bonds had ruined many an investor, large and small. He did not want investors to have the same experience with the aviation industry. Far too many of the early railroads had lacked the qualities required for an investment as Keys defined them: principally the assurance of a reasonable return with minimal risk of loss. In his experience, there was a vast difference between a well-established industry with a demonstrated record of sound financial performance and an infant industry in the pioneering stage of its development. He wanted to make clear that there was risk involved in these ventures, and he did his best to secure the backing of prominent men in the business and financial communities. The use of common equity only was, he said, "admirably suited to the needs of a new and rapidly growing industry. There are no bonds outstanding against any of our companies.... We are financing aviation pioneering wholly on a common stock basis. In no prospectus of ours does the word "investment" occur. This is no business for widows and orphans; it is a businessman's proposition.... We have a greater degree of freedom, of maneuver ability, because of our one-class capital. The business is not underwritten by "investors"; it is underwritten by businessmen, ready ... to take a business risk."[61] These certainly were Keys's intentions, but he was not always able to control how the stocks in his companies were described or who they were ultimately sold to.

From June 1928 to May 1929, Keys helped arrange $82 million in financing through the stock market for eight of the companies he had organized or was involved with.[62] The amount raised represented about a third of all stock underwritings for aviation companies during this period. All of these offerings proved successful, a reflection of the demand for aviation stocks in the market and the strong banking syndicates that led the underwritings. As both James J. Hill and Edward R. Harriman had done, Keys built strong working relationships with several investment bankers, principally Grayson Murphy, senior partner of the investment banking firm of G.M.-P. Murphy & Company; James C. Willson, head of the Lexington, Kentucky investment banking company James C. Willson & Company; J. Cheever Cowdin, vice president at the investment banking firm of Blair & Company; and Harry Knight of Knight, Dysart & Gamble. All these men, together with their banking firms, were actively involved in underwriting aviation securities during 1928–1929; their partnership with Keys was successful and lucrative.[63]

In working with Cowdin, Knight, Murphy, and Willson, Keys took on the role of promoter for his ventures. He would often craft the prospectus for a new issue for one of his companies, describing the business of the company in straightforward terms, never overpromising but putting a positive tone to the company's plans with a few key words or phrases. These letters, often quoted verbatim, would go into the prospectuses the banking syndicate would put out before issue of the new stock.[64] Keys had the ability to attract and persuade men of standing in the business community to participate as directors in his companies. Keys used his own standing as a reputable and knowledgeable financier and president of one of the largest aviation companies in America. He built on the widespread interest in aviation and, no doubt, the desire of many individuals to get in on the ground floor of an industry that seemed destined for large profits in the future.[65] For his corporate boards Keys tried to get a mix of men from the aviation industry, representatives of the investment banking community involved in the syndications for his companies, and prominent business figures involved in banking, manufacturing, or transportation. He also endeavored to get a geographic representation on his boards, including men from the Midwest and the West Coast when he could.

Managing the Curtiss-Keys Group

By the spring of 1929, Clement Keys was serving as chairman of the board, chairman of the executive committee, president, or director of eighteen aviation companies covering engineering and manufacturing, sales and service, transportation, and finance. Collectively, the companies Keys was involved with had come to be called the Curtiss-Keys group, although there was no formal group structure. Managing these diverse companies was a challenge, but at a time when many large corporations were beginning to experiment with a multidivisional structure as a means of coordinating both product and geographic diversification, Keys chose to maintain all his companies as separate entities.[66] Keys and his associates did not own these companies outright, but controlled them through a series of interlocking directorates. He did not want to create a holding company structure, nor did he want to merge these various companies into a single entity. His reasons were severalfold.

In the first place, his group of companies had grown up through a process of evolution that was still continuing. Keys's situation was quite different from the normal business process whereby holding companies and mergers were put together by purchasing existing concerns and combining them into a larger entity in order to eliminate duplication and create economies of scale. Only two of the eighteen companies in his group were more than two years old; they were still in an "infant," startup stage, in an industry that was itself in its infancy. No one could predict with any certainty what the aviation and air transport industries would look like even a few years later. Keys did not believe that molding together companies involved in different activities at such an early stage in their development would necessarily lead to better management. "Merger," he believed, "tends to submerge the individual unit, while this partial method, consisting of companies connected loosely, if at all, seems to enhance rather than lessen the importance of the individual units."[67] He wanted to allow each of the companies in his group the freedom to use their initiative, to change their financial structure, or to expand or contract as the needs of the business would dictate over time.[68]

Nor did Keys believe that the Curtiss Aeroplane & Motor Company itself could form the basis of an aviation holding company. When the possibilities for commercial expansion of aviation emerged after 1926, Curtiss Aeroplane was still a comparatively small company specializing in the production of military aircraft. The idea of expanding this small organization to take on the administration of not only the highly technical manufacture of military and commercial airplanes, but also a nationwide operating company, airlines, and airports was, Keys thought, "appalling."[69] At the time, Keys later wrote, he and the other Curtiss directors agreed that "it would probably be disastrous to try to make this little organization, settled in its ways and definite in its tasks, undertake these other great jobs."[70] It was also clear to Keys that financing the expansion into commercial aviation in any number of areas would require a far greater amount of capital and financing activity than the Curtiss Aeroplane & Motor Company could deal with. It would be better, he and the other directors decided, to have Curtiss Aeroplane concentrate on what it could do best, which was to build airplanes and engines. And even in this area Keys was willing to bring in other parties who were willing to provide capital for expansion, as he had done with the Curtiss-Robertson Aircraft Company. There was another reason that Keys chose to expand through setting up separate companies. As he explained to a reporter from the *Wall Street Journal* in the spring of 1929, he wanted to bring a wide cross-section of businessmen into aviation, which he thought ultimately would be critical to the industry's success. As he explained,

> If we had organized one big company and done all our financing
> in one big stock, we might have had twenty or thirty or a few more

leaders of the country invested in that concern as directors and active stockholders. In the method that we have followed there are more than 200 men, nearly all well known in aviation, in banking, in transportation or in manufacturing, who are interested in one or more of our companies. Three years ago, when the foundations were laid for expansion, there was more need for public education about aviation than there was for anything else. Particularly was this need evident among businessmen, because the interest and the active backing and the work of businessmen are essential to any industry that is going to amount to anything and the best way to accomplish such a result appears to me to be the method we have followed.

Of course, the widespread interest of the great group of businessmen in the affairs of our companies has broadened and deepened the foundations upon which all these companies rest.... We do not claim to have invented or discovered anything new in corporate administration, and we do not recommend this group system of finance as a substitute for the well-known conventional merger, consolidation, or holding company. Probably, in the long run, when we have time to get around to it, partial consolidation will find its place in our scheme of things, manufacturing companies grouping closely together, the transport companies possibly merging entirely, and the finance companies becoming part of a larger whole. To mass all three kinds of companies together would not seem to me to tend toward good management in anyone of them, but even on that point there is room for difference of opinion. In the meantime, we are not even thinking of such things, being too busy in the job of carrying on.[71]

Getting on with the job was putting increasing demands on Keys's time and energy. He oversaw the activities of all his companies from the offices of C. M. Keys & Co. on 39 Broadway where he had moved in September 1928, much as Edward Harriman had done from the offices of E. H. Harriman & Company. Taking up most of the far wall of his office was a map of the United States, through which Keys could visualize airline routes across the country. From this office on Broadway, Keys conducted his corporate meetings, gave interviews, consulted with his counsel, held meetings with James C. Willson, J. Cheever Cowdin, and other financiers, and dealt with the many issues and problems that came across his desk. He was a prolific correspondent, keeping in touch with the men doing the day-to-day running of the companies

A drawing of Clement Keys that appeared in the April 1929
issue of *Forbes* profiling Keys as one of the "Big Three"
in the aviation industry. (Courtesy Donald Furin)

through a constant stream of letters and frequent phone calls. Despite the growing complexity of his group of companies, he did not build a corporate staff, preferring instead to work through a small group of key individuals from the Curtiss Aeroplane & Motor Company whom he had known and worked with for years. Keys relied heavily on J. A. B. Smith, who had started as Treasurer of the Curtiss Aeroplane & Motor Company. Keys appointed Smith as treasurer or director of ten other companies in the group. Similarly, Frank

Russell, Casey Jones, and his brother, C. Roy Keys, took positions of responsibility at other companies.

Keys believed in delegation, getting the most capable managers he could find to run operations on a daily basis, agreeing on a strategy and, for the most part, letting them get on with it much as James J. Hill had done with his railroad empire. He saw his own role as one of contributing ideas, suggestions, and questions. With his attention to detail and his ability to digest prodigious amounts of information, he requested and received regular reports on the operating and financial details of the companies he was involved with. This enabled him to keep track of progress on the development of engines and airplanes, the status of contracts with the Army and Navy, and monthly performance figures for the transportation companies. He appears to have been a voracious reader, keeping in touch with the aviation industry through intensive reading of aviation and financial journals and using a clipping service covering his areas of interests. He made regular visits to Washington to meet with contacts in the military and other branches of government, passing the information he gained on to his colleagues.[72] Once TAT was underway, he corresponded regularly with Paul Henderson and Charles Lindbergh, setting up a system of regular reports from Henderson and Lindbergh's Technical Committee.

But invariably, even though he was removed from the day-to-day management of the companies in his group, questions and decisions arose that only he could opine on, prompting Chester Cuthell's admonishing letter to Keys on his trip west in the summer of 1928. Keys recognized the problem. That fall he tried to address the issue of operating two TAT offices with Paul Henderson, and wrote to J. A. B. Smith about his need for more support. "I feel the necessity," he wrote Smith, "of building up an organization here to take care of my work, which is much too heavy for me. The work itself is not so heavy, but the time taken up in interviews makes it impossible for me to be efficient in any part of my work and I am going to be obliged to make an organization here."[73] Nonetheless, it does not appear that Keys made much progress in building up a corporate staff to assist him in his work. Smith, for example, continued to maintain his office at Curtiss's Garden City factory, although he was effectively acting as a group treasurer for the entire Curtiss-Keys group of companies.

Keys maintained an active business life apart from his many aviation companies, which imposed even more demands on his time and attention. The business of C. M. Keys & Co. grew steadily in the bull market of the late 1920s. The firm continued to deal in railroad, utility, and municipal bonds and to provide investment advice to clients large and small. Keys's involvement with his aviation companies brought C. M. Keys & Co. much new business. As Keys became involved with other aviation companies, he arranged for them to engage C. M. Keys & Co. as their fiscal agent. By 1929, his firm was acting as fiscal agent for the Curtiss Aeroplane & Motor Co., Curtiss Flying Service, Curtiss Airports Corporation, National Air Transport, Transcontinental Air Transport, and North American Aviation. In June 1929, these companies had placed $14.5 million on call with C. M. Keys & Co. , with the major portion, amounting to $14,171,000, coming from North American Aviation and its subsidiaries.[74]

Keys's involvement with the financing of the many aviation companies he helped organize brought a steady stream of stock sales and trading to C. M. Keys & Co. Keys's firm never participated in the syndicates underwriting the stocks in these companies for public sale. He did, however, arrange the sale of large blocks of shares to business associates, his regular investor clients, and friends who wanted to participate in this booming industry. He also regularly bought shares through his company for his own account and that of his wife Indiola. These transactions were profitable for C. M. Keys & Co. and for Keys personally. As an example, C. M. Keys & Co. earned a profit of $161,000 for participating in the selling group for the stock of the Curtiss Flying Service and $184,000 for participation in the sale of stock and options for the Douglas Aircraft Company.[75] Between 1924 and 1933, Keys and his wife invested $9,037,642 in aviation stocks and gained a profit of $2,161,000.[76] These investments were highly speculative; Keys lost as well as made money on the stocks he purchased, though he was fortunate to be able to ride the bull market that created the demand for aviation stocks. He was generous with others as well, ensuring that the key employees among his companies and close business associates received allocations of shares.

As Keys rose in prominence in the business and Wall Street communities, he was asked to serve as a director on the boards of other companies, again adding more demands on his time. He had served as a director of the Willys-Overland Company and the B. F. Goodrich Company from time to time, and he became a director of the Skinner Organ Company and the National Cash Register Company. In May 1928 Keys joined Howard Coffin as a director of the Montauk Beach Development Company, organized by Carl Fisher, the developer of Miami Beach, to build an elite summer resort in the Long Island town of Montauk. Keys immediately became involved in refinancing the company's outstanding bonds through C. M. Keys & Co., working with Richard

Hoyt, also a director, and his firm Hayden, Stone & Co. A few months later Keys joined Hoyt again when in January 1929 he was asked to become a director of the Commercial National Bank & Trust Company, a new commercial bank set up at 60 Wall Street, joining a prestigious group of directors from the business and financial communities that included Hoyt.[77]

The Harriman of Aviation

In 1925, Colonel Harold Hartney, former commander of the 1st Pursuit Group in World War I and an aeronautical advisor, wrote an article for the magazine *The Independent* titled "A Chance for a Harriman." Hartney stated that just as the railroads had prospered through the efforts of the Hills and Harrimans and others, air transportation needed the same. "What is needed is another Harriman," he wrote, "a pioneer, organizer, financier, transportation expert! . . . The Harriman of commercial aviation will make a fresh start, put aside a great many delusions about aircraft, see clearly that aviation has, in point of great speed and ability to operate in any direction, great advantages in supplementing railway and motor transportation, and will set to work . . . another pioneer who can get the picture whole and organize and win through with a project that must in the end mean as much to the United States as the development of railway and motor transportation."[78]

Keys had done this. By the spring of 1929, he had come to be recognized as one of the leading figures in the growing aviation industry. In an article in *Forbes* magazine in April 1929 entitled "Why Aviation's Future Is Strictly Business," the journalist Earl Reeves profiled Keys. Reeves began his profile with a comparison to the great men of the railroad era:

> The Harrimans, Hills, and Goulds of the aviation industry are now beginning to be discernable on closer analysis. In the public mind, Wright, Curtiss, and Lindbergh are the names which have stood for aeronautical achievement. But after the pioneer comes always the builder. Of the "empire builders of the air" three now stand forth head and shoulders above all contemporaries:
>
> They are:
>
> C. M. Keys, ex-newspaperman
> Richard F. Hoyt, a banker
> William Boeing, a lumber baron of the far Northwest[79]

Two months later, Howard Mingos wrote an article for the *New York Herald Tribune* making the connection explicit; he titled his article "The Harriman of Aviation." "Just as E. H. Harriman, the financier of steel and steam dominated the expansion of our railroads into vast systems," Mingos wrote, "so now is C. M. Keys, financier of aviation, guiding the construction and development of our highways of the sky."[80]

> Today C. M. Keys is the E. H. Harriman of aviation. Harriman had a vision of railroad expansion. He had the confidence of the money powers. They gave him credit without limit. He took their money and built vast railroad systems. He speeded up business and industry. He quickened the country's growth and he paid dividends. He was the dominant figure in the age of steel and ocean. Keys has his dominion in the air.
>
> His vision is one of winged machines carrying people and things through space faster than any bird in its maddest flight. He is building up a network of air transport lines and aerial service systems. He plans to have them soon sprawling all over the country. Like Harriman he has the confidence of money. Wall Street supports any project he launches. Shareholders show faith in anything he manages. More than a hundred leading bankers, capitalists, industrialists and business men are associated with him in his enterprises. Keys has financed more than $80,000,000 worth of airlines, airplane plants, flying services, holding companies and investment trusts. Yet he is just beginning.[81]

Mingos went on to describe Keys's vision for aviation, his belief that every community in America would have access to air service, that feeder lines across the country would link into the trunk line systems, that one day airplanes would leave the main terminals not just once or twice a day, but every hour, and that soon airplanes would be carrying passengers at the unheard of speed of 200 miles per hour and more. Keys was in every phase of aviation, with his group of aviation companies producing a full range of aircraft for commercial and private uses, aircraft engines, propellers, aircraft instruments, providing flying training and flying services as well as transportation. "So," Mingos wrote, "with his air transport lines extending from coast to coast and linked together in a vast system of aerial service companies and feeder routes, Keys is insuring their future by providing sources of equipment and trained men to carry on the development. It is that sort of long range vision that has made him the most interesting personality in the industry. And the manner in which he is bringing all the various elements virtually under the same financial control is winning him recognition as the most important person in all aviation."[82]

Merging the Curtiss and Wright Groups

Air transportation and the aviation industry had made dramatic strides since Hartney's 1925 article. In an article reviewing developments in the aviation industry, the business magazine *Forbes* noted "the astounding progress the industry has recorded in the last year or two. Huge companies are being formed to build air terminals, to finance individual projects or to construct an ever increasing number of airplanes."[83] The structure of the aviation industry was changing rapidly.

That there would be mergers and consolidations in the aviation industry was not unexpected; the railroads, the automobile industry, and most recently the radio industry had all gone through a period of mergers. The stronger and more successful companies saw a clear benefit to mergers. There were, potentially, benefits to be obtained through economies of scale and the pooling of research and development efforts, the costs of which would inevitably expand as the technology of aviation advanced. With the rapid growth of air transportation, "it appeared to make sense for a manufacturer to assure himself a share of the commercial market by a corporate relationship to one or more airlines. From that it was a logical step to envision an integrated organization operating its own airlines and building its own airframes, engines, and components."[84]

William Boeing, seeing the potential for creating a large integrated organization from the companies he had created, responded positively to the suggestion that he form a holding company to raise the capital he needed for expansion.[85] In October 1928, Boeing established the Boeing Airplane and Transport Corporation, which controlled the Boeing Airplane Company and Boeing Air Transport. Boeing then proposed to Frederick Rentschler, president of Pratt & Whitney, a major supplier of airplane engines to the Boeing Airplane Company, that Pratt & Whitney join Boeing Airplane and Transport Corporation. Rentschler agreed and brought the Chance Vought Company into the negotiations. In mid-December a new holding company, United Aircraft & Transportation Corporation (UATC), was organized with the help of the National City Bank of New York, combining Boeing Airplane and Transportation Corporation, Pratt & Whitney, and the Chance Vought Company. In addition, the holding company owned 70 percent of the stock of Pacific Air Transport, which flew the mail from Seattle to San Francisco. William Boeing became chairman of the new corporation and Frederick Rentschler was appointed president. The combined securities of these companies had a market value of $150 million representing the largest consolidation to date in the aviation industry.[86] The companies involved in United Aircraft & Transportation all had strong links; Boeing Air Transport and Pacific Air Transport flew Boeing airplanes with Pratt & Whitney engines, while both the Boeing and the Chance Vought military airplanes also used Pratt & Whitney engines. The United Aircraft & Transportation Corporation reflected the business view that military sales and air transport would be the backbone of the demand for aircraft.[87]

Then, on March 6, 1929, Sherman Fairchild, president of Fairchild Aviation Corporation, Robert Lehman of the investment banking firm of Lehman Brothers, and W. Averell Harriman, son of E. H. Harriman and head of W. A. Harriman and Company, the investment bank, formed the Aviation Corporation (AVCO), a combined holding company and investment trust. AVCO quickly raised $35 million through an issue of common stock. AVCO acquired the Fairchild Aviation Corporation and its manufacturing subsidiaries, and over the next several months took a controlling interest in several airline companies including the Aviation Corporation of the Americas, the holding company that controlled Pan American Airways. Although located in different parts of the country, in combination these air transport ventures made AVCO in theory the largest airline operator in the United States.[88]

By the late spring of 1929, the aviation industry had aligned into four main groups with a constellation of around sixty smaller and still independent companies surrounding them. The leading aviation groups were now:

1. The United Aircraft & Transport Corporation
2. The Aviation Corporation
3. The Curtiss-Keys Group
4. The Wright-Hoyt Group

The Wright-Hoyt Group of companies, the smallest of the four groups, was linked through the management of Richard F. Hoyt, chairman of four principal companies in the group: Wright Aeronautical Corporation, Keystone Aircraft Corporation, Travel Air Company, and the Aviation Corporation of the Americas. During the 1920s, Hoyt, a partner in the investment banking firm of Hayden, Stone & Company, was active in promoting aviation. He helped organize Pan American Airways and numerous other aviation companies, acquiring interests in Keystone Aircraft and Travel Air. The Wright-Hoyt Group had perhaps the loosest affiliation of the four leading aviation groups. Of greater interest to analysts of the aviation industry were the common interests between the Wright-Hoyt Group and the Curtiss-Keys Group.[89]

Keys and Hoyt likely met shortly after the end of World War I. They began corresponding and meeting to discuss issues of mutual concern in the early 1920s, with Keys writing to Hoyt in August 1921 saying "it seems to me that there are some problems ahead of us which it would be safer and better for us to face together, rather than separately."[90] Over the next several years, they continued to meet and to correspond on business matters relating to the Curtiss Aeroplane & Motor Company and Wright Aeronautical Corporation, as well as sharing views on aviation matters and potential business opportunities. But it was not until the expansion of commercial aviation following the passage of the Air Commerce Act and Lindbergh's flight to Paris that they formed a close working relationship in several aviation ventures. In 1925 Keys brought Hoyt into National Air Transport as an investor, and during 1928 into Transcontinental Air Transport as a director and member of TAT's Executive Committee, as well as the National Aviation Corporation, the Aviation Credit Corporation, and the Aviation Corporation of California, while Hoyt sought Keys's participation in the Aviation Securities Corporation of New England and in the Aviation Corporation of the Americas as an investor and board member. Both men were directors of the newly formed Bendix Aviation Corporation, in which the General Motors Corporation held a 25 percent interest.[91]

Keys saw the merger and consolidation movement within the industry as a positive force, helping to eliminate duplicate facilities and reducing overheads.[92] Like other leaders in the aviation industry, Keys fully expected that there would be a readjustment within a few years. He wanted to ensure that his own companies would be able to weather such a readjustment. Companies that integrated the production of airplanes, engines, and accessories appeared to have an advantage. A combination with Hoyt's group of companies, complementary with Keys's group in many ways, made sound business sense.

The Curtiss Aeroplane & Motor Company had a strong position in military aircraft producing trainers, pursuits, attack, and observation aircraft, while Hoyt's Keystone Aircraft Company was building large bombers for the Army Air Corps and developing a large trimotor airliner. In the commercial aircraft field, the Keys Group had the Curtiss-Robertson Airplane Manufacturing Company building the Curtiss Robin, while the Hoyt Group could potentially bring in the Travel Air Company, providing single-engine aircraft for the private flyer. A merged group could offer a full range of military and commercial aircraft. There would be a similar benefit in the manufacture of airplane engines. While the Curtiss Aeroplane & Motor Company had developed a small air-cooled radial engine, the company had continued to specialize in large, water-cooled engines for the military. Keys could not ignore the growing trend toward air-cooled radial engines for military and commercial airplanes. Wright Aeronautical would bring a full range of air-cooled radial engines through its Whirlwind series and the more powerful Wright Cyclone.[93]

There were two potential challenges that a merged Keys and Hoyt group would face. First, both the Keys and Hoyt groups were loosely knit groupings of individual companies. In contrast to the more tightly controlled United Aircraft & Transport Corporation, Keys had been deliberately following a policy of developing small companies to cover different aspects of aviation rather than forming a more unified holding company structure under the Curtiss Aeroplane & Motor Company. This approach conformed to his belief that at this stage of the aviation industry's development, smaller companies could better respond to change than a single large entity.[94] Coordination of a common policy was assured through Keys's participation on the boards of these companies and their management by his close associates. Hoyt's companies were, if anything, even more loosely connected. Establishing economies of scale in research and development and in manufacturing would be a challenge.

Second, both Keys and Hoyt had invested heavily in the private aircraft market through manufacturing companies like Curtiss-Robertson and Travel Air, through development of small radial engines like the Whirlwind and the Curtiss Challenger, and through Keys's expansion of the Curtiss Flying Service into sales and service.[95] They believed that the private airplane owner would be a key component of the demand for aircraft in the future, at least if one looked at the statistics of private aircraft sales, rather than from the air transport companies. Not all analysts, however, agreed that the private airplane market had a robust future. By the spring of 1929, there was a general recognition that the manufacturing of airplanes for the private flyer was expanding too rapidly, and that production was getting ahead of demand.[96]

Moreover, the assumption that the public would adopt the airplane as a means of transportation was by no means a certainty. As one analyst put it, "hopes and predictions based upon the extensive use of airplanes for private use appears to be farfetched for it does not seem possible to eliminate the personal risk in aviation as has been fairly well accomplished in the automobile field."[97] If demand for private aircraft was not sustainable, what would happen to this investment?

When Keys and Hoyt first broached the idea of a merger isn't precisely clear, but it probably took place in the late spring of 1929. As good businessmen and financiers, the logic of a merger would have appeared evident to both of them. As rumors of a pending consolidation of the two groups spread in the market, both men strenuously denied them; Hoyt flatly denied that any negotiations leading to a merger of Curtiss and Wright had been opened.[98] News that the two iconic names in American aviation would, in fact, merge came two days later, on June 27, 1929, with the announcement of the formation of the Curtiss-Wright Corporation to take over the Curtiss Aeroplane & Motor Company, Wright Aeronautical Corporation, and their associated companies through an exchange of stock. It was ironic that given the Wrights' precedence in developing the airplane and their longstanding animosity to Glenn Curtiss, the Curtiss name should come first in the merger. The *Wall Street Journal* reported that the merger would bring into being the largest aviation company in the world with tangible assets of more than $70 million and a market value of $220 million.[99] But not all their interests were included in the merger. North American Aviation, the National Aviation Corporation, and several other aviation financing and investment companies remained outside the group, as did the air transport companies. Keys and Hoyt were adamant that the airlines they were involved with would continue to remain independent of the new Curtiss-Wright Corporation. As Keys explained, neither he nor Hoyt "thinks that transport and manufacturing and sales can be run by the same group of executives. It requires a different kind of training and a different kind of mind to carry on the two types of operations."[100]

Keys explained that the new Curtiss-Wright Corporation would function more as a holding company than as a true merger. There were no immediate plans for absorbing the manufacturing units. The more important task of the consolidation, Keys said, was to fill out and complete the product line. "In general," Keys stated, "it may be said that our object will be to make an almost complete line of planes and motors for every use. In some cases, no doubt, the filling of the line as time goes on may involve the purchase of other companies, but it is just as likely as not to come about through additions to the products now made under our control."[101] By combining existing airplane designs with the greater range of engines that would be available through Wright Aeronautical, Keys argued that the combined group could relatively easily increase the potential markets available to the group's existing airplane designs at lower cost and with less duplication of effort.

In general, the merger was well received in the market. The securities firms were quick to promote the stock of the new corporation, particularly those firms that were most closely involved. In a joint special circular, Hayden, Stone & Company, Bancamerica-Blair Corporation, and James C. Willson & Company waxed eloquent:

> Bringing together in one holding corporation the interests repre-
> sented by the two oldest names in American aviation—Curtiss
> and Wright—will mean not only the association of eleven compa-
> nies whose outstanding securities on June 26 had a market value of
> more than $220,000,000, but will give new and powerful impetus to
> all those policies and ideals for which the Curtiss-Keys and Wright-
> Hoyt groups have stood in the public mind. . . . Even the most coldly
> analytical investor cannot fail to catch something of the glamour
> which clings to this latest development in American aeronautics. To
> the public it means still greater assurance of safety in air travel. To the
> investor it means the kind of unified program which ultimately brings
> sound and profitable achievements.[102]

On August 15, 1929, the new Curtiss-Wright Corporation formally began operations. At the first organization meeting of the Curtiss-Wright Corporation, Richard Hoyt was elected chairman of the Board and Keys became president. The companies brought into the new Curtiss-Wright Corporation were as follows:

Curtiss Aeroplane & Motor Company
Curtiss Airports Corporation
Curtiss Flying Service, Inc.
Curtiss Aeroplane Export Corporation
Curtiss-Caproni Corporation
Curtiss-Robertson Airplane Manufacturing Company
Wright Aeronautical Corporation

Keystone Aircraft Corporation
Moth Aircraft Corporation
Travel Air Company
New York and Suburban Air Lines, Inc.
New York Air Terminals, Inc.

The new corporation had no debt and no need for financing. All the companies within the group had a strong liquid position. There was every expectation that the benefits to be gained through the merger would soon start to show in profitability.

In the summer of 1929, Clement Keys reached the pinnacle of his career. In ten years he had gone from a position as the vice president for finance for the Curtiss Aeroplane & Motor Corporation to the presidency of the first transcontinental airline and the largest airplane and engine manufacturer in the world. The path to success had by no means been an easy one, but he had taken on the challenges as they came, first to rescue Curtiss Aeroplane and ensure its survival, then the struggle for government policies that would promote and not destroy the American aviation industry, and then the promotion of commercial aviation and his main interest, air transportation. By 1929, the breadth of his interests, activities, and plans, which stretched from manufacturing to services to air transport in America and around the world, had no parallel among his peers. He had attained his position through his own intelligence, drive, persistence, and, above all, through a consistent vision and faith in the future of aviation and a commitment to being a systems-builder like the older generation of railroad men he so admired. A contemporary journalist described Keys in these words:

His supreme skill, his fine execution of finance and his sheer ability make him today the figure around which aviation revolves and gains momentum.[103]

His position as a leader of the aviation industry would not last long.

7

TAILSPIN

Nemesis

IN GREEK MYTHOLOGY, TYKHE, DAUGHTER OF ZEUS, WAS THE GODDESS OF Fortune bestowing success and prosperity on mortal man. There were some who received the blessings of Tykhe that were thought to be undeserving of her good fortune; who by their actions or arrogance violated the sense of order and divine law. For these few, Nemesis, the goddess of retribution, intervened in the affairs of men to humiliate those who were deemed to have received inordinate success, thereby serving to check "the extravagant favors of Tykhe."[1] Having reached the pinnacle of success, it would be Keys's fate to be cast down through his own actions which, at the time, seemed in his eyes to be the normal and proper course of business, but which from another perspective appear to be tinged with arrogance. For Keys, Nemesis, retribution for his hubris, would come through his own company, C. M. Keys & Co.

When Keys rescued the Curtiss Aeroplane & Motor Corporation from bankruptcy in 1920, as we have noted, he did so initially using his own money and credit. During the subsequent years of struggling to preserve the company, Keys often provided badly needed working capital funds to Curtiss Aeroplane from his own resources. As the fortunes of Curtiss Aeroplane improved and its working capital position turned positive, these funds were placed in the call market on Wall Street through C. M. Keys & Co. As he expanded the number of aviation-related companies he became involved with, Keys continued this practice of having C. M. Keys & Co. act as fiscal agent, placing their surplus funds in the call market. Keys chose to use his own firm as fiscal agent rather than a bank so that the surplus funds of these companies could gain the advantage of the much higher rates of interest to be found in the call market. At the beginning of September 1929, C. M. Keys & Co. was managing more than $16 million for the Curtiss group of companies, National Air Transport, Transcontinental Air Transport, and North American Aviation. The money placed on call with C. M. Keys & Co. from North American Aviation and its subsidiaries amounted to $14,171,767.[2]

As with any broker, C. M. Keys & Co. had an obligation to pay interest on the money that was placed with it on call. Typically, a brokerage firm on Wall Street would borrow money on call from a bank or corporation to finance the firm's inventory of stocks and bonds and to finance margin loans to the brokerage's customers. Margin lending allowed a customer of the brokerage to purchase stocks by putting up less than the full purchase price of the stock, the margin then being 25 percent for small investors and as little 10 percent for large investors; the brokerage firm then lent the client the balance of the purchase price. The call loans the brokerage firm borrowed would be secured with collateral from the brokerage firm's inventory or with the stocks the brokerage's customers had purchased through margin lending. The brokerage firm could earn a spread by charging its margin loan customers a higher rate of interest than the rate at which the brokerage borrowed the funds. While call loans were collateralized, a lender could gain added protection through the brokerage firm's careful selection of customers that it would, in turn, lend to. While there was no formal contract between the lender and the brokerage firm, the brokerage firm had an implicit fiduciary responsibility to protect the funds placed with it on call.

In this respect, Keys seems to have made an error in judgment; instead of taking the funds from Curtiss Aeroplane and the other companies where C. M. Keys & Co. was acting as fiscal agent and simply placing them in the call market with reputable parties, during 1928 and well into 1929 Keys apparently used these funds to finance the brokerage business of C. M. Keys & Co., taking the money placed on call and lending it to the firm's brokerage clients. As he later admitted in 1928 and 1929, he had used the funds to make large advances to assist other people.[3] Keys was simultaneously serving in an executive capacity with the firms that were providing the call loans to C. M. Keys & Co. and as the firm's senior partner, but he does not seem to have deemed this a conflict of interest, nor does he appear to have shared with the other executives or directors of the lending firms exactly what he was doing with the money entrusted to C. M. Keys & Co.

The temptation to use the money placed on call with C. M. Keys & Co. to finance margin lending for the firm's clients was likely overwhelming, given the interest rates that could be earned and the fact that these loans were secured by the stocks they were financing, stocks that steadily rose in value as the stock market accelerated through 1928 and into 1929. The bull market that had begun in the mid-1920s was in full force; after March 1928, stocks continued a steady climb upward, in some cases with spectacular rises. Concerned with the rise in the stock market, the Federal Reserve Bank raised interest rates over the course of the year, sending rates in the call market up to 10 percent. The public's demand for securities "was insatiable," and despite a few negative voices decrying this orgy of speculation, the common belief was that the stock market would continue to improve.[4] With customers clamoring to buy stocks and with call loan rates at 9 percent or better, it would have been hard for any broker to turn away business, and over that fateful summer of 1929 almost all brokerage firms, including C. M. Keys & Co. , did not.

As previously related, over the summer of 1929 Keys went to Europe. On his return in early September 1929, he found that his business associate, James C. Willson, who had underwritten so many of the companies that Keys had organized, had encountered severe financial difficulties. The exact nature of these difficulties remains unclear. It may be that Willson simply became overextended, but for whatever reason, out of a sense of loyalty Keys and two other friends of Willson agreed to attempt to save him. On October 2, 1929, Keys and F. H. Hardy of Chicago advanced $632,000 to the Willson Holding Company, which controlled James C. Willson & Co., Willson's brokerage firm. Willson's holding company was apparently heavily indebted to the Guaranty Trust Company, the brokerage firm of Spencer, Trask & Co. and Bancamerica-Blair Corporation. In return, Willson provided collateral in the form of shares in various companies, options to purchase shares in several Curtiss companies, and a 75 percent interest in warrants to purchase the shares of North American Aviation at $12.50 a share up to December 1, 1931.

To finance this loan to Willson, Keys used some of the funds from North American Aviation and its subsidiaries that had been placed on call with C. M. Keys & Co. In his capacity as president of North American, Keys would have known the future capital and cash requirements of the company and its subsidiaries, and he apparently drew funds from subsidiaries that had no immediate need for funding, such as Intercontinent Aviation, Inc. It would seem that Keys did not inform the other directors of North American Aviation of what he was doing. He used North American Aviation's money as if it was his own, no doubt believing that Willson's financial predicament was temporary and would soon be resolved. Keys had no way of knowing what was shortly to befall him and the rest of the stock market, but his loyalty to his friend and his desire to help his clients would cost him dearly. The total amount he had lent out to Willson and to other clients of C. M. Keys & Co. is uncertain, but it may have been as much as $6 million.[5]

Back to Business

Keys found much to deal with on his return from Europe. In his absence over the summer, steady progress had been made in completing the merger of the Curtiss and Wright groups. The exchange of stock of existing companies for shares of the new Curtiss-Wright Corporation had gone so well that the incorporation and organization of the new company had been completed on the date originally set in the merger plan. Under Richard Hoyt's direction, a detailed survey of all twelve companies in the new corporation had been undertaken, listing all facilities, production capacity, staffing, and unfilled orders. The survey showed that at the end of June, on a pro forma consolidation, the Curtiss-Wright Corporation was in sound financial condition. The group had strong liquidity, with more than $26 million in cash and call loans compared to $2.9 million in current liabilities, and capital and surplus of $76 million. Gross sales were $13 million, with a net profit after tax of $2.7 million. The group had a healthy backlog of business with Curtiss Aeroplane and Wright Aeronautical, both of which showed healthy gains in sales.[6] More good news on the order front came in August when the War and Navy Departments awarded Curtiss-Wright contracts for seventy-two airplanes for the Army Air Corps and thirty-six two-place fighters for the Navy; Wright Aeronautical was awarded contracts for one hundred and thirty-two aircraft engines for the Navy.[7]

The outlook for Curtiss-Wright seemed promising, and the market reflected this bright prospect. The weekly business magazine *Barron's*, responding to a reader's inquiry on Curtiss-Wright, commented that "the companies controlled by Curtiss-Wright Corporation form the outstanding manufacturing group in the industry today . . . operation of the 12 companies at present in the group under one coordinative management should result in larger profits as well as the quicker and more efficient development of an important industry."[8] *Barron's* recommended the Curtiss-Wright A shares as "…an outstanding investment in an infant industry with unusual possibilities for future growth."[9] On September 14, 1929, Keys, Richard Hoyt, and the other directors of the Curtiss-Wright Corporation attended a day-long meeting at the Westchester Biltmore Country Club in Rye, New York, with the senior managers of all twelve of the companies in the corporation to discuss plans for coordinating efforts among the companies, particularly in the area of research and development. The meeting reaffirmed the earlier decision that Curtiss-Wright would continue to act as a holding company and would not consolidate the subsidiary companies. The executives of these companies were to be retained, and the companies would continue to maintain and publish their own separate balance sheets. Coordination would come through the appointment of experienced men to key positions within Curtiss-Wright and several of the component companies.

Keys took the opportunity to reiterate Curtiss-Wright's management philosophy. "The administration of the manufacturing companies is not being materially changed," he said, "The Curtiss-Wright Corporation itself is not a manufacturing company and it is believed that the high morale of the separate manufacturing organizations, Wright, Curtiss, Keystone, Travel Air, Curtiss-Robertson, and Moth, can best be maintained under their own separate organizations."[10]

The Great Crash of 1929

During the summer of 1929, trading in aviation stocks was somewhat lackluster compared to other segments of the market. The lack of movement in aviation securities was not considered a bad sign. As one commentator noted, the lack of public interest "has given a healthy breathing space to this class of security which has, frankly, been somewhat overemphasized during the last year."[11] And, too, there was a general recognition within the aviation industry that production of airplanes and engines, particularly small airplanes, had significantly outpaced demand and that the industry was in a state of "overexpansion and overproduction"; production in the first six months of 1929 equaled 75 percent of the total production in 1928.[12]

Writing in the magazine *Aeronautics* in October, B. C. Forbes, the editor of *Forbes*, warned of the dangers of overexpansion. The danger for the aviation industry, he warned, was in letting production get too far ahead of the actual growth in demand. In this situation, a period of consolidation within the industry would actually be healthy for its long-term development. "Any tendency toward cutting down too-rapid expansion in the aviation field would be a favorable, rather than an unfavorable, factor," said Forbes.[13] The industry appeared to have come to the same realization, with many companies scaling back their production, but Forbes predicted that a period of readjustment was more than likely. He was critical of the excessive promotion of aviation stocks, noting that "the temptation to promote securities of questionable economic prospect has been especially marked in the aviation field. The only result that seems logical to the experienced economist and market analyst is a future period of readjustment, during which the weaker sisters must present a discouraging picture and during which many of the smaller and less necessary promotions will either disappear or be taken over by the larger and stronger concerns."[14]

The market had come to accept that a period of readjustment in the aviation industry was inevitable; concerns over the government's plan to reduce air mail contracts and possible cuts in the Army and Navy appropriations for aviation added to the sense of uncertainty surrounding the outlook for the industry.[15] As the stock market began a steady decline during the month of September, with the Dow Jones Industrial Index dropping from 381 to 343 at the end of the month and then to 325 in the first week of October, many of the leading aviation stocks experienced declines of 20 percent or more. Collectively, the seven leading firms in the industry had seen their combined market value drop 56 percent from their market highs earlier in 1929.[16] Then the collapse came, first on "Black Thursday," October 24, 1929, a day of real panic on the markets, followed by "Black Tuesday," October 29, 1929, which saw 880 stock issues lose $8 to $9 billion in value and the Dow Jones Industrial Index close at 230.[17] While the market managed a slight recovery over the next few days, aviation stocks had been hard hit across the board. Curtiss-Wright common had dropped 43 percent since mid-September, Transcontinental Air Transport 40 percent, National Air Transport 38 percent, and North American Aviation 47 percent, with AVCO and UATC dropping 47 percent and 28 percent, respectively.[18]

On October 31, 1929, Keys gave an interview to the *Wall Street Journal* in which he tried to inject a more rational view of what had just taken place in the stock market, making a clear distinction between the small, poorly capitalized companies that had sprung up during the aviation boom and the long-established, well-financed leaders of the industry. Keys began the interview by saying that he "believed that the wildcat stages of the industry which existed in 1929 have passed and that the future promises further expansion on a sound basis."[19] He went on to say that "after a complete study of all phases of aviation, I have no hesitation in stating that the reaction in the trade has been grossly exaggerated in the public mind. Most of it is, in fact, a reaction from a boom which played no part in the plans and policies of the old companies in the business."[20]

Throughout the industry, Keys said, a wholesome cleaning out of loose ends was underway. Small concerns that had been able to start up transport operations or manufacturing because of the availability of easy financing, and that were operating below their true actual costs, were going out of business. These operations had made it difficult to establish a business that would pay a profit, but as they fell away, the future prospects for the leading firms were becoming brighter. The leading transport companies, Keys said, "provided ample reserves of capital and have been building for the time when, through the growth of volume of traffic and the betterment of equipment, a fair margin would be established. All of them welcome the passing of the wildcat stage as the first step toward the goal."[21] While admitting that profits in the industry would likely be lower than the estimates made earlier in the year, Keys nonetheless believed that they would be satisfactory. "The future of aviation in all its phases," he insisted, "continues to promise sound growth."[22]

Keys may have been sanguine about the future of aviation in the aftermath of the stock market crash, but the impact on his investment business and C. M. Keys & Co. had been devastating. As noted earlier, Keys had relied heavily on call loans from his aviation companies to finance part of his brokerage and investment business. Prior to the Crash, C. M. Keys & Co. probably had more than $7 million in call loans from various Curtiss companies and $8.5 million from North American Aviation with an estimated $5 million in liabilities from other sources, making approximately $20 million in total liabilities. With the collapse in stock values that occurred during October, many of the corporate lenders to the New York call market called in their loans, forcing a mad scramble among brokerage houses to find alternative financing. Margin calls went out to investors, demanding that they put up more cash as collateral for the stocks they had purchased on margin. Many investors, however, already overextended, could not come up with the required cash, which added to the selling pressure in the market as brokers sold off the shares they had been holding as collateral.[23]

In the space of a few weeks, the volume of call loans in the New York call market dropped by half. C. M. Keys & Co. faced the same problem of needing to repay call loans as the market collapsed. By calling in margin loans from his customers, selling assets, and putting out margin calls, Keys managed to repay all the money he had borrowed from North American Aviation, with interest, and to reduce his overall call loan funds by $4.4 million in October and by another $1.2 million in November, but he still held call loans amounting to $7.4 million from other Curtiss companies, Pitcairn Aviation, and Transcontinental Air Transport, and an additional $1.7 million from subsidiaries of North American Aviation, which he appears to have borrowed in order to pay back North American Aviation itself. He reduced his total indebtedness to $6,548,000 by the end of December, though he still had the $1.7 million owed to North American Aviation subsidiaries.[24]

Keys now had a real concern as to whether he would be able to meet his obligations and repay these loans. Extrapolating from data he submitted a few years later, it appears that Keys had lent from $4 million to $5 million to the customers of C. M. Keys & Co., in all likelihood for margin loans, and like other brokerage firms, was having trouble getting paid back. During December Keys advised Chester Cuthell, his legal counsel, and J. Cheever Cowdin of Bancamerica-Blair Corporation, his closest advisors and members of the Executive Committee of North American Aviation, of his situation and the fact that he had borrowed money from North American's subsidiaries that he was not in a position to immediately repay. In lending these funds to his clients, Key had done nothing illegal, but arguably he had violated his fiduciary responsibility as a financial advisor to North American Aviation, which one would assume would include a responsibility to safeguard the company's funds.

For whatever reason, Cuthell and Cowdin decided not to bring the matter to the attention of the other members of the North American Aviation Executive Committee but instead agreed with Keys's plan to repay these loans over time. Keys had ordered an audit of C. M. Keys & Co.'s position as of December 31, 1929, which showed that he had sufficient assets in the company and from his own resources to cover all his liabilities. He immediately instructed his staff to begin liquidating assets and paying off the call loans. For the next two years, while dealing with all the other problems that beset him as the Depression deepened, Keys had C. M. Keys & Co. continue to liquidate assets, as he did himself from his own portfolio, most often with heavy losses, in order to meet his obligations as they came due. This constant burden added to Keys's mental and physical strain. In the background to all the other challenges he faced was the worry that he would not be able to pay back what he owed, as well as the fact that his own fortune was rapidly being eroded.[25]

Outwardly at least, in the months following the Crash, Keys remained reasonably confident for the future in regard to both the general economy and the aviation industry. The aviation industry was passing through its first great readjustment, which Keys considered to be a wholesome development. The plans that many companies in the industry had developed during 1929 constituted, he thought, "a rather dangerous period of hotly competitive building and expansion, which will be much sounder if it is spread out over a longer period of time and not executed with such great haste."[26] In an interview with the *Wall Street Journal*, Keys said: "I believe that the greater part of the inevitable deflation of aviation is over and that all well-founded and well-financed industries in the trade can now see their way clear to a sound and constructive growth in 1930."[27]

Keys elaborated on this theme of slower but sounder growth in early 1930 in articles for the aviation magazines *Airway Age* and *Aviation*. Commenting for the magazine *Aviation* on the outlook for 1930, Keys said the keynotes for the industry going forward were caution and courage. The first part of the year would likely see the curtailment of expansion plans as manufacturing companies worked to reduce inventories from the overproduction of 1929, conserved working capital, and pushed through greater efficiencies. Still, Keys said, "there seems to be plenty of courage and vision left in this trade after the great recession of 1929."[28] The hope of the commercial manufacturers was the growth in private ownership of airplanes by business and individual pilots. "One may look forward," he said, "with some confidence to the development of a very substantial volume of new business in 1930 compared with any previous year."[29]

In the first few months of 1930, the stock market staged a recovery, with the prices of many stocks regaining the ground they had lost in the Crash. Even aviation stocks, which had continued their decline until the end of 1929, managed a comeback of sorts, with Curtiss-Wright common climbing from its December low of $6½ a share to $10⅝ a share by March 1. For Keys and many other men with long experience in the stock market, there was a sense of having gone through it all before. Keys had lived through the Panic of 1907, the depression of 1913–1914, and the depression of 1920–1921; in each instance, the stock market had faltered, followed by a business recession, and then recovery. It seemed to many in early 1930 that this pattern was simply repeating itself.[30]

Tailspin

Not Single Spies but in Battalions

As the year went on, the American economy did not take the path that so many had expected when the year began. By the middle of the year, it was evident that the economy was in a steep decline; during 1930 both the index of industrial production and the index of employment fell by 20 percent.[31] As evidence of the worsening condition of the economy continued to accelerate, the market steadily declined with "an air of inevitability"; the Dow Jones Industrial Index dropped from its high of 297 to 164 in December of 1930.[32]

Aviation stocks followed the market. Although they did not reach their previous pre-Crash levels, during the spring of 1930 aviation stocks managed a comeback of sorts. Curtiss-Wright common rose from $7¾ a share in January to a high of $14¼ in April, shares in Curtiss-Wright Flying Service increased from $5⅛ to $11⅞ a share, and North American Aviation, from $5¼ to $15⅝ a share. United Aircraft & Transport Corporation rose from $49⅞ a share to $91½, while even AVCO managed to climb from $4½ to $9 a share in early May.[33] When the market turned, aviation stocks sank, and they sank quickly. By the beginning of September, Curtiss-Wright common had dropped to $6⅝ a share, North American Aviation to $9⅛, and United Aircraft & Transport Corporation to $60¼. The sector continued what seemed to be an inexorable decline, with Curtiss-Wright closing the year at $3 a share, North American at $6, United Aircraft at $28⅞, and AVCO at $3½—nearly all ending the year well below where they began.[34] Stock prices reflected the market's view of the value of the underlying assets and future prospects for each company and the sector, as well as what had befallen the companies during the course of 1930. For North American Aviation, the year proved to be one of mixed fortunes; for Curtiss-Wright it was a disaster.

In the aftermath of the Crash, North American Aviation did not initially alter its strategies, continuing to invest in securities and directly in companies. The company took advantage of the sharp decline in aviation stocks to add to its securities portfolio. In March 1930, Keys reported that North American Aviation had recorded a net profit of $1,908,819 for 1929 on revenues of $2,301,697, with profits from investment activities (interest, dividends received, and profits on securities sold) amounting to $1,625,117, and profits from wholly owned companies totaling $676,580, a 70 percent/30 percent split between investment earnings and operating earnings.[35] North American was still interested in investing directly in sound companies with good earnings prospects. When Thomas Morgan, president of Sperry and a newly appointed director of North American Aviation, brought to the Executive Committee an opportunity to purchase the Ford Instrument Company, Keys and the other members of the Executive Committee decided to go ahead. Ford Instrument held exclusive patents and contracts with the U.S. Navy and the Army for fire control systems for naval guns and antiaircraft guns; this was a narrow market, but one that had few other competitors. On February 15, 1930, the Ford Instrument Company accepted an offer from North American Aviation to purchase all of its common stock for $2.9 million, becoming a wholly owned subsidiary of North American.[36]

But the North American annual report showed that the open market securities on its books at the end of the year had a cost of $14,688,998 and a market value of $6,813,040.[37] Accounting conventions of the day gave Keys some leeway in how to treat the loss in value in North American Aviation's stockholdings. They did not have to be written off immediately, and the loss in value could be taken through the profit and loss account. Stocks that could be readily liquidated and that "can be used for the current needs of the business may be included in current assets." in which case it was recommended that they be valued at "cost or market, whichever is lower." However, if the investments were not easily liquidated or not intended for the current needs of the business, then they did not have to be included in current assets and could therefore be carried at cost "unless there has been a radical and permanent change in the conditions which makes a revaluation advisable."[38] One could have well argued that a collapse in value of some 54 percent was a radical change, but Keys appears to have believed that it was not permanent and that the value of the stocks in North American Aviation's portfolio would recover. He therefore chose not to write down the value of these securities, thereby avoiding posting a horrendous loss.

The market, or at least the New York Stock Exchange (NYSE), appeared to agree with Keys's approach, for on April 9 the board of governors of the Exchange approved the listing of North American Aviation's stock. Moving from the Curb Market, where North American Aviation's stock had been traded, to the NYSE was a measure of both recognition and approval, even though the company was still thought of as a speculative investment. Writing a few months later in response to an investor's inquiry, the *Wall Street Journal* commented that North American Aviation could not be considered "as at present ranking with those few concerns which combine a strategic position in the industry with an already developed substantial earnings power."[39] Still, the paper noted, North American Aviation had attractive possibilities; its two principal subsidiaries, Sperry Gyroscope and Eastern Air Transport, were both doing well,

A photo of Keys after he was appointed chairman of
Transcontinental & Western Airlines in August 1930.
By this point, he was under increasing strain from
his own financial problems and the challenges facing
his many companies as the Depression deepened.
(The Hatfield Collection, Museum of Flight)

and considering that most aviation companies were operating at a loss, North
American Aviation's record of profits to date was considered satisfactory.

When North American Aviation issued its semiannual report in July, Keys
announced a change in the company's strategy that was more a tacit recognition of the fundamental change the collapse in the stock market had imposed
on the company's business. The new policy, Keys said, was "to enlarge its controlling and operating functions, rather than its investment function."[40] For the

first six months of the year, North American Aviation had earned $267,821 from interest, dividends, and the sale of securities, compared to $896,860 in income from wholly owned subsidiaries, a near reversal of the contributions during 1929. North American had been established as a form of investment trust to provide capital for long-term investments in aviation companies, but with the investments it had made in manufacturing and transport companies as opposed to stock investments it had come to be more of an operating holding company itself, providing direct management and direction of its subsidiaries. As the stock market, aviation stocks, and the economy all continued to decline in the second half of the year, few new investment opportunities came North American's way. The value of North American's stock portfolio continued to decline, but as Keys had said, this was not the focus of the members of the Executive Committee. But if there was one issue that concerned the other members of North American's Executive Committee, it was the condition of the Curtiss-Wright Corporation, North American Aviation's largest stockholding; Curtiss-Wright was not doing well, and its losses were growing alarmingly.

When the Curtiss-Wright Corporation released its first annual report for the year 1929 at the end of March, it was not a surprise that the Corporation had posted a net loss of $668,532. It was generally known that the aviation industry was in a slump, and so the financial papers reported the Curtiss-Wright results in a rather matter of fact way.[41] As the *Wall Street Journal* noted, the great amount of organizational expenses involved in merging the Curtiss and Wright groups and expanding their activities, combined with the decline in sales in the final quarter of the year, offset the profits of the three largest companies in the group, Curtiss Aeroplane & Motor Company, Wright Aeronautical, and Keystone Aircraft.[42] Keys, who signed the annual report in his capacity as president of Curtiss-Wright, explained to stockholders that the officers and directors of the corporation had decided to write off all establishment, construction, and development costs in 1929 rather than amortize these costs over time. The approach was more conservative and had the advantage of giving a clearer statement of the corporation's assets. "In view of the consolidation of the large group of companies," Keys wrote, "it has been considered better policy to get as much as possible of the necessary development costs of your industry written off and behind us, rather than burden the future with the slow but inevitable charges for this work."[43]

There had been extraordinary expenses. There were development expenses at Wright Aeronautical of $1,501,000 and $1,421,268 at the Curtiss Aeroplane & Motor Company.[44] In addition, there were the costs of expanding the operations of the renamed Curtiss-Wright Airports Corporation and the Curtiss-Wright Flying Service, amounting to $2,471,013, which were also written off.[45] At the end of the year, sixteen airports were under construction or in partial operation around the country, and there were forty-two flying fields and flying schools. The development of airports and flying services, which had been central to Keys's strategy for expanding commercial aviation, had been slow to develop and were operating well below their projected earnings capacity. Curtiss-Wright ended the year with a strong government order book of more than $10 million, but some of the group companies had overexpanded commercial aircraft production in expectation of increasing sales, leaving the group with an unsold inventory of 585 airplanes.[46]

In contrast, United Aircraft & Transport Corporation (UATC) posted net income for 1929 of $8,966,032. UATC had been the beneficiary of large government orders for airplanes and engines and, disregarding the frenzied market for private aircraft in the first part of the year, had maintained a more disciplined manufacturing policy, building airplanes only on the basis of firm contracts and not in expectation of future orders. As a result, compared to Curtiss-Wright, UATC ended the year with a minimal amount of unsold aircraft in inventory. Its airlines, Boeing Air Transport, Pacific Air Transport, and Stout Air Services, all posted substantial increases in the volume of mail they carried, which also contributed to the Corporation's strong revenues. Although UATC had made some investments in airport facilities and opened an aeronautical school, it was far less exposed to property and less dependent on flying and related services than Curtiss-Wright. UATC's consolidated fixed assets, including land and buildings, machinery and equipment, and aircraft, amounted to $11 million, compared to $45 million at Curtiss-Wright. UATC had roughly the same proportion of sales to governments as Curtiss-Wright at 40 percent of total revenues, but the key difference between the two companies was the mix of commercial business, with Curtiss-Wright more exposed to the private flying segment of the market than UATC.[47]

Keys tried to put a brave face on the situation. The aviation industry, he said, "is passing through a normal period of strain, disturbance, and readjustment, due largely to too-rapid a growth and to overproduction; and partly due to economic conditions in the country as a whole."[48] But he sought to reassure the stockholders that the stronger and more important companies in the industry, which were taking steps to deal with the situation, would continue to prosper. "Your company has taken," he said, "all necessary steps to meet the situation."[49] Plans were underway to consolidate all engine manufacturing at

Wright Aeronautical and to concentrate the manufacture of airplanes in a further bid to cut manufacturing costs. Some good news came shortly after the annual report was released when it was announced that Curtiss Aeroplane & Motor Company had won a contract from the Army Air Corps to provide fifty Curtiss attack aircraft and one hundred D-12 engines in a contract worth $1.2 million, while Keystone Aircraft received a contract for seventy-three Keystone bombers worth $2.2 million.[50]

Something happened over the next few months to change Keys's position within Curtiss-Wright. Despite the fact that he retained the title of president, he effectively removed himself from active management of the corporation. Whether this was by his design or was forced upon him is unclear, although it appears to have been his choice. When the Curtiss-Wright Corporation interim report was released at the end of August, announcing an even larger net loss, it was Richard Hoyt as chairman of Curtiss-Wright who signed the report and gave a statement to the press, not Keys. The few mentions of this development in Keys's correspondence from the period provide enigmatic hints, but no clear indication of what was going on. In July Keys wrote to his brother Roy saying that he wanted to see him soon to discuss Roy's future plans. The two brothers had decided that it would be best if Roy, who was then vice president and general manager of Curtiss Aeroplane & Motor Company's Buffalo factories, left Curtiss-Wright. As Keys explained to his brother "after my own practical withdrawal from Curtiss-Wright affairs, I think you would be much too heavily handicapped going on with them."[51] Roy Keys resigned from the Curtiss Aeroplane & Motor Company in September.

A month later, in a personal letter to a broker he knew discussing North American Aviation and its prospects, Keys said that the other North American directors "get peevish about Curtiss-Wright now and again, because we are not having much to do with the management of it, but I have made it plain that I am not going to assume again the burden of operating that company until I am thoroughly on my feet again physically, and perhaps not even then."[52] Keys was under increasing strain, and this was, apparently, having an impact on his health. In April he had gone to St. Catherine's Island in Georgia planning to get a good rest, only to become involved in the takeover battle for National Air Transport. At the end of the month, J. Cheever Cowdin sent a telegram to Keys saying "Hope you are getting full of health."[53] It may well be that with all the pressures on Keys from his involvement with North American Aviation, Transcontinental Air Transport, his China project, and all his other directorships, not to mention his travails at C. M. Keys & Co., he realized that continuing in an active managerial role at Curtiss-Wright was beyond his capability, and so he arranged with Richard Hoyt to quietly withdraw from that responsibility.

The Curtiss-Wright Corporation announced a loss of $1,620,920 for the first quarter of 1930. In June, Richard Hoyt gave a statement to the press saying that the corporation would likely have a loss of about the same amount for the second quarter. "However," he said, "I expect a decided improvement in the third quarter, and the deficit in that period will be substantially reduced. We are making no effort to show earnings at present, and I do not think much can be expected from our income statement until 1931. We will spend the rest of the year 'getting our house in order', looking forward to a constructive program for 1931."[54] But the actual results were far worse. In early September, Curtiss-Wright issued its interim report showing that for the first six months of the year the corporation had a net loss of $5,351,661, some $2 million more than Hoyt had estimated in June. All areas of the corporation were suffering; the manufacturing companies had a loss of $1,850,000, the flying service and airports reported a loss of $1,680,000, and there were additional write-offs for developmental expenses of $2,030,000.[55]

In explaining the results, Hoyt noted that Curtiss-Wright was a new company in an industry still in its pioneering phase, which required the expenditure of large sums of money on development. The aviation industry was coping with the effects of overproduction during 1929, which the current market could not absorb. Despite the loss, Curtiss-Wright was making progress. The inventory of unsold aircraft had been reduced from 585 at the beginning of the year to 300 at the end of June by slashing prices, and the program of consolidation of facilities was continuing, as was a program of general reduction in staffing and production. Hoyt admitted that progress toward putting the Curtiss-Wright Airports Corporation and the Curtiss-Wright Flying Service on a profitable basis was a slow process. "We are confident," Hoyt said, "that this branch of our activities will steadily reduce its operating loss until within a reasonable period its business becomes profitable."[56] There was a tremendous amount of work still to be done to complete the planned adjustments in organization and production, but Keys would play a less and less active role as time went on.

Great Expectations

In January 1931, Keys sent a handwritten note to J. Cheever Cowdin asking for his assistance with his financial problems. Outlining what he had accomplished during the previous year, Keys told Cowdin that an audit by Price,

Waterhouse & Co. of C. M. Keys & Co. as of December 31, 1929, had found that Keys's company had $14,564,000 in debts. "I went to work to clean up that position," Keys wrote, explaining to Cowdin that "during 1930, by steady sales and by turning away business, I have accomplished this: 1) Have reduced from $6,844,000 to $1,350,000 open credit balances; 2) Have reduced from $7,720,000 to $2,546,000 other borrowed funds. This will be reduced again by December 31 to $1,846,000 or less, fully secured. To accomplish this, no personal loans or other direct or indirect obligations have been received, no payments deferred, no future debt piled up. I have, however, done nothing much else but this, and now want help to complete the job, and get back to the constructive side of life."[57]

Keys had spent the year grinding out cash to repay the debts he owed. He had been forced to liquidate large amounts of stock from his own portfolio to raise funds to repay these outstanding loans. He was fortunate that the prices of many of the stocks he owned recovered during the first half of the year, which enabled him to sell most at a profit. Ironically, United Aircraft's takeover of National Air Transport enabled him to sell his NAT shares at a profit of $1,226,940. In total, Keys sold $3,814,678 worth of stocks from his own account and that of his wife.[58] It appears, from Keys's handwritten note to Cowdin, that he applied these funds to repayment of the debts of C. M. Keys & Co. Given the steady decline in the stock market during the second half of the year, Keys had been remarkably fortunate in what he had achieved; as he told Cowdin, he had managed to meet all the calls on C. M. Keys & Co. as they came due. But C. M. Keys & Co. still had $3,896,000 in debts outstanding, including $860,132 owed to subsidiaries of North American Aviation, and it had fewer resources available to ensure repayment. So far Keys had managed to restrict knowledge of his financial situation to J. Cheever Cowdin and Chester Cuthell, but his day of reckoning, his meeting with Nemesis, was not far off.

As the New Year of 1931 dawned, many could look back on the year just passed with disbelief and to the year ahead with no little trepidation. The optimistic prognostications of a turnaround in the economy issued at the beginning of 1930 had proved to be wildly off the mark. As the business weekly *Barron's* put it, describing the statistics for 1930, the year read "like an after-battle tale of casualties—so many perished and so many more wounded.... It has been a continued story of loss and retreat."[59] The financial sector was hit as well. The volume of new stock and bond issues fell to the lowest level since 1924, which depressed the income of investment banks and brokers, who had to bear the additional burden of inventories of unsold stocks and bonds. Pushed to the wall, a number of small and medium-sized brokers failed, and worse, banks began to fail in large numbers, with 256 banks failing in November and 352 in December.[60] Yet, as *Barron's* reported, "the twelve-month went out with a generally hopeful feeling, based mainly on confidence that after some 18 months of reaction all these various price and production and profit indices cannot now—in the light of experience—sink much farther toward zero."[61]

The aviation industry was in its own depression, with a continuing slump in sales causing a sharp decline in production. During 1930 the industry built 1937 commercial airplanes compared to 5357 during 1929.[62] Manufacturers managed to reduce their excess inventory of airplanes built in 1929 only by slashing prices by as much as 40 percent. One of the few bright spots amidst the gloom was the change in the fortunes of the air transport industry with passage of the Watres Act in April. The Act allowed the passenger air lines to carry the air mail, while the number of passengers carried more than doubled compared to 1929. Another was the fact that despite the sharp decline in sales and heavy losses on the part of some holding companies and manufacturers, surprisingly few companies actually failed. While a number of mergers and absorptions had taken place, which no doubt saved a number of companies, including Transcontinental Air Transport, only nine companies had been forced into bankruptcy.[63] As one observer noted, "the capital structure of the American aeronautical industry has withstood the strain astonishingly well."[64] Earnings had dropped for much of the industry, but most of the larger companies had managed to maintain strong capital ratios and positive working capital, partly through sharp cutbacks in production and new capital investment. It had been a year of retrenchment, of economy, of driving efficiencies in manufacturing and operations; the overproduction of 1929 was a thing of the past.

With the needed retrenchment and readjustment under their belts, the leaders of the industry were cautiously optimistic for the coming year. Richard Hoyt said, "As we approach 1931 I feel that the industry in general is in a far sounder position than it was a year ago, and faces the coming year with renewed confidence and a keener understanding of the problems to be met."[65] Keys stated his belief that "all the major units in the industry, both in manufacturing and in transportation, have met the business readjustment squarely and are in a strong position either to stand still or to go forward boldly in 1931 as circumstances may dictate."[66] A number of men, Hoyt included, still believed that private flying would rebound, once the manufacturers had produced planes that were easier to fly and more economical and the infrastructure for flying had been more developed.

Keys was now more skeptical. "It is too early to be optimistic," he said, "because it cannot be said that the tide has definitely turned in manufacturing, selling, or operations. There has been no great flood of private buying of aeroplanes or engines. This is due partly to the fact that this element of the business was tremendously overestimated in 1928–1929, and partly to the fact that buying of all expensive merchandise has been greatly curtailed in the United States by economic causes."[67] Frederick Rentschler, reflecting on the sources of demand for the companies in United Aircraft & Transport Corporation, was even less sanguine. The industry had suffered losses from the effects of overproduction, forced liquidation, and continued research and development. These losses would have to be met, and Rentschler believed only through government support for the industry, in the form of "extension of equipped airways, contract air mail routes and the programmed military and naval construction."[68]

For Keys, the challenges that the companies in the Curtiss-Wright group, in North American Aviation, and his air transport ventures had faced in 1930 and would continue to face in 1931 were qualitatively different and an order of magnitude greater than anything he had encountered before. He had sustained the Curtiss Aeroplane & Motor Company through its lean years, but Curtiss Aeroplane had been a small company, with only several million dollars in assets, a relatively small staff, two factories, and a single customer, the military. Curtiss-Wright and North American were composed of multiple companies, with total assets close to $100 million, a broad range of customers, plants, products, and markets, and a far larger staff in the midst of an ever-deepening depression. Strategies had to be more clearly defined and operating decisions loomed larger, including which factories to close, how to consolidate production, how to implement cost reductions and drive efficiency, and how to adjust the different business to the new market conditions. The business was becoming a matter of grimly hanging on.

Meeting the challenges that these companies now faced called for a different set of skills and experience than Keys had acquired in his long years in business. Keys was at his best when he took on the role of entrepreneur, innovator, financier, and systems builder. These were his strengths: his ability to process large amounts of information; to identify the key issues or problems; to come up with ideas or a vision for a way forward; and his ability to attract investment capital to make his vision a reality. In Alfred Chandler's terms, Keys was better at the formulation of policies and procedures than in their implementation, and he was better at making strategic decisions involving long-term plans for an organization than in tactical decisions governing day-to-day operations.[69] He was, to use Thomas Hughes's phrase, a "systems conceptualizer" who could combine technological, business, finance, and economic issues and clearly articulate the concept that emerged from his deliberations.[70] Not every entrepreneur or systems conceptualizer can successfully make the transition to management, shifting from the strategic to the tactical.[71] It was difficult for Keys. He was the first to admit that he lacked the necessary experience. As he would write some years later, after the fact, "I was never fool enough to believe that I was qualified, either by temperament or by training, to run such enterprises, being neither an engineer, sales executive, factory executive, accountant, nor chief pilot."[72] Unlike other entrepreneurs, however, Keys did not hesitate to seek out capable managers to run the companies he was associated with and in delegating considerable responsibility to them. As the year went on, he would become more and more reliant on these men.

Shifting the Burden

"I am worn out." Keys wrote these words in a letter to Charles Lawrance toward the end of March describing the work he had been doing on the management situation at TWA, adding: "I am not yet able to work continuously and am leaving for the South next weekend. I hope to be gone at least a month but may not be able to do so."[73] The continual strain Keys was under was affecting his health. Early in the year William Langhorne Bond, who was being considered for a post as operations manager of China National Aviation Corporation, had an interview with Keys, whom he recalled as looking "harassed and ill...I doubt if he really knew why I was there."[74] Keys had hoped to spend time at St. Catherine's Island off the Georgia coast where he could get some much needed rest, but it was not to be. There were, as always, incessant demands on his time and energy.

Keys had finally begun to take steps to ease at least some of the burdens he was carrying. At a special meeting of the Board of Directors of North American Aviation, he had proposed a number of changes in the organizational structure of the company. As he explained in the annual report, "the executive work of your company, involving continuous contact with all of [its subsidiary companies], has assumed such proportions that it is necessary to increase the official organization, which has remained almost without change since incorporation."[75] Keys proposed, and the Board approved, adding several formal positions to the senior management, creating the position of chairman of the board, a vice-chairman of the board in addition to the president, two or more vice-presidents, and a designated secretary and treasurer.[76] Keys relinquished the office of president of North American Aviation to become chairman of the board, but kept his position as chief executive officer. In his place Keys promoted Thomas Morgan, president of the Sperry Gyroscope Company, to be president of North American Aviation. J. Cheever Cowdin was elected as vice-chairman, in effect the assistant chief executive officer of the company. Since much of the work in the Executive Committee related to Eastern Air Transport, Keys brought in Thomas Doe, Eastern's president, to be first vice president of North American Aviation. Keys finally brought in more administrative support, appointing John Sanderson, who had been acting as treasurer of Intercontinent Aviation, as secretary and treasurer with two assistants, and appointing Charles Kraemer as vice president. Then, a month later, Keys arranged for a permanent reduction in the number of directors of the company from sixty to forty-five, which was still a very large number compared to other American companies, but further simplified his administrative burden.[77]

All things considered, North American Aviation had a reasonably good year. Net income for 1930 was $1,911,110 compared to $1,908,819 for 1929.[78] The income account brought out the dramatic shift in the company's mix of business from a pure investment company to a mixed holding company/investment company; 95 percent of North American Aviation's net income came from the net earnings of its operating subsidiaries, principally Sperry Gyroscope, Ford Instrument Company, and Eastern Air Transport, and a mere 5 percent came from the parent company's investment activities. North American Aviation's subsidiaries had performed satisfactorily through a difficult period. Although the Watres Act had changed the rate structure for the air mail service Eastern Air Transport provided, the healthy 49 percent increase in mileage flown helped offset this and, at least in part, the increased costs due to the introduction of passenger services. Earnings declined during the year, but North American's operations remained profitable. For the first six months of 1931, the company had net income of $540,410, which was just over half the profit earned over the same period in 1930, but still it was a profit.[79]

The market value of North American Aviation's portfolio of marketable securities had continued to decline during the year with the general decline of the stock market. Securities carried at a cost of $15,269,186 had a market value of $3,918,946 at year end, though a brief rally in the first few months of 1931 raised this value to some $4.6 million at the end of June.[80] The annual report noted that net income for the year was based on carrying these investments at cost and not at their market value. Under the accounting conventions of the day, there was no formal requirement that the portfolio be written down to its market value and the resulting loss taken through the profit and loss statement. As these assets were considered to be long-term investments, the issue revolved around the question of permanent impairment, which was a subjective judgment. Since there was broad sentiment at the time that the economy would recover during the latter part of the year, there seemed to be no immediate pressure on Keys to recommend reducing these investments to their market value. The financial press at the time, at least, did not raise the issue, though a number of North American Aviation directors expressed their concern when Keys attempted to institute a profit-sharing plan for executives and senior managers. Several directors protested against introducing the plan, which Keys wanted to implement as an inducement to retain senior managers, at a time when North American was not paying dividends and was holding a portfolio of securities that had experienced such a sharp decline in value.[81]

While the news from North American Aviation was reasonably good under the circumstances of a deepening depression, Curtiss-Wright posted the largest loss of any company in the aviation industry. Releasing its annual report on March 27, 1931, the corporation reported a net loss of $9,012,919.[82] In comparison, the Aviation Corporation posted a loss of $4,703,601, while United Aircraft & Transport Corporation had net income of $3,302,206, a sharp decline from 1929 but a far better performance than its rivals.[83] Once again the Curtiss-Wright annual report went out under Richard Hoyt's name, not Keys; it did not make for pleasant reading. In his report to the stockholders, Hoyt said that the program of liquidation and consolidation of business activities and facilities announced in the semiannual report had been continued "aggressively," although these measures had not been sufficient to reverse the losses.[84] The cost of sales and administrative expenses had been $6.7 million greater than revenues. On top of the operating loss, the corporation had taken an additional $2.7 million charge for inventory write-downs and development expenses. The only positive news coming out of these figures was that, despite the size of the loss, the corporation had managed to reduce expenses compared to the prior year. Hoyt reported that "the progress that we have made in placing our operations on the most economical basis possible is evidenced by the fact that the general and administrative expenses of our companies are now less than half similar expenses at this time last year."[85]

Curtiss-Wright had taken steps to consolidate operations and streamline the organization. The Curtiss-Robertson Airplane Manufacturing Company in St. Louis and the Travel Air Manufacturing Company in Wichita, Kansas, had been renamed the Curtiss-Wright Airplane Company to continue production of small aircraft. The New York plant of the Loening Aeronautical Engineering Corporation, a division of the Keystone Aircraft Corporation, had also been closed, and production was moved to the Keystone plant in Bristol, Pennsylvania. Engine production at the Curtiss Aeroplane & Motor Company's plant in Buffalo was being transferred to the Wright Aeronautical factory in Paterson, New Jersey. Similarly, the Curtiss Aeroplane factory at Garden City, Long Island, which Keys had been instrumental in getting started back in World War I, was shut down, and its production was shifted to Buffalo where, in addition, a second production plant was shut down. The Curtiss Aeroplane factory in Buffalo would now be responsible for the production of military airplanes and experimental work. The inventory of surplus airplanes had been steadily worked down from 585 at the beginning of 1929 to 151 by the end of the year, but as Hoyt commented in the annual report "competition in the commercial field is still exceedingly keen, and it is not anticipated that the Curtiss-Wright Airplane Company can do better than meet expenses under present conditions."[86] Engine sales from Wright Aeronautical and military airplane sales from Curtiss Aeroplane and Keystone Aircraft were carrying the company, and thanks in part to new military orders, the Curtiss-Wright group of companies had $12.2 million in unfilled orders, a 20 percent increase over the prior year.[87]

The most serious drain on the corporation's financial position continued to be from the Curtiss-Wright Airports Corporation and the Curtiss-Wright Flying Service. The Flying Service, Hoyt admitted, "has been one of the most difficult problems and in the past has contributed a very considerable proportion of the total operating loss. Flying equipment and personnel were provided far in excess of the actual demand for flying service."[88] The Flying Service had been conceived, and expanded, based on a projection of a continuing strong demand for flying that had been evident in 1929. Keys and many others had had great expectations for an explosion in private flying; the expansion of the Flying Service into a national network of sales and service centers, distributors for Curtiss-Wright products, and air taxi services had been a key component of his plan for developing commercial aviation. The Depression was having a severe negative impact on the plan, which was now rapidly unraveling. Not nearly enough people had the money for flying lessons, and much less to purchase a small airplane. Sales of private airplanes were roughly a third of what they had been in 1929, and Curtiss-Wright's own commercial sales were practically at a standstill.[89]

Despite a near-fourfold increase in the number of passengers carried in charter and air taxi services, Hoyt stated that "the revenue to be derived from taxi and charter flights has not equaled the cost of the service."[90] The Flying Service had been forced to reduce its fleet of airplanes from 350 to 288, and Hoyt admitted that even this figure was well in excess of requirements. The situation at Curtiss-Wright Airports was similar, if not quite as bad; the revenues being generated from operating airports and concessions were simply too small in relation to the original investment. The Curtiss-Wright Corporation was now encumbered with a large number of properties across the country that were not making money. As a precaution, after a careful analysis of the corporation's assets the Curtiss-Wright directors agreed to set up a contingency reserve of $6 million to provide for the probable excess of book value over real value.[91]

Although the losses began to decline, they continued. For the first quarter of 1931, the Curtiss-Wright Corporation reported a loss of $1,088,124, followed by a loss of $375,030 in the second quarter.[92] The market's view of Curtiss-Wright had changed radically. Its market value had declined from a high of $239 million in 1929 to a mere $13.6 million as of November 9, 1931.[93] Whereas Curtiss-Wright common stock had a leading position at the time the merger was announced in 1929, purchase of the stock was now seen as "a radical speculation."[94] With a large proportion of its original business devoted to the commercial aviation field, Curtiss-Wright had been "forced to make an about-face, from the manufacturing viewpoint, and concentrate almost wholly on the military field."[95] As the *Wall Street Journal* commented in an analysis of the corporation and its painful reorganization,

> while the substantial reductions have been made in the operating deficits shown, it is too early to predict with any degree of certainty the ultimate results of this reorganization from the commercial viewpoint. The commercial demand currently is very small in the airplane industry, and is likely to remain so for some time. . . . Curtiss-Wright's commercial business now consists almost entirely of the operation of

The Grand Central Air Terminal in Glendale, California, one of the many airports in which the Curtiss-Airports Corporation and the Curtiss-Wright Flying Service had an interest. By 1931, both companies were facing steep declines in revenues from overambitious expansion. (2002-01-15_image_076_01, Folder 2, Box 30, Curtiss-Wright Photos, Museum of Flight)

flying schools and flying services. The latter branch has been operating at a deficit since incorporation of the company, and it is doubtful if the profitmaking abilities of the military units can offset the deficits of the Curtiss-Wright Flying Service in the remaining months of this year.[96]

What Keys thought about this change in fortune or about the growing necessity to dismantle the edifice he had so laboriously created, he did not say. He did, however, finally take steps, as he had at North American Aviation, to remove himself from close involvement with the operating companies in the Curtiss-Wright group. The reorganization of these companies, the consolidation of manufacturing plants, and the combination of production facilities were tasks that were beyond him at this stage. Curtiss-Wright had little choice but to continue the program of reorganization and retrenchment, cutting excess capacity and expenses in the commercial field, and trying to hang on to sufficient military business to reduce losses. At the end of June, Richard Hoyt announced that Thomas Morgan had been elected president of the Curtiss-Wright Corporation, replacing Keys, who was now made chairman of the Executive Committee.[97] In making this change, Hoyt may have realized that Keys was perhaps not the best person to oversee a sharp cutback in the operations in which he had invested so much effort. To what extent this move was forced on him remains unclear, for nothing was said at the time, but if it was forced, Keys took it with good grace. He fell back onto the pattern that had served him in good stead, providing ideas to good operating managers and letting them get on with their jobs.

Keys Leaves Aviation

The end, when it came, came quickly. All year long Keys had worked to repay the call money C. M. Keys & Co. had borrowed from various Curtiss-Wright and North American Aviation subsidiaries. C. M. Keys & Co. had begun the year with some $985,000 in call loans, and it had been a struggle to reduce this total. Keys liquidated what assets he could, but by the end of the summer of 1931, C. M. Keys & Co. still had close to $800,000 in call loans outstanding.[98] With personal incomes continuing their decline and many of his clients sharply cutting back on their investment activity, there had not been much traditional investment business to support the company. Many investors were choosing to leave their funds in only the safest of investments. The stock market continued its steady decline. Volumes on the New York Stock Exchange were half of what they had been in 1929, and the bond markets were little better.[99] New issues of common stock and bonds were down substantially from those of prior years. As a result, the earnings of investment banks and brokerage houses dropped sharply. The number of brokerage offices shrank and several medium and large brokerage firms failed during the course of the year.[100]

Then, in September, Sterling fell into crisis, and the United Kingdom abandoned the gold standard, sending shock waves through financial markets around the world as the international currency system neared collapse. The American economy suffered a sharp contraction from what President Hoover called "a concatenation of catastrophes from abroad," experiencing a record number of bankruptcies and declines in almost every industrial sector.[101] The contribution of corporate profits to National Income was actually negative for the year, implying sharp cuts in corporate dividends. The financial system approached the crisis point as both foreign and domestic investors rushed to withdraw their deposits from banks as the panic spread. During the last four months of the year, 1360 American banks failed across the country, compounding the failure of 932 banks up to August.[102] The Dow Jones Industrial Index declined from 140 to 87 between September 1 and October 5, rose to 116 during early November, and then dropped again, ending the year at 74.[103] The real estate market was similarly hard hit, with values collapsing and a growing number of foreclosures and defaults on bank loans. These events, all happening more or less concurrently, had a devastating impact on thousands of individual investors. As one contemporary observer wrote, "In the mind of the average American, 1931 was the year of the Great Depression, for it was in the past 12 months that it really affected us who are just ordinary people, not international bankers, not financiers of any sort, not great executives, and not derelicts who are chronically on the verge of unemployment in all years."[104]

Keys's clients began to default on their obligations to C. M. Keys & Co. During the heady days of the stock market boom, C. M. Keys & Co. had taken on many clients on open account, providing a line of credit for the purchase of securities, which were then pledged as collateral for the loan. Now the value of the underlying collateral was collapsing, and many of these individual clients who were by now heavily overextended could not come up with the funds to repay their loans. During 1930–31, C. M. Keys & Co. had liquidated approximately $20 million in accounts, basically collapsing its business. Keys now ordered whatever collateral was left on these accounts to be liquidated, but in the market's steady decline in the last months of the year, only a fraction of their value could be realized. This was not nearly enough to repay what was owed, leaving C. M. Keys & Co. with outstanding claims against its clients of over $1 million with little hope of recovery.

James C. Willson still owed Keys more than $330,000 and was unable to repay this debt. A number of securities in the investment portfolio of C. M. Keys & Co. had gone into default that fall, notably $2 million in First Mortgage Bonds of the Montauk Beach Development Corporation, which was on its way to bankruptcy.[105] By December C. M. Keys & Co. was effectively insolvent, its assets insufficient to pay all its liabilities. Keys was forced to the realization that he would be unable to repay the call loans due to North American Aviation's subsidiaries. As of December 31, 1931, C. M. Keys & Co. owed $766,967 to the Sperry Gyroscope Company, Intercontinent Aviation,

Inc., Aviation Exploration, Inc., and the Ford Instrument Company. Against these call loans C. M. Keys & Co. could put up collateral consisting of shares in Intercontinent Aviation, Transcontinental Air Transport, and Douglas Aircraft Company options worth approximately $112,750, and Montauk Beach Development Corporation bonds with a face value of $1.4 million, but worth only a fraction of this amount.[106] Keys had kept his financial condition and the call loan problem hidden from the other directors of North American Aviation, but now he had no alternative but to admit what he had done and take the consequences.

Keys's extensive papers make no record of what transpired at the meeting of the North American Aviation Executive Committee when Keys informed them of his indebtedness. One can imagine a mix of reactions; shock, disbelief, dismay, anger, sense of betrayal, and perhaps, finally, sadness and pity. So many others had faced ruin in the Crash and Depression that seemed to be spiraling ever downward, and now a trusted colleague, a leader in the industry, a man respected for his vision and insight, had been brought down by these same forces and his own errors of judgment. The Executive Committee ordered an immediate audit of North American Aviation's call loans and an investigation of the financial condition of C. M. Keys & Co., which concluded that Keys's company was in desperate straits requiring a substantial amount of money to remain in business.[107]

The Committee held a number of informal meetings to decide how to proceed. There was, apparently, surprise that North American Aviation's funds had been placed on call with C. M. Keys & Co., and shock at the magnitude of the call loans.[108] It was clear that Keys had violated the trust of his fellow directors, and for that he had to take responsibility. On December 29, 1931, Keys wrote a letter of resignation as a director and officer of North American Aviation and its subsidiaries, which was submitted to the Executive Committee on January 5, 1932. There may have been a "Gentlemen's Agreement" surrounding Keys's resignation; nothing was ever said publicly about the reasons for his action, and there was no contradiction of his own public explanation. It may be that his fellow directors said, in effect, resign from all your positions, walk away, and we will preserve your reputation. Resignation was sufficient humiliation.

On January 8, 1932, Keys issued the following statement to the press;

On account of health, and also the pressure of other business, I am retiring from my aviation activities. I hope, in time, to resume some of these activities, but I don't know when. In the meantime, I believe they are all in very good hands, well organized, and ready to go forward as the trade revives.[109]

Keys was chairman or director of twenty-five companies in the aviation industry. He resigned from all of them, from the Curtiss-Wright Corporation, from Transcontinental & Western Air, from Eastern Air Transport, from every company he had been associated with. He was four months shy of his fifty-sixth birthday.

8

A SUMMING UP

Aftermath

For the first few months after his resignation, Keys was mentally and physically exhausted. He had withdrawn completely from aviation, no longer spoke to the press, and was overwhelmed with his own financial problems. He left no record of what transpired that year or how he spent his time. There were several other expressions of regret at his resignation from associates he had worked with at a number of companies, the most heartfelt coming from Richard Robbins, who wrote Keys to tell him that his letter of resignation from Transcontinental & Western Air had been accepted "with sincere regret and appreciation of the loss to us of your wise counsel and unceasing efforts on behalf of Transcontinental & Western Air, Inc."[1] Robbins then put down his own personal thoughts, which provides an apt and cogent summary of Keys's achievement over the past fifteen years:

> Air transportation owes its present state of development to the foresight and organizing ability of such men as yourself. Transcontinental Air Transport and this company have been fortunate in being guided through the crucial pioneer period of the last few years by you. Your work in bringing together companies representing all phases of aviation industry and focusing their expansion and efforts toward the

achievement of a definite purpose has been an heroic task. The pioneering of our own transcontinental airline constitutes a work which in time to come will be ranked with the spanning of the continent by the railroads. Credit for conceiving and developing the first air-rail transcontinental passenger service is due you in a great measure. It took courage, perseverance and ability of a large order to carry forward this undertaking. We, who have had the privilege of your advice, will miss you at the council table. I personally regret and feel the loss of your resignation. It is of some comfort to know that we may still come to you with our problems.[2]

By the end of 1932, Thomas Morgan, president of Curtiss-Wright and North American Aviation, had dismantled much of what Keys had built up with such confidence and expectation during the heady days of 1928 and 1929. After the stock market crash and the consequent collapse, Curtiss-Wright had to pursue a program of consolidating manufacturing facilities, cutting expenses, and getting rid of unprofitable operations. The Curtiss-Wright Airports Corporation and the Curtiss-Wright Flying Service, a central part of Keys's vision, were now floundering. Despite previous cuts in facilities and operating expenses, there weren't sufficient sales and service business to support the Curtiss-Wright Flying Service. Nor were the majority of Curtiss-Wright airports operating near their potential. Thomas Morgan decided that the Curtiss-Wright Flying Service had to cease operations, while the Curtiss-Wright Airports Corporation had to be reduced to a skeleton operation.[3]

Following Keys's resignation from North American Aviation, the Board appointed a special committee "to investigate any or all actions of officers or members of the Board of Directors or Executive Committee of North American Aviation and report to the Board recommendations on these points." Five members of the Board were appointed to the committee, none of whom had connections with Keys.[4] A new Executive Committee was elected, which included only two men associated with Keys, Leonard Kennedy and J. Cheever Cowdin, who somehow managed to survive the fact that he and Chester Cuthell had hidden Keys's financial problems from the Board. The directors' wrath appears to have fallen on Chester Cuthell, who resigned as a director of North American, and a month later North American terminated its contract with Cuthell's law firm, Cuthell, Hotchkiss & Mills.[5]

Summing Up

The Special Committee recommended that North American undertake a recapitalization to deal with the problem of the steep drop in value of the securities in the company's investment portfolio. At the annual meeting on March 9, 1932, the stockholders approved the recapitalization plan resulting in a loss of close to half the value of the company. At the same meeting, the stockholders approved a reduction in the number of directors from forty-five to twenty-four, a far more manageable number, but reducing the broad diversity of directors that Keys had wanted for the company.[6] Harold Talbot was elected chairman of the board and chairman of a reconstituted Executive Committee, with Morgan remaining as president.[7]

With North American Aviation's income under pressure Morgan sought to cut back on operations that weren't making a profit. North American steadily withdrew from Keys's investments in foreign airlines through Intercontinent Aviation. On May 6, 1932, Intercontinent sold the stock of Compañia Nacional Cubana de Aviación Curtiss S.A. to Pan American Airways for $500,000.[8] Nearly a year later, on April 1, 1933, Pan American purchased Intercontinent Aviation's interest in China Airways Federal, Inc. and through this purchase acquired China Airways Federal's 45 percent interest in the China National Aviation Corporation, for $282,258 in cash and shares of Pan American stock, giving Pan American a foothold on the continent of Asia.[9]

A change in ownership at North American Aviation brought what might be called the Keys era to an end. In January 1933, the General Motors Corporation acquired North American Aviation through its subsidiary, the General Aviation Corporation. The acquisition built on an existing link between General Aviation and North American through their participation in Transcontinental & Western Air. General Motors gained a larger presence in the air transportation industry and a possible outlet for the General Aviation Corporation's airplanes. Through North American's portfolio of aviation securities, General Motors gained a stake in several of the larger aviation manufacturing concerns, notably Curtiss-Wright and Douglas Aircraft.[10] In March the directors of North American and General Aviation agreed to give General Aviation a 43 percent interest in North American. Through these transactions, General Motors became the dominant participant in Transcontinental & Western Air. North American Aviation was now almost exclusively an operating holding company. The new management brought North American Aviation's investment activities to a halt.[11]

Restoring His Financial Health

When Keys resigned from North American Aviation, his own company, C. M. Keys & Co., was effectively insolvent. His first task was to prevent his company from going into bankruptcy and then, as best he could, to restore his financial position. He could not have picked a more challenging time or circumstance. His partner, Morris Sinsabaugh, was bankrupt; his clients owed the firm over $1.1 million with little prospect that these debts could be recovered; the advances he had made to his clients in 1928 and 1929 had resulted in losses of over $4 million; and the American economy was entering what would be an "Annus Terriblus."[12]

Keys's first step toward financial recovery was to come to an agreement with North American Aviation on the borrowed money he was unable to repay. This amounted to $766,000, with an additional $77,000 owed to Transcontinental Air Transport.[13] He offered to provide what collateral he had available for these loans. In early February 1932, North American's Executive Committee reached an agreement with C. M. Keys & Co. whereby the various subsidiaries took over this collateral, consisting of shares in Intercontinent Aviation and Transcontinental Air Transport, and options in the stock of Douglas Aircraft Company.[14] Later that same month, Keys entered into a Creditors Agreement with all the creditors of C. M. Keys & Co. to arrange for the orderly liquidation of his debts. From January to April Keys sold what securities he had available at C. M. Keys & Co. and from his own account to raise more money to repay his debts.[15] In June Keys legally ended his partnership with Morris Sinsabaugh and reestablished C. M. Keys & Co. under his own name, but with a substantially smaller business and smaller staff at his offices at 39 Broadway.[16] It would take Keys more than six years to work out his financial problems, a period that Keys later referred to as being "completely submerged for six years in a long business battle to restore my own affairs to a proper basis."[17] He would not be completely free of debt until 1939. While working on his own affairs, he tried as best he could to help rebuild the finances of his clients who had invested in the aviation sector on his advice.[18]

Clement Keys's nephew recalled that his uncle had the ability to move on from whatever challenge he had completed. After resigning his positions in the aviation industry, he took a matter-of-fact view of what had happened; he had helped establish the aviation industry, but it was now up to someone else to take the industry to the next stage.[19] In the following years, Keys turned his energies to serving as a corporate director and as a member of protective committees for bondholders and stockholders in bankruptcies. Keys also became involved in several corporate reorganizations where he could use his financial expertise. Keys maintained C. M. Keys & Co. through the 1930s, registering with the new Securities and Exchange Commission as a broker and dealer in securities. Keys and his wife lived a quieter life under more constrained circumstances; there do not appear to have been any more trips to Europe, or leases of properties at high-end resorts.

There was, however, one final venture in aviation in Keys's future. The genesis of this venture was a meeting Keys had in Toronto in 1944 with John Gundy, a Canadian banker who was interested in starting an airline after the war was over. Keys suggested that Gundy consider an airline that would run from Montreal via New York, Havana, Panama, and Lima, Peru, to Santiago, Chile, connecting North America with the western and southern parts of Latin America. Keys had considered and surveyed this route back in 1928.[20] After the end of World War II, in January 1946, Keys arranged $4 million in financing from a group of American, Canadian, and Peruvian backers and organized Peruvian International Airways (PIA). In December of that year, the United States and Peru signed a commercial air agreement that allowed two U.S. carriers and the newly formed Peruvian airline to operate routes between Peru and the United States. Keys hired Lt. General Harold George, who had led the Air Transport Command during World War II, to be president and chairman of the board of PIA. The airline purchased four surplus Douglas C-54 transport planes and set about converting them to civilian passenger configuration. PIA received a certificate from the Civil Aeronautics Board to operate a route between New York and Lima against bitter opposition from Pan American Airways and its Latin American affiliate, Pan American Grace.[21]

The airline had much going for it—a president experienced with global air transport operations, strong financial backing, adequate capital, and a promising route structure—but problems developed all too quickly. Peruvian International Airways began operating the route from Lima to Havana via Panama on May 14, 1947. Three months later, a service to Santiago was introduced, and finally, on September 26, 1947, PIA inaugurated its service to New York beginning the first direct air service between Santiago and New York. By the end of the year, the airline was running three flights a week to New York, but it was far from breaking even. Competition along its routes had increased when Panagra (Pan American Grace Airways) introduced more modern Douglas DC-6s along the route, followed by Braniff which initiated a service with the DC-6 from Santiago to Houston, Texas, via Havana. A plan to increase the capital to $9.5 million came to nothing that fall, and in February 1949 the airline ceased operations.[22]

Clement Keys died of a stroke at his home on West 55th Street in Manhattan on January 12, 1952, just four months shy of his seventy-sixth birthday. The *New York Times* carried a long obituary outlining his career and achievements, but sadly his passing went little noticed in the aviation press.[23]

A Summing Up

There remains the question of what American aviation would have been like during the 1920s had Keys remained a teacher of classics, a journalist, or even an investment banker and not gone into aviation. Few human endeavors rely solely on the shoulders of one individual. But it can be argued that Keys was clearly one of perhaps a dozen or so men who played pivotal roles in developing the industry during that decade. One thinks of Major General Mason Patrick and Rear Admiral William Moffet and their contributions to the development of military aviation; Harry S. New, Paul Henderson, and Walter Folger Brown at the Post Office; Herbert Hoover as secretary of Commerce and William P. McCracken, as assistant secretary of Commerce for Air, in government; the few congressmen and senators who were strong supporters of aviation legislation, like Representatives Clyde Kelly and Florian Lambert and Senator Hiram Bingham; Charles Lindbergh, for stimulating America's interest in flying; in the business of aviation, there were Henry Ford, Howard Coffin, William E. Boeing, Richard Hoyt, Frederick Rentschler, Harris Hanshue, Graham Grosvenor, and Averell Harriman of the Aviation Corporation, and Keys. Had Keys not been involved, the American aviation industry might well have developed more or less along the lines that it did—but perhaps not.

Keys was a catalyst. He stimulated the growth of the aviation industry as a business, first in fighting for an end to the military's destructive competitive bidding system and then in supporting more rational government policies toward commercial aviation. In forming National Air Transport, Transcontinental Air Transport, and Eastern Air Transport, Keys advanced aviation as a transportation system. He formed North American Aviation as a means of raising capital for aviation ventures. Finally, his successful rescue of the Curtiss Aeroplane & Motor Corporation led, in time, to the merger of the Curtiss and Wright groups. His name, less well known today, became associated with progressive, successful ventures. Founding companies like National

A photograph of Clement Keys in his later years. His last aviation venture, Peruvian International Airways, ceased operations in early 1949. (Courtesy C. Roy Keys Jr.)

Air Transport and Transcontinental Air Transport made news around the country, helping to stimulate the flow of capital into aviation. Writing an article for the *New York Herald Tribune* in June 1929 titled "The Harriman of Aviation," Howard Mingos called Keys "the most interesting personality in the industry" and "the most important person in aviation."[24]

Keys fits both Thomas Hughes's definition of a "systems builder" and Joseph Schumpeter's definition of an entrepreneur. He had in large measure one of the characteristic motivations of the entrepreneur: a willingness to accept a challenge. Keys sought what Schumpeter called "the joy of creating, of getting things done, or simply exercising one's energy and ingenuity"; he was the type of man who "seeks out difficulties, changes in order to change, delights in ventures."[25] Keys understood from the very beginning that unless aviation manufacturing was to be conducted solely in government-owned factories and air transportation run as a public utility or the recipient of massive subsidies, neither of which he thought possible in the United States, then the aviation industry and air transportation would have to pay their own way. And that meant making a profit and providing a return to shareholders. Keys saw nothing wrong with this. As he said in his 1929 speech, "the greatest incentive to progress and the greatest contributions to human comfort and welfare has always been the desire of men seeking to make money. It is old-fashioned, in this day, to sneer at the motives of business or to condemn the desire of mankind to get forward in a material way, because all modern economists agree and even the people in general subscribe to the theory that material progress in a nation makes not only for the welfare of a few, but makes for the happiness and general welfare of the nation as a whole."[26]

Unlike so many in Congress in that era and later during the New Deal, Keys recognized that for aviation to become a real business, molding it into a "new combination," to use Schumpeter's term, invariably meant separating the invention from the inventor. Whereas many members of Congress bemoaned the fact that "big business" and "Wall Street" were coming to dominate aviation and sought to preserve space for the independent inventor or operator, Keys viewed this process as inevitable. As Schumpeter noted, "as long as they are not carried into practice, inventions are economically irrelevant. And to carry any improvement into effect is a task entirely different from the inventing of it, and a task, moreover, requiring entirely different kinds of aptitudes."[27] Thomas Hughes makes this same point in his analysis of the evolution of large technological systems where the initial inventor–entrepreneur gives way to the manager–entrepreneur and then the financier–entrepreneur at different stages of the system's development.[28]

Keys exercised what Schumpeter called "entrepreneurial leadership."[29] In Schumpeter's view, the entrepreneur marshaled the means of production, including capital, into his new combinations. This form of leadership had, Schumpeter argued, little of the glamour of other forms of leadership, but it required "fulfilling a very special task which only in rare cases appeals to the imagination of the public. For its success, keenness and vigor are not more essential than a certain narrowness which seizes the immediate chance and *nothing* else."[30] With regard to marshalling the means of production and the special tasks involved, Schumpeter and other classical economists put a premium on the significance of raising capital; financing was "fundamentally necessary" to the formation of new combinations.[31] Keys became adept at the special task of raising financing for his aviation ventures. He was one of the few men, along with William Boeing, Richard Hoyt, Frederick Rentschler, and Averell Harriman, who succeeded in bridging the world of aviation and the world of finance.

Keys's insistence on financing his companies with equity instead of debt was financially prudent. In coming up with this approach, he benefited greatly from his study of the railroads. His understanding of the causes of so many of the railroad failures of the 1890s and early 1900s—inept management, poorly planned expansion, excessive capitalization, and above all, overreliance on financing with debt with the associated heavy burden of fixed interest charges—helped him determine the characteristics he needed for the companies he organized. A study of the "negatives" helped him focus on the "positives." While several of his aviation ventures encountered severe difficulties during the Depression, none failed due to an inability to meet fixed charges on debt, as they had none. In all likelihood, this helped Curtiss-Wright get through the Depression. Keys was extremely fortunate that the Lindbergh boom and the demand for aviation stocks gave him the opportunity to issue stocks to finance his ventures. Had this surge of interest in aviation stocks not happened, Keys might never have been able to expand to the extent he did.

Keys wanted to organize companies that had strong, reputable backers who were, in the main, experienced businessmen. He needed to add financiers like J. Cheever Cowdin, James C. Willson, and G. M.-P. Murphy because he needed their connections to the investors on Wall Street and elsewhere, but he wanted men who understood business problems and were willing to take a business risk. Keys also sought out businessmen from across the country to be directors of his companies so as to diversify the support for and interest in aviation among the broadest possible community of investors. Having notable businessmen on his boards would be equally reassuring to investors. What seems like a natural, certain progression in hindsight was actually, for much of the decade of the 1920s, a highly speculative investment in an unproven industry. It was not, Keys said repeatedly, an investment for "widows and orphans." However, much as Keys would have liked to believe that the stocks in his companies were only sold to businessmen who could afford to take the risk, he could not completely control distribution of the underwritings. Just as was the case in later speculative markets, many investors who bought aviation stocks probably could not afford their subsequent losses.

CHAPTER EIGHT

246

In Schumpeter's schema of "new combinations," it may be seen that Keys contributed to the introduction of new products and services, the opening of new markets, and the carrying out of new forms of organization. With regard to new products and services, one thinks first of the Curtiss D-12 engine, which captured world records and brought world attention to American aviation products. The D-12 existed because Keys rescued the Curtiss Aeroplane & Motor Corporation from bankruptcy and pushed development of this powerful engine. An argument can be made that if Keys had walked away from Curtiss there would have been no D-12 engine, no Curtiss racing planes to win the 1923 and 1925 Schneider Trophy races, nor the hundreds of Curtiss fighter, attack, and observation aircraft built for the Army Air Corps and Navy during the 1920s and 1930s. The Curtiss D-12 played an important role not just in American aviation, but also in subsequent liquid-cooled engine development. As one historian of the D-12 noted, the D-12 was the "American link between the French highlight in engine building of World War I and British supremacy in aero engines during World War II."[32] The D-12 had a profound influence on the British Air Ministry, which turned to Rolls-Royce to develop a similar engine.[33] This became the Rolls-Royce Kestrel, the first Rolls-Royce aluminium Monoblock engine, which led in turn to development of the Rolls-Royce Merlin of World War II fame, a scaled-up version of the Kestrel.[34]

Keys's role in the development of American air transportation was equally important. In helping to organize National Air Transport, building Transcontinental Air Transport, and developing Pitcairn Aviation into Eastern Air Transport, Keys, like William Boeing, Frederick Rentschler, and Harris Hanshue, helped introduce new transportation services to the American public. In helping to form National Air Transport, Keys contributed to successfully transferring the air mail service from the government to private enterprise.

Keys considered Transcontinental Air Transport his greatest achievement, even though this bold experiment in all-passenger transportation had to be rescued through air mail contracts. In many ways it *was* his greatest. For air transport to develop, someone had to take the first step. Keys was willing to do this, even though he realized that with neither suitable aircraft nor the required infrastructure, early experiments were fraught with risk. But TAT made a start. The airline cut twenty-four hours off the time it took to cross the continent, demonstrating the airplane's potential to dramatically shrink distance and speed connections between people. With TAT Keys achieved what the earlier railroad builders had not: a single system spanning the continent instead of a hodgepodge of connecting railroad systems limited by geography and regulation. American air transport could, conceivably, have gone the way of the railroads, with a multitude of smaller regional airlines developing. TAT showed that single systems were not only possible, but also a far more efficient use of this new technology.[35] Keys's work in organizing TAT exemplifies his often expressed belief that nine-tenths of aviation was on the ground. If it was to succeed, aviation had to be based on disciplined execution and careful organization.

From his first days at the Curtiss Aeroplane & Motor Corporation, Keys, to a greater extent than his contemporaries, pushed the opening of new markets for aviation products around the world. He sustained the Curtiss Aeroplane Export Corporation and, when financing became available, its reorganization. During the 1920s, Curtiss salesmen traveled through Europe, Asia, and Latin America in search of new markets for Curtiss products, with much success, capturing sales in markets that the European nations had previously dominated. Similarly, he recognized the important role air transport could play in other countries with less developed systems of transportation, especially China. For his role in setting up the first successful commercial airline in China, he is considered by one historian as the father of Chinese commercial aviation.[36]

In creating new combinations, the entrepreneur often encounters resistance or reluctance to adopt what is new and untried; overcoming the inertia of fixed habit takes hard work. Keys's adaptation of the investment trust vehicle was an innovative means of addressing the problem of financing new and speculative ventures that would be less able to attract capital from more traditional banking sources. Linking these investment trusts with his companies, as he did with North American Aviation, was a means of ensuring continuity of management. Keys was, however, slow to adopt the holding company structure for his aviation group, the merger between the Curtiss and Wright groups taking place some months after other aviation groups had formed. As Thomas Morgan noted in a 1932 report to Richard Hoyt, "the Curtiss–Wright combination, consisting largely of several new companies in this young industry, a number of which having been in keen and bitter competition, created an unusual and complicated administrative problem."[37] Keys did not have time to address this issue before the Crash of 1929, and subsequently he was overwhelmed with his many problems.

Keys's role as a promoter of aviation and his vision must be added to this list of achievements. Keys was a tireless advocate for aviation, though most often not in the public eye. Unlike Brigadier General William Mitchell, Keys did not seek out publicity to advance his cause. During the 1920s, he wrote numerous articles about aviation, and as the decade wore on and aviation became more popular, he gave frequent speeches to various public forums. But most of his effort was devoted to countless private meetings with congressmen, government and military officials, journalists, businessmen, financiers, and potential investors. By his own count he testified before more than twenty congressional, governmental, or military committees or investigations. During the early 1920s, Keys was a fierce critic of the competitive bidding system, which he believed would destroy the aviation industry. His position as president of Curtiss Aeroplane gave him access to senior military and governmental officials, and he took advantage of this access to present his arguments, as he did in his extensive correspondence with Major General Mason Patrick, head of the Army Air Service. In the latter half of the decade, he devoted his effort to promoting air transport, speaking with confidence and candor about not just the problems that had to be addressed, but also how to address them. It was, he always said, a job of hard work, but he conveyed his belief and confidence that, in time and with perseverance, the problems would be solved, and air transport would come to play a significant role in the life of the nation.

Keys's vision for aviation and the range of his involvement were arguably broader than those of any of his contemporaries. While some individuals were just as involved in certain aspects of aviation, few could match the diversity of Keys's activities. Keys played a role in airplane and engine manufacturing, the development of air transportation of mail and people, the buildup of the infrastructure to support aviation in the form of airports, air services, and training facilities, the promotion of aviation exports, and the formation of airline companies abroad. Keys had tremendous faith that aviation would ultimately come to rank as one of the leading forms of transportation not just in America but throughout the world, though at times the course of this development frustrated him.

There were flaws in his vision. With the benefit of hindsight, it can be seen that he put too much emphasis on private flying as being one of the driving forces behind the growth of the aviation industry, but then many others did too at the time. The idea that the airplane would rival the car as a form of transportation captured the imagination. Even if the Depression had not happened, it is doubtful that the reality would ever have lived up to Keys's expectations. Moreover, the structure he built up through the Curtiss-Wright Flying Service and the Curtiss-Wright Airports Corporation would likely have become substantially smaller. Keys was also slow to realize the growing dominance of the air-cooled radial engine; he clung to the belief that the liquid-cooled engine, in which Curtiss had a dominant technological lead, was best suited to the needs of the airlines, when there was ample evidence that it was not. The solution came through the merger of Curtiss and Wright Aeronautical, but one wonders at the opportunities that might have been lost along the way.

Where Keys was less successful was in making the transition from financial entrepreneur to manager. The entrepreneur's role is, in the main, a temporary one; as Schumpeter noted, "everyone is an entrepreneur only when he actually 'carries out new combinations', and loses that character as soon as he has built up his business, when he settles down to running it as other people run their businesses."[38] The manager's focus is on "discipline, structure, and organization," and the manager's role is to monitor and coordinate the producing and distributing units within the organization to ensure the optimum allocation of resources for the future.[39] Keys had little experience of management, as he himself admitted. As he wrote to Thomas Morgan years later, "I was never fool enough to believe that I was qualified, either by temperament or training, to run such enterprises, being neither an engineer, sales executive, factory executive, accountant, nor chief pilot. All I did was have a good idea now and then. . . . The rest of it was a matter of choosing men."[40]

Keys chose to manage his companies through delegation, and he was fortunate in being able to attract, for the most part, capable subordinates, but the structure of the organization he was building was not his real focus. His reluctance to build a support structure around him as the range of his activities expanded dramatically, despite requests to do so from his associates, is curious. His removal as president of Transcontinental Air Transport and his appointment as chairman is evidence of the difficulty he had taken up when he accepted an active management role. Clearly, his financial problems in the first years of the Depression appear to have been a constant drain on his time and energy, yet only belatedly did he change the management structure at North American Aviation and bring in more staff to ease his burden.

Keys's decision to retain individual companies rather than create a divisional structure within the Curtiss-Wright group reflected his own preferences and management style, and played to his strengths as an entrepreneur and financier. But it made coordination that much more difficult and the attainment of economies of scale that much more challenging, especially during the desperate years of the Depression. Thomas Morgan noted that within the newly merged Curtiss-Wright group of companies, "there were comparatively few men of mature business experience in the management to shoulder the burden and shape the affairs of the corporation into a stabilized and smooth running organization."[41] It is interesting to note that within a few years of his departure from Curtiss-Wright, the corporation moved to a fully divisional structure. By character and temperament, Keys was better placed in a senior executive role as a director or chairman where his ability to generate ideas could have been put to good use through a chief executive and a stronger administrative structure.

The story of Keys's rise from modest beginnings to the pinnacle of his industry and his subsequent expulsion through his own failure in judgment reads like a Greek tragedy. Why he let his judgment fail him and allowed the commingling of his companies' funds with his own investment banking business remains something of a mystery. One can only point to his excessive confidence in the financial markets, an excess of hubris, and as he tried to explain, the habits of many years. That he should have abandoned his fiduciary responsibilities was uncharacteristic and in the absence of a more detailed explanation may simply remain unexplainable. His financial problems proved to be a distraction of time and energy away from his other responsibilities, at a time when all the companies he was involved with needed wise counsel.

Shortly after Keys died, the magazine *U.S. Air Service* published a eulogy by a man who knew him. His tribute provides a fitting end to this story of his life:

> During the era immortalized by Lindbergh's flight to Paris, Mr. Keys was one of the most active and influential men in the aviation industry. He accumulated great wealth and was proportionately a sufferer due to the equally immortalized market crash in 1929. He seemed before the crash to be a Miracle Man. He had the magic touch. Crowds followed him from place to place to hear him on any aspect of life as it was then lived. He was delightfully interesting in conversation in a quiet way. He had a great fund of information. A luncheon with him when there were only three or four present, became a high spot in one's experience. He had the story-teller's art. He must have been a dynamo to do the things he had done, handling so many large companies with such apparent ease and success, but in conversation one had to be alert to hear all he said, because of his softspoken words, and in appearance he was rather frail.

Summing Up

Mr. Keys was hard hit by the '29 crash, but we have heard many authentic references to the splendid way he took it. Also have been told that he made many personal sacrifices to care for friends whom he felt had invested because of his advice—and lost. We shall long remember various conversations with C. M. Keys, some during the time when he was riding the crest of the wave, others when he was in the trough.[42]

APPENDIX

Clement M. Keys Principal Aviation Interests
and Offices Held 1930–1931[1]

Manufacturing Companies

Curtiss-Wright Corporation: President and Director
Curtiss Aeroplane & Motor Company: Director
Curtiss-Wright Airplane Company (Delaware): Director
Curtiss-Wright Airplane Company (Missouri): Director
Curtiss-Caproni Corporation: Director
Keystone Aircraft Corporation: Director
Wright Aeronautical Corporation: Director
Bendix Aviation Corporation: Director
Douglas Aircraft Company, Inc.: Director
Ford Instrument Company: Director
Sperry Gyroscope Company: Chairman

Aviation Services

Curtiss-Wright Flying Services, Inc.: Director
Curtiss-Wright Airports Corporation: Director
Curtiss-Wright Export Corporation: Director
Intercontinent Aviation, Inc.: Chairman
Aviation Exploration, Inc.: Chairman and President
New York Air Terminals, Inc.: Director

Investment Companies

Aviation Corporation of California: Director
Aviation Credit Corporation: Director
Aviation Securities Corporation, Chicago: Director
National Aviation Corporation: Director
North American Aviation, Inc.: Chairman
Condor Corporation: Chairman

Airlines

China Airways Federal, Inc.: Chairman
Compania Nacional Cubana de Aviacion Curtiss: Director
Eastern Air Transport, Inc.: Chairman
National Air Transport, Inc.: Director (until May 1930)
Transcontinental Air Transport, Inc.: Chairman
Transcontinental & Western Air, Inc.: Chairman

NOTES

Preface

1. Roger E. Bilstein, *The American Aerospace Industry: From Workshop to Global Enterprise* (New York: Twayne Publishers, 1996), p. 21.
2. Ibid., p. 22.
3. Ibid.
4. Rudolf Modley, Editor, *Aviation Facts and Figures 1945* (New York: McGraw-Hill Book Company, 1945), p. 7.
5. Department of Commerce: Bureau of Foreign and Domestic Commerce: *Statistical Abstract of the United States 1934* (Washington, DC: U.S. Government Printing Office, 1935), Table 759: Vehicles and Aircraft: Production, p. 745.
6. Ibid., Table 429, Civil Aeronautics, p. 377.
7. William M. Leary, *The Dragon's Wings: The China National Aviation Corporation and the Development of Commercial Aviation in China* (Athens, GA: University of Georgia Press,1976), pp. 7–8.
8. Naomi R. Lamoreaux and Kenneth L. Sokoloff, Eds., *Financing Innovation in the United States 1870 to the Present* (Cambridge, MA: MIT Press, 2007), pp. 185–86.
9. Ibid., p. 237, note 19.
10. William M. Leary, Editor, *Encyclopedia of American Business History and Biography: The Airline Industry* (New York: Bruccoli Clark Layman, 1992).
11. Edward M. Young, "Clement M. Keys," in Leary, Ed., *Encyclopedia of American Business History and Biography*, pp. 258–66.

Introduction

1. William M. Leary, *Aviation's Golden Age: Portraits from the 1920s and 1930s* (Iowa City: University of Iowa Press, 1989), p. xi.
2. Roger E. Bilstein, *The American Aerospace Industry* (New York Twayne Publishers, 1996), p. 22.
3. Rudolf Modley, *Aviation Facts and Figures 1945* (New York: McGraw-Hill Book Company, 1945), pp. 7–8.
4. Ibid., pp. 66–67.
5. R. R. Doane, "Aeronautical Finance in 1930," *Aviation*, 30, no. 1 (January 1931): 45; John B. Rae, *Climb to Greatness: The American Aircraft Industry, 1920–1960* (Cambridge, MA: MIT Press, 1968), p. 2.
6. Roger F. Bilstein, *Flight in America: From the Wrights to the Astronauts*, Rev. Ed. (Baltimore, MD: Johns Hopkins University Press, 1994), p. 74.

7. Modley, *Aviation Facts and Figures 1945*, pp. 66–67, 116; U.S. Department of Commerce, *Fifteenth Census of the United States: Manufactures:1929*, Volume II: Reports by Industries, (Washington, DC: Government Printing Office, 1933), p. 1189.

8. Donald M. Pattillo, *Pushing the Envelope: The American Aircraft Industry* (Ann Arbor: University of Michigan Press, 1998), p. 60.

9. Bilstein, *The American Aerospace Industry*, p. 22; Maurer Maurer, *Aviation in the U.S. Army 1919–1939* (Washington, DC: U.S. Government Printing Office, 1987), pp. 191–95; Archibald D. Turnbull and Clifton L. Lord, *History of United States Naval Aviation* (New Haven, CT: Yale University Press, 1949), pp. 257–61.

10. John B. Rae, *Climb to Greatness: The American Aviation Industry, 1920–1960* (Cambridge, MA, 1968), pp. 30–35; Bilstein, *Flight in America*, pp. 48–59, 75.

11. Rae, *Climb to Greatness*, pp. 39–46.

12. Bilstein, *Flight in America*, pp. 74–76.

13. *The Aviation Industry* (New York: Punchon & Co., 1928), p. 9.

14. Pattillo, *Pushing the Envelope*, pp. 64–67.

15. Thomas P. Hughes, "The Evolution of Large Technological Systems," in Weibe Bijker, Thomas P. Hughes, and Trevor Pinch, *The Social Construction of Large Technological Systems: New Directions in the Sociology and History of Technology* (Cambridge, MA: MIT Press, 2021), p. 45.

16. Hughes, "The Evolution of Large Technological Systems," p. 48.

17. Thomas P. Hughes, *Networks of Power: Electrification in Western Society, 1880–1930* (Baltimore, MD: Johns Hopkins University Press, 1988), pp. 18–19; Hughes, "The Evolution of Large Technological Systems," pp. 46, 50–51.

18. Hughes, "The Evolution of Large Technological Systems," pp. 46, 50–51.

19. Ibid., pp. 46, 50–51; Thomas P. Hughes, *American Genesis: A Century of Invention and Technological Enthusiasm, 1870–1970* (Chicago: University of Chicago Press, 2004), pp. 2–6.

20. Hughes, "The Evolution of Large Technological Systems," p. 46.

21. Schumpeter, *The Theory of Economic Development: An Inquiry into Profits, Capital, Credit, Interest, and the Business Cycle* (Cambridge, MA: Harvard University Press, 1934), pp. 65, 74.

22. Schumpeter, *The Theory of Economic Development*, p. 92. See also the discussion of the role of the entrepreneur in Anthony J. Mayo and Nitin Nohria, *In Their Time: The Greatest Business Leaders of the Twentieth Century* (Boston: Harvard Business School Press, 2005), pp. xxi–xxiii, xxix.

23. Schumpeter, *The Theory of Economic Development*, p. 86.

24. Margaret B.W. Graham, "Entrepreneurship in the United States, 1920–2000," in David S. Landes, Joel Mokyr, and William Baumol, eds., *The Invention of Enterprise: Entrepreneurship from Ancient Mesopotamia to Modern Times* (Princeton, NJ: Princeton University Press, 2010), p. 402.

25. Ibid.

26. Pattillo, *Pushing the Envelope*, pp. 62–63.

27. Earl Reeves, *Aviation's Place in Tomorrow's Business* (New York: B.C. Forbes Publishing Company, 1930), p. 79.

28. Reeves, *Aviation's Place in Tomorrow's Business*, pp. 79–89.

29. Reeves, *Aviation's Place in Tomorrow's Business*, p. 95.

30. C. M. Keys, "The Business End of Aviation" (May 8, 1929), Folder 5, Box 22, C. M. Keys Papers, National Air and Space Museum (hereafter CMKP, NASM).

31. C. M. Keys to H. G. Wells, May 3, 1927, Folder 29, Box 20, CMKP, NASM.

Chapter 1

1. Pauline Keys Ladd, *The Story of the Evans and Lewis Families*, ND. Diane C. Coates, Diocese of Huron, to the author, March 29, 2004; Descendants of Andrew Keys, genealogical chart provided to the author by C. Roy Keys III.

2. Conversation with Gordon Keys, a cousin, August 27, 2005.

3. St. Paul's Church, Lindsay, Ontario: *Parish and Home*, No. 24 (August 1893); *Honors Classics in the University of Toronto, by a group of classical graduates, with a foreword by Sir Robert Falconer* (Toronto: University of Toronto Press, 1929), pp. 18–19; *The Canadian Post*, June 30, 1893.

4. John Bartlett Brebner, *Canada: A Modern History* (Ann Arbor: University of Michigan Press, p. 362.

5. *Honors Classics in the University of Toronto*, pp. 27, 36; A. B. McKillop, *Matters of Mind: The University in Ontario 1791–1951* (Toronto: University of Toronto Press, 1994), p. 225.

6. I am indebted to my niece, Ms. Antonia Young, PhD in Classics and Art History at the University of California, Berkeley, for these observations on study of the Classics.

7. This information was provided in a letter to the author from Lagring Ulanday, University Archives, University of Toronto, April 22, 2003.

8. Kim Beattie, *Ridley: The Story of a School*, Volume I (St. Catherine's, Ontario: Ridley College, 1963), p. 108.

9. Ibid., p. 85.

10. Claude T. Bissell, *University College: A Portrait 1853–1953* (Toronto: University of Toronto Press, 1953), p. 121.

11. C. M. Keys, "Hosts and Guests of Winter in City's Birdland," *New York Times*, January 12, 1902, p. SM5.

12. Howard Mingos, "The Harriman of Aviation," *New York Herald Tribune*, June 2, 1929; *The New Yorker*, June 16, 1928.

13. Earl Reeves, *Aviation's Place in Tomorrow's Business* (New York: B. C. Forbes Publishing Company, 1930), pp. 92–93.

14. Mingos, "The Harriman of Aviation."

15. Lloyd Wendt: *The Wall Street Journal: The Story of Dow Jones & the Nation's Business Newspaper* (Chicago: Rand McNally, 1982), p. 77; *The Wall Street Journal*, August 27, 1945, p. 4; *The New York Times*, August 27, 1945.

16. Thomas P. Hughes, *Networks of Power: Electrification in Western Society, 1880–1930* (Baltimore: Johns Hopkins University Press, 1988), p. 5.

17. Thomas P. Hughes, *American Genesis: A Century of Invention and Technological Enthusiasm* (Chicago: University of Chicago Press, 2004), p. 185; Chandler, *The Visible Hand* (Cambridge, MA: Belknap Press, 1977), pp. 167, 170.

18. Mingos, "The Harriman of Aviation."

19. Mary A. O'Sullivan, *Dividends of Development: Securities Markets in the History of US Capitalism* (Oxford: Oxford University Press, 2016), p. 36.

20. See, for example, "Railroad Affairs," *The Wall Street Journal*, May 4, 1901, p. 1; October 23, 1901, p. 1; December 25, 1901, p. 2.

21. See, for example, "Current Railroad Earnings," *The Wall Street Journal*, January 18, 1902, p. 2; March 10, 1902, p. 2; November 11, 1902, p. 3.

22. Charles R. Geisst, *Wall Street: A History* (New York: Oxford University Press, 1997), p. 123.

23. Thomas F. Woodlock, *The Anatomy of a Railroad Report* and *Ton-mile Cost* (New York: S. A. Nelson, 1900), p. 11.

24. Emory R. Johnson, *American Railway Transportation*, Revised Edition (New York: D. Appleton and Company, 1908), pp. 1–3.

25. Ibid., p. 301.

26. Ibid.

27. E. G. Campbell, *The Reorganization of the American Railroad System* (New York: Columbia University Press, 1938), pp. 223–25.

28. *The Wall Street Journal*, October 25, 1905, p. 1.

29. Albro Martin, "James J. Hill," in Robert L. Frey, editor, *The Encyclopedia of Business History and Biography: Railroads in the Nineteenth Century* (New York: Bruccoli Clark Layman, 1988), pp. 169–78; see also Albro Martin, *James J. Hill & the Opening of the Pacific Northwest* (New York: Oxford University Press, 1976).

30. Lloyd J. Mercer, "Edward Henry Harriman," in Frey, *Encyclopedia of Business History and Biography: Railroads in the Nineteenth Century*, pp. 155–63. See also Maury Klein, *The Life and Legend of E.H. Harriman* (Chapel Hill: University of North Carolina Press, 2000).

31. "W. E. Hazen, Former Wall Street Journal Industrial Editor, Dies," *The Wall Street Journal*, August 22, 1940, p. 7.

32. C. M. Keys, "A 'Corner in Pacific Railroads," *The World's Work*, 9, no. 4 (February 1905): 5816–22.

33. C. M. Keys, "The Newest Railroad Power," *The World's Work*, 10, no. 2 (June 1905): 6302–13.

34. C. M. Keys, "The Contest for Pacific Traffic," *The World's Work*, 10, no. 4 (August 1905): 6503–9.

35. C. M. Keys, "As Many Railroad Methods as Railroad Kings," *The World's Work*, 10, no. 5 (September 1905): 66652–59.

36. *The New York Times*, October 20, 1906, p. BR686.

37. Robert J. Rusnak, *Walter Hines Page and The World's Work 1900–1913* (Washington, DC: University Press of America, 1982), p. 44.

38. Walter Hines Page, "On a Tenth Birthday," *The World's Work*, 21, no. 3 (January 1911): 13910.

39. Serano S. Pratt, *The Work of Wall Street: An Account of the Functions, Methods, and History of the New York Money and Stock Markets* (New York: D. Appleton and Company, 1912), p. 66.

40. Clement M. Keys, "The Country Investor and His Mortgages," *The World's Work*, 18, no. 1 (May 1909): 11539.

41. Howard Mingos, "The Harriman of Aviation," *New York Herald Tribune* (June 2, 1929), pp. 12–13.

42. Clement M. Keys to Walter Hines Page, February 8, 1911, Garden City, Long Island Container 596, MS AM 1090 Page, Walter Hines 1855–1918, recipient, Letters from Various Correspondents, American Period, Houghton Library, Harvard College Library, Harvard University, Cambridge, MA.

43. Vincent P. Carosso, *Investment Banking in America: A History* (Cambridge, MA: Harvard University Press, 1970), p. 85.

44. *The Literary Digest*, October 26, 1912, p. 753.

45. *The Literary Digest*, January 25, 1913.

46. C. M. Keys to C. N. Griffis, October 3, 1947, File 15 (C. N. Griffis, 1947), Box 28, Clement M. Keys Papers, National Air and Space Museum.

47. *The Magazine of Wall Street*, 14, no. 5 (September 1914): 435–49.

48. *Los Angeles Times*, September 26, 1914, p. II9.

49. Susan B. Carter, Ed., *Historical Statistics of the United States: Earliest Times to the Present, Millennial Edition, Volume 4, Part D: Economic Section* (New York: Cambridge University Press, 2006), Table Cj831–837 Corporate Security Issues: 1910–1934.

50. Walter Hines Page, "The Solving of Aerial Navigation," *The World's Work*, 16, no. 4 (August 1908): 10519–520.

51. Mingos, "The Harriman of Aviation."

52. Owen Thetford, *British Naval Aircraft Since 1912*, Fifth Revised Edition (London: Putnam, 1982), pp. 84, 405; Louis Casey, *Curtiss: The Hammondsport Era* (New York: Crown Publishers, 1981), p.191; Timothy Wilson, "Broken Wings: The Curtiss Aeroplane Company, K Boats, and the Russian Navy, 1914–1916," *Journal of Military History*, 66, no. 4 (October 2002): 1064; Thetford, *British Naval Aircraft Since 1912*, pp. 84, 404–5; *The New York Times*, March 8, 1915; Alexis Mehtidis, *Italian Military Aviation in World War I 1914–1918* (Tiger Lily Publications LLC, 2005), p. 11.

53. *The New York Times*, December 12, 1915; *The Wall Street Journal*, December 16, 1915.

54. C. R. Roseberry, *Glenn Curtiss: Pioneer of Flight* (New York: Doubleday and Company, 1972), p. 406.

55. *New York Times*, December 31, 1915, p. 14; *The Wall Street Journal*, December 31, 1915, p. 2.

56. *Aerial Age Weekly*, 2, no. 19 (January 24, 1916): 448; *The Wall Street Journal*, January 14, 1916, p. 7.

57. *Aerial Age Weekly*, 2, no. 19 (January 24, 1916): 448.

58. Mingos, *The Harriman of Aviation*, pp. 12–13.

59. *Wall Street Journal*, January 11, 1916, p. 6.

60. This event was covered in a speech by C. M. Keys at American Legion Post 501 Dinner, April 17, 1951, Folder 23, Box 27, Clement Melville Keys Papers, Accession Number XXXX.0091, National Air and Space Museum, Smithsonian Institution (hereinafter CMKP, NASM); *Wall Street Journal*, December 19, 1916, p. 4; *Commercial & Financial Chronicle*, December 23, 1916.

61. *The Wall Street Journal*, December 28, 1916, p. 7.

62. *Aerial Age*, 14, no. 26 (March 12, 1917): 747.

63. Ibid.

64. *Wall Street Journal*, March 3, March 5, 1917.

65. Arthur Sweetser, *The American Air Service: A Record of Its Problems, Its Difficulties, Its Failures, and Its Final Achievements* (New York: D. Appleton and Company, 1919), pp. 43–45; *Aerial Age Weekly*, 5, no. 3 (April 2, 1917): 74–75.

66. *Aerial Age Weekly*, 5, no. 3 (April 2, 1917): 74–75.

67. Ibid., p. 74.

68. Juliet A. Hennessy, *The United States Army Air Arm: April 1961 to April 1917* (Washington, DC: Office of Air Force History, 1985), pp. 197–98; George van Deurs, *Wings for the Fleet: A Narrative of Naval Aviation's Early Development, 1910–1916* (Annapolis, MD: United States Naval Institute 1966), p. 197.

69. Sweetser, *The American Air Service: A Record of Its Problems, Its Difficulties, Its Failures, and Its Final Achievements*, p. 46.

70. Ibid., pp. 46–47.

71. Clement M. Keys to J. C. Hunsaker, Federal Aviation Commission, October 2, 1934, Folder 22, Box 27, CMKP; B. C. Forbes, *Automotive Giants of America:*

Men Who Are Making Our Motor Industry (New York: B. C. Forbes Publishing Company, 1926), pp. 279–95; The Allpar.com website, devoted to owners of Chrysler, Dodge, Plymouth, and DeSoto cars, contains a short biography of John Willys which gives the production statistics for Willys-Overland in 1910 and 1915; *The Wall Street Journal*, May 7, 1917, p. 5.

72. Sweetser, *The American Air Service*, p. 48.

73. Ibid., pp. 67–68.

74. Brigadier General J. D. Cormack, Department of Aeronautical Supplies, Ministry of Munitions of War in U.S.A, to Howard E. Coffin, Aircraft Production Board, September 26, 1917, Box 9, Entry 87, Army Air Corps Central Decimal Files, 1917–1938, Record Group 18, National Archives and Records Administration (RG-18, NARA).

75. Clement M. Key to Howard E. Coffin, Chairman, Aircraft Production Board, November 22, 1917, Box 9, Entry 87, RG 18, NARA.

76. *The Wall Street Journal*, June 18, 1917, p. 5.

77. *The Wall Street Journal*, June 22, 1917, p. 2; *Aerial Age Weekly*, 5, no. 16 (July 2, 1917): 522–23.

78. *Aerial Age Weekly*, 5, no. 17 (July 2, 1917): 562; *The Wall Street Journal*, July 24, 1917, p. 6.

79. John Willys, Willys-Overland Company to Howard E. Coffin, Aircraft Production Board, July 14, 1917, Box 9, Entry 87, RG 18, NARA.

80. R. M. McFarland, "History of the Bolling Aeronautical Mission," October 29, 1919, Box 2, Entry 111, General Records of the Chief of the Air Service, RG 18, NARA.

81. U.S. Congress. Senate. Committee on Military Affairs. *Aircraft Production: Hearings Before the Subcommittee of the Committee on Military Affairs*, 65th Congress, 2nd Session, May 29–August 15, 1918, Volumes I and II (Washington, DC: Government Printing Office, 1918), pp. 69–71, 75–76; A. B. Gregg, "Spad-Single Seat Pursuit Plane," October 1919, p. 1647, Box 4, Entry 111, RG 18, NARA.

82. A. B. Gregg, "Spad-Single Seat Pursuit Plane," p. 1651; Col. G. W. Mixter and Lt. H. H. Emmons, *United States Army Aircraft Production Facts* (Washington, DC: Government Printing Office, 1919), p. 46.

83. Ray Wagner, *American Combat Planes of the 20th Century: A Comprehensive Reference* (Reno, NV: Jack Bacon & Company, 2004), p. 38.

84. Clement M. Keys to Howard Coffin, Chairman, Aircraft Production Board, November 22, 1917, Box 9, Entry 87, RG18, NARA.

85. Ibid.

86. "Brief Resume of the Bristol Fighter," Office of the Chief of the Air Service, 1917–1927, Correspondence, Reports, Studies and other Records Relating to the Bureau of Aircraft Production in World War I, 1917–1921, File 452.1–Planes-Curtiss, Box 32, Entry 110, RG 18, NARA.

87. A. B. Gregg, "Bristol Fighter," September 1919, p. 1632, Box 2, Entry 111, General Records of the Chief of the Air Service, RG 18, NARA.

88. Ibid., p. 1633; *Aerial Age Weekly*, 18, no. 9: 464.

89. *The Wall Street Journal*, July 25, 1918, p. 8.

90. "Brief Resume of the Bristol Fighter."

91. Gregg, "Bristol Fighter; A. B. Gregg, "History of the SE-5- Single Seat Pursuit Plane," N.D., File 452.1 SE-5, Box 37, Entry 110, RG 18, NARA; "Brief Resume of the Bristol Fighter"; *Aerial Age Weekly*, 8, no. 9 (November 11, 1918): 464.

92. Jacob Vander Meulen, *The Politics of Aircraft: Building an American Military Industry* (Lawrence: University Press of Kansas, 1991), pp. 36–37; Sweetser, *The American*

Air Service: A Record of Its Problems, Its Difficulties, Its Failures, and Its Final Achievements, pp. 215–17.

93. "Supply of Aircraft," RG 45, NARA; Archibald D. Turnbull and Clifford L. Lord, *History of United States Naval Aviation* (New Haven, CT: Yale University Press, 1949), pp. 107–18.

94. *The Wall Street Journal*, May 11, 1918, p. 6; undated memorandum in the records of the Aircraft Production Board, Box 9, Entry 87, RG18, NARA.

95. Erie Finance Corporation to Clement M. Keys, December 31, 1917, Folder 28, Personal Securities (1017, 1931), Box 18, C. M. Keys, CMKP.

96. J. F. Prince, Treasurer, Curtiss Aeroplane & Motor Corporation, to C. M. Keys, February 6, 1918, Folder 1, J. F. Prince (1918), Box 11, Curtiss Aeroplane and Motor Co., CMKP.

97. Resolution of the Aircraft Production Board, December 7, 1917, Box 9, Entry 87, Aircraft Production Board, RG18, NARA.

98. Curtiss Aeroplane and Motor Corporation, Agreement Relative to Advances by War and Navy Departments on Account of Current Contracts, January 18, 1918, Box 118, Entry 5, General Correspondence Initiated in the Office of the Chief of Naval Operations, 1917–1925, Records Group 72, Bureau of Aeronautics, NARA.

99. Clement M. Keys to J. F. Prince, Treasurer, Curtiss Aeroplane & Motor Corporation, January 7, 1918, Folder 1, J. F. Prince (1918), Box 11, Curtiss Aeroplane & Motor Co., CMKP.

100. J. F. Prince, Treasurer, Curtiss Aeroplane & Motor Corporation, to Clement M. Keys, January 25, 1918, Folder 1, J. F. Prince (1918), Box 11, Curtiss Aeroplane & Motor Co., CMKP.

101. Clement M. Keys to J. F. Prince, wire of February 15, 1918, Folder 1, J. F. Prince (1918), Box 11, Curtiss Aeroplane & Motor Co., CMKP; *The Wall Street Journal*, February 14, 1918, p. 1; February 20, 1918, p. 6.

102. Curtiss Aeroplane & Motor Corporation, February 19, 1918, Folder 3, Publicity (1918–1928), Box 11, Curtiss Aeroplane & Motor Co., CMKP; *The Commercial and Financial Chronicle*, February 23, 1918, p. 824.

103. *The Wall Street Journal*, February 25, 1918, p. 8.

104. C. M. Keys, Curtiss Aeroplane & Motor Corporation, to the War Credits Board, April 16, 1918, Folder 29, J. E. Kepperley (January–June 1918), Box 9, Curtiss Aeroplane & Motor Co., CMKP.

105. C. M. Keys to William Moss, Comptroller, Curtiss Aeroplane & Motor Corporation, July 2, 1918, Folder 22 (M. W. Moss, July 1918–1919), Box 10, Curtiss Aeroplane & Motor Co., CMKP.

106. Major H. S. Brown, Bureau of Aircraft Production, to W. C. Potter, July 12, 1918, Box 16, Entry 22, RG18, NARA.

107. Peter M. Bowers: *Curtiss Aircraft 1907–1947* (London: Putnam, 1979), pp. 83–84.

108. C. M. Keys to J. N. Willys, October 2, 1918, Folder 16, J. N. Willys (1918), Box 13, Curtiss Aeroplane & Motor Co., CMKP.

109. *America's Munitions 1917–1918:* Report of Benedict Crowell, Assistant Secretary of War, Director of Munitions, p. 256; *The Wall Street Journal*, November 15, 1918, p. 7.

110. Clement M. Keys to Mr. G. C. Westervelt, Superintendent, Constructor of Aircraft, U.S. Navy, April 9, 1918, Folder 15, George Conrad Westervelt (1918–1928), Box 25, CMKP.

111. Clement M. Keys to John Willys, President, Curtiss Aeroplane & Motor Corporation, August 30, 1918, Folder 16; B. A. Guy (July 1918–1919), Box 13, CMKP, NASM.

112. Committee on Stock Listing, New York Stock Exchange, A-7697: Curtiss Aeroplane & Motor Company, Inc., October 7, 1927, Folder 8- Listing Statement [NYSE; legal-size documents] (1927–1928), Box 29, CMKP, NASM.

113. Clement M. Keys to Howard Coffin, Chairman, Aircraft Production Board, August 23, 1917, Box 9, Entry 87, Aircraft Production Board, Record Group 18, Records of the Army Air Forces, NARA.

114. Clement M. Keys to Commander Conrad Westervelt, February 25, 1918, Folder 15, Westervelt, George Conrad (1918–1928), Box 25, CMKP.

115. J. E. Kepperley to C. M. Keys, April 2, 1918, Folder 29, J. E. Kepperley (January–June 1918), Box 9, Curtiss Aeroplane & Motor Co., CMKP.

116. Clement M. Keys to J. E. Kepperley, Curtiss Aeroplane & Motor Corporation, April 20, April 29, 1918, Folder 29; J. E. Kepperley (January–June 1918), Box 9, Curtiss Aeroplane & Motor Co., CMKP; Westervelt, Richardson, and Read, *The Triumph of the NCs* (New York: Doubleday, Page & Company, 1920), p. 109.

117. Clement M. Keys to J. E. Kepperley, April 29, 1918, CMKP.

118. F. G. Coburn, Bureau of Construction and Repair, to Chief of Naval Operations (Aviation), June 21, 1918, Box 118, Entry 5, General Correspondence Initiated in the Office of the Chief of Naval Operations, 1917–1925, Record Group 72, NARA.

119. Clement M. Keys to the Curtiss Aeroplane & Motor Corporation, April 17, 1918, Folder 15, Box 9, CMKP.

120. Bowers, *Curtiss Aircraft 1907–1947*, p. 140; Hugo T. Byttebier, *The Curtiss D-12 Engine* (Washington, DC: Smithsonian Institution Press, 1972), p. 17.

121. Byttebier, *The Curtiss D-12 Engine*, p. 17.

122. U.S. Congress, House, Subcommittee No. 1 (Aviation) of the Select Committee on Expenditures in the War Department, *Hearings on War Expenditures*, Vol. 3, 66th Congress, First Session, November 14, 1919 (Washington, DC: Government Printing Office, 1920), pp. 3611–612.

123. *The Curtiss Fuselage*, 2, no. 9, November 16, 1918; *The Wall Street Journal*, November 20, 1918; Memorandum: "Curtiss Aeroplane & Motor Corporation Contracts for Flying Boats," December 24, 1918, Box 118, Entry 5, General Correspondence Initiated in the Office of the Chief of Naval Operations, 1917–1925, RG 72, NARA.

124. John N. Willys, Curtiss Aeroplane & Motor Corporation, to W. C. Potter, Aircraft Production Board, November 12, 1918, Box 16, Entry 22, RG 18, NARA.

125. Bowers, *Curtiss Aircraft 1907–1947*, p. 74.

126. *Hearings on War Expenditures*, pp. 3611, 3615.

127. Clement M. Keys to J. C. Hunsaker, Federal Aviation Commission, October 2, 1934, Folder 22, Howell Committee (1934), Box 27, CMKP.

128. Clement M. Keys to J. C. Hunsaker, Federal Aviation Commission, October 2, 1934.

Chapter 2

1. *Aerial Age Weekly*, 8, no. 18 (January 13, 1919): 898–903; 8, no. 19 (January 20, 1919): 941–44.

2. *Aerial Age Weekly*, 8, no. 19 (January 20, 1919): 941.

3. Robert Everett, "New York to Bagdad via the Air Line," *Nation's Business* (November 1918): 22–23, 36–37; Willard Hart Smith, "The Wings of Tomorrow," *The Forum*, 61, no. 2 (February 1919): 197–209; "Putting the Airplane to Work," *The Forum*, 61, no. 3 (March 1919): 317–27.

4. *Aviation and Aeronautical Engineering*, 5, no. 11 (January 1, 1919): 676.

5. *Aerial Age Weekly*, 8, no. 19 (January 20, 1919): 943.

6. Clement M. Keys to W. C. Potter, Bureau of Aircraft Production, War Department, December 14, 1918, Box 16, Entry 87, RG 18, NARA.

7. *The New York Times*, March 9, 1919, p. 25.

8. *Aerial Age Weekly*, 9, no. 20 (July 28, 1919): 922.

9. Curtiss Aeroplane & Motor Company, Inc., "House of Representatives Select Committee of Inquiry into the Operations of the United States Air Services: Testimony of C. M. Keys, president" (1925), p. 21; *Aerial Age Weekly*, 9, no. 19 (July 21, 1919): 888.

10. R. S. Pearce, "Does the Public Want to Fly?," *Illustrated World*, 32, no. 2 (October 1919): 220.

11. Pearce, "Does the Public Want to Fly?," p. 221.

12. Dr. W. F. Durand, "An Analysis of the Need for Civil Aviation," *Automotive Industries*, 41, no. 18 (October 30, 1919): 874.

13. *The New York Times*, June 13, 1919, p. 17; June 17, 1919, p. .17; Maurer Maurer, *Aviation in the U.S. Army 1919–1939* (Washington, DC: U.S. Government Printing Office, 1987), pp. 44–45.

14. Maurer, *Aviation in the U.S. Army 1919–1939*, pp. 10–13.

15. Clement M. Keys, "Memorandum Concerning the Purchase of 2,727 Planes & 4,608 Motors from the Government for $2,720,000," CMKP, Folder 9 (Frear Committee), Box 9, CMKP, NASM.

16. U.S. Congress, House, Subcommittee of the Committee on Military Affairs, A United Air Service. Hearing, 66th Congress, 2nd Sess. on H.R. 10380 and H.R. 9804, December 9, 1919, p. 245.

17. U.S. Congress, House, Subcommittee of the Committee on Military Affairs. A United Air Service. Hearing, 66th Congress, 2nd Sess. on H.R. 10380 and H.R. 9804, December 9, 1919, p. 245.

18. Clement M. Keys, "Memorandum Concerning the Purchase of 2,727 Planes & 4,608 Motors from the Government for $2,720,000."

19. *Aerial Age Weekly*, 9, no. 12 (June 2, 1919): 588.

20. Keys, "Memorandum Concerning the Purchase of 2,727 Planes & 4,608 Motors from the Government for $2,720,000"; *The New York Times*, April 27, 1919, p. 58.

21. Irving Brinton Holley Jr., *Buying Aircraft: Material Procurement for the Army Air Forces* (Washington, DC: U.S. Government Printing Office, 1962), p. 84.

22. Ibid., p. 85.

23. Ibid., pp. 85–86.

24. John Morrow, "The Outlook for Three War Brides," *The Magazine of Wall Street*, 24, no. 6 (July 5, 1919): 528.

25. Robert Everett, "Peace Fleets for the Air Lanes," *Nation's Business* (May 1919), p. 30.

26. *Aircraft Journal* (May 24, 1919), pp. 5, 20; *The New York Times*, May 23, 1919, p. 13.

27. *Aircraft Journal* (May 24, 1919), pp. 5, 20.

28. Ibid., pp. 479–80; "Report of American Aviation Mission," Reprint from *Aircraft Journal*, August 23, 1919, pp. 5–6, 14–15, 20–21; *The New York Times*, January 5, 1919, p. 25; R.E.G. Davies, *A History of the World's Airlines* (London: Oxford University Press, 1964), pp. 14–17; Robin Higham, *Britain's Imperial Air Routes 1918 to 1939: The Story of Britain's Overseas Airlines* (Hamden, CT: Shoe String Press, 1961), pp. 19–26, 28–29; Reorganization of the Army, p. 480.

29. "Report of American Aviation Mission," Reprint from *Aircraft Journal*, August 23, 1919, pp. 11–12.

30. Ibid., p. 18.

31. Ibid., p. 20.

32. *The Literary Digest*, 62, no. 4 (July 26, 1919): 15.

33. Ibid.; *The Literary Digest*, 62, no. 10 (September 6, 1919), p. 19.

34. U.S. Congress, Senate, Subcommittee of the Committee on Military Affairs, Reorganization of the Army, 66th Congress, 1st Session, August 7–November 5, 1919 (Washington, DC, 1919), pp. 477–95.

35. U.S. Congress, House, Subcommittee No. 1 (Aviation) of the Select Committee on Expenditures in the War Department. Hearings on War Expenditures Volume 3, 66th Congress, 1st Session, November 14, 1919 (Washington, DC, 1920), pp. 3607–41; U.S. Congress. House. Subcommittee of the Military Affairs Committee. A United Air Service. Hearing on H.R. 10380 and H.R. 9804, 66th Congress, 2nd Session, December 4, 1919 (Washington, DC, 1920), pp. 242–53.

36. U.S. Congress, Senate, Subcommittee of the Committee on Military Affairs. Reorganization of the Army. 66th Congress, 1st Session, August 7–November 5, 1919 (Washington, DC, 1919), p. 489.

37. A United Air Service, p. 252.

38. U.S. Congress, House, Subcommittee of the Military Affairs Committee. A United Air Service. Hearing on H.R. 10380 and H.R. 9804, 66th Congress, 2nd Session, December 4, 1919 (Washington, DC, 1920), p. 251.

39. Reorganization of the Army, p. 485.

40. A United Air Service, p. 246.

41. Clement M. Keys to the Manufacturer's Aircraft Association, August 13, 1919, Folder 12 (Manufacturer's Aircraft Association, Inc., 1917–1931), Box 19, CMKP, NASM; *The Wall Street Journal*, October 15, 1919, p. 15; October 25, 1919, p. 2.

42. Poor's and Moody's Manual Consolidated 1922, Industrial Section (New York, 1922), p. 1024.

43. G. C. Newton, Deputy Commissioner, Office of the Commissioner of Internal Revenue, Treasury Department, Washington, to the Curtiss Aeroplane & Motor Corporation, February 5, 1920. Copy in black 3-ring binder, Box TY5: Curtiss-Wright Tax and Legal Documents, 1916 to 1946, Wright Collection, Museum of Flight, Seattle, Washington.

44. Clement M. Keys to the Directors of the Curtiss Aeroplane & Motor Corporation, April 5, 1920, Folder 3 (Finance Committee 1920–22), Box 9, CMKP, NASM.

45. G. C. Newton, Deputy Commissioner, Office of the Commissioner of Internal Revenue, Treasury Department, to the Curtiss Aeroplane & Motor Corporation, March 30, 1920. Copy in black 3-ring binder, Box TY5: Curtiss-Wright Tax and Legal Documents, 1916 to 1946, Wright Collection, Museum of Flight, Seattle, Washington.

46. Clement M. Keys to the Directors of the Curtiss Aeroplane & Motor Corporation, April 5, 1920; see C. M. Keys deposition of August 5, 1920, in "Curtiss Aeroplane and Motor Corporation (General Settlement), October 15, 1920, Folder 25 (C. W. Cuthell, 1919–1927), Box 8, CMKP, NASM.

47. Clement M. Keys to the Directors of the Curtiss Aeroplane & Motor Corporation, April 5, 1920.

48. C. M. Keys Deposition, August 5, 1920.

49. Clement M. Keys, "Memorandum of 1st Meeting of Financial Committee of Curtiss Aeroplane & Motor Corporation," April 19, 1920, Folder 3 (Finance Committee 1920–22), Box 9, CMKP, NASM.

50. Ibid.; C. M. Keys Deposition, August 5, 1920.

51. Clement M. Keys, "Memorandum of the 1st Meeting of the Financial Committee"; George E. Holmes to Clement M. Keys, April 27, 1920, Folder 3 (Finance Committee 1920–22), Box 9, CMKP, NASM.

52. "Memorandum of 1st Meeting of Financial Committee."

53. Clement M. Keys, "2nd Meeting of Finance Committee," April 22, 1920, and "3rd Meeting of Financial Committee," April 26, 1920, Folder 3 (Finance Committee 1920–22), Box 9, CMKP, NASM.

54. "3rd Meeting of the Finance Committee."

55. Clement M. Keys, "Memo of Observations," May 1, 1920, Folder 3 (Finance Committee 1920–22), Box 9, CMKP, NASM.

56. *New York Times*, June 12, 1920, p. 17.

57. Ray Wagner, *American Combat Planes of the 20th Century* (Reno, NV: Jack Bacon & Company 2004), p. 33.

58. Ibid., p. 117.

59. Maurer, *Aviation in the U.S. Army*, p. 45; *The New York Times*, June 7, 1920.

60. C. M. Keys, Curtiss Aeroplane & Motor Corporation, to R. N. Denham, Liquidation Division, Air Service, June 10, June 15, 1920, Folder 25 (C. W. Cuthell, 1919–1926), Box 8, CMKP, NASM.

61. "Curtiss Aeroplane and Motor Corporation (General Settlement)."

62. *Automotive Industries*, 63, no. 21 (November 18, 1920): 1049.

63. C. M. Keys, Deposition of August 5, 1920.

64. *The Wall Street Journal*, September 27, 1920; *The New York Times*, September 26, 1920, p. 25.

65. Vincent Curcio, *Chrysler: The Life and Times of an Automotive Genius* (New York: Oxford University Press, 2000), p. 265.

66. Clinton L. Mosher, "Money Miracle Saved Aviation," *Brooklyn Eagle*, March 4, 1928; "House of Representatives Select Committee of Inquiry into Operations of the United States Air Services: Testimony of C. M. Keys," Curtiss Aeroplane & Motor Company, p. 10.

67. Mingos, "The Harriman of Aviation"; Harry Bliss, "Clement M. Keys: He Bought a Company Nearly Insolvent," *Air Transportation* (July 27, 1929), p. 51; "Testimony of C.M. Keys," p. 10.

68. Clement M. Keys to J. C. Hall, February 1, 1923, Folder 7 (Reorganization 1923), Box 11, CMKP, NASM.

69. "Testimony of C. M. Keys," p. 10.

70. Ibid., p. 11.

71. Ibid., p. 3.

72. Ibid., p. 11.

73. *Aviation and Aeronautical Engineering*, 5, no. 11 (January 1, 1919): 675.

74. "Testimony of C. M. Keys," p. 10.

75. See contract between C. M. Keys, Curtiss Engineering Corporation, and the Curtiss Aeroplane and Mortgage Corporation, dated August 6, 1920, and mortgage between Curtiss Aeroplane and Motor Corporation and C. M. Keys and Curtiss Engineering Corporation, dated August 6, 1920, both in Folder 14, Box TY4: Curtiss Contracts, 1915 to 1942, Wright Collection, Museum of Flight, Seattle, Washington.

76. Clement M. Keys to Newton D. Baker, December 8, 1924, Folder 6 (Newton D. Baker, 1924), Box 4, CMKP, NASM; Curtiss Aeroplane & Motor Company, Inc.: House of Representatives Select Committee of Inquiry Into Operations of the United States Air Services: Testimony of C. M. Keys, President, No Date, pp. 10–11.

77. Copy of the minutes of the meeting of the War Claims Board, August 4, 1920, included in "Curtiss Aeroplane & Motor Corporation (General Settlement) Supporting Papers," Folder 25 (C. W. Cuthell, 1919–1927), Box 8, CMKP, NASM.

78. Copy of the minutes of the meeting of the War Claims Board, August 4, 1920.

79. C. M. Keys deposition of August 5, 1920; Major General Charles T. Menoher, chief of Air Service, to the Disbursing Officer, Air Service, August 20, 1920, in "Curtiss Aeroplane & Motor Corporation (General Settlement) Supporting Papers."

80. Clement M. Keys to Newton D. Baker, December 8, 1924.

81. C. M. Keys to J. C. Hall, Curtiss Aeroplane & Motor Corporation, February 1, 1923, Folder 7 (Reorganization), Box 11, CMKP. NASM.

82. C. M. Keys to S. S. Bradley, Manufacturers' Aircraft Association, September 15, 1920, Folder 12 (Manufacturers' Aircraft Association 1917–1931), Box 19, CMKP, NASM.

83. C. M. Keys to Frank G. Allen, State Savings Bank & Trust Co., October 30, 1920, Folder 2 (Frank G. Allen, 1920–1921), Box 2, CMKP, NASM.

84. Clement M. Keys, Curtiss Aeroplane & Motor Corporation, to Frank G. Allen, State Savings Bank & Trust Co., November 29, 1920, Folder 2 (Frank G. Allen: 1920–1921), Box 2, CMKP, NASM.

85. Frank G. Allen, State Savings Bank & Trust Co. to C. M. Keys, November 27, 1920, Folder 2 (Frank G. Allen, 1920–1921), Box 2, CMKP, NASM.

86. C. M. Keys to John J. Raskob, General Motors Corporation, January 2, 1923, Folder 32 (John J. Raskob, 1922–1924), Box 21, CMKP, NASM.

87. As of June 30, 1920, Curtiss had approximately sixty-six Orioles and seventy-seven Seagulls completed or under construction, nine Eagles, and 758 surplus aircraft. For the Curtiss Aeroplane & Motor Corporation inventory position, see "Memorandum in Connection with the Value of Patents of Curtiss Aeroplane & Motor Corporation," Folder 25 (C. W. Cuthell, 1919–1927), Box 8, CMKP. NASM. See also Keys's deposition of August 5, 1920, in "Curtiss Aeroplane & Motor Corporation (General Settlement) Supporting Papers."

88. Clement M. Keys, "Value of Racing Planes," *Aero Digest*, 7, no. 4 (October 1925): 568.

89. Hugo Byttebier, *The Curtiss D-12 Aero Engine, Smithsonian Annals of Flight No.7* (Washington, DC: Smithsonian Institution Press, 1972), pp. 26–28, 30–32; Thomas G. Foxworth, *Speed Seekers* (New York: Doubleday and Company, 1975), pp. 62–64.

90. Bowers, *Curtiss Aircraft 1907–1947*, pp. 224–25; Byttebier, *The Curtiss D-12 Aero Engine*, pp. 36–39.

91. Byttebier, *The Curtiss D-12 Aero Engine*, p. 36.

92. Keys, "Value of Racing Planes," p. 535.

93. "House of Representatives Select Committee of Inquiry into Operations of the United States Air Services: Testimony of C. M. Keys," p. 6; Byttebier, *The Curtiss D-12 Aero Engine*, pp. 44–45.

94. C. M. Keys to Frank G. Allen, February 19, 1921, Folder 2 (Frank G. Allen: 1920–1921), Box 2, CMKP, NASM; Moody's Analysis of Investments: Industrials 1920, p. 933; Annual Report and Condensed Balance Sheet and Profit and Loss Account of the Curtiss Aeroplane & Motor Corporation, December 31, 1921, Folder 31 (Curtiss Aeroplane & Motor Co., 1917–1929), Box 25, CMKP, NASM.

95. *Moody's Manual of Investments: Industrials 1921* (New York, 1921), p. 1182.

96. E. H. Batson, Acting Deputy Commissioner, Office of the Commissioner of Internal Revenue, Curtiss Aeroplane & Motor Corporation, March 21, 1921, copy in black three-ring binder in Box TY 5: Curtiss-Wright Tax and Legal Documents, 1916–1946, Wright Collection, Museum of Flight.

97. C. M. Keys to Frank G. Allen, May 14, 1921, Folder 2 (Frank G. Allen, 1920–1921), Box 2, CMKP, NASM.

98. Ibid.

99. C. M. Keys to Frank H. Russell, Curtiss Aeroplane & Motor Corporation, May 12, 1921, Folder 10 (Frank H. Russell, 1920–1921), Box 11, CMKP, NASM.

100. C. M. Keys, "The Curtiss Aeroplane and Motor Corporation after the War," *Aerial Age Weekly*, 13, no. 13 (June 6, 1921): 295.

101. Ibid.

102. Ibid.

103. Ibid.

104. Ibid.

105. Keys, "The Curtiss Aeroplane and Motor Corporation after the War," p. 295.

106. Wagner, *American Combat Planes*, p. 41.

107. Ibid., pp. 146–47.

108. C. M. Keys to Frank H. Russell, June 26, 1921.

109. Foxworth, *The Speed Seekers*, pp. 196–98.

110. Bowers, *Curtiss Aircraft*, pp. 228–29; Foxworth, *The Speed Seekers*, pp. 194–98.

111. *The New York Times*, January 27, 1922, p. 2; April 22, 1922, p. 9.

112. Annual Report and Condensed Balance Sheet and Profit and Loss Account of the Curtiss Aeroplane and Motor Corporation and Subsidiaries December 31, 1921, Folder 31 (Curtiss Aeroplane and Motor Co., 1917–1929), Box 25, CMKP, NASM.

113. C. M. Keys to Frank G. Allen, State Savings Bank & Trust, October 20, 1922, Folder 3 (Allen, Frank G. 1922), Box 2, CMKP, NASM.

114. C. M. Keys, Memorandum, February 3, 1922, Folder 18 (Meetings 1921–1925), Box 10, CMKP, NASM.

115. Byttebier, *The Curtiss D-12 Engine*, p. 3.

116. Foxworth, *The Speed Seekers*, p. 208.

117. *Aviation*, 12, no. 25 (June 19, 1922): 719.

118. *Aviation*, 13, no. 16 (October 16, 1922): 483.

119. *Aviation*, 13, no. 17 (October 23, 1922): 541.

120. Ibid., p. 588.

121. C. M. Keys to Frank G. Allen, State Savings Bank & Trust Co., December 11, 1922, Folder 3 (Frank G. Allen, 1922), Box 2, CMKP, NASM.

122. C. M. Keys to Frank G. Allen, December 11, 1922.

123. C. M. Keys to C. Roy Keys, Curtiss Aeroplane & Motor Corporation, Buffalo, December 13 and December 18, 1922, Folder 13 (Buffalo Office-C. Roy Keys 1918–1930), Box 8, CMKP, NASM.

124. C. M. Keys to Frank G. Allen, December 11, 1922.

125. Ibid.

126. Ibid.

127. Frank H. Russell to Major General Mason Patrick, chief of Air Service, War Department, Washington, March 2, 1923, File 452.1 Pursuit Planes, Box 995, Entry 166, Record Group 18, Army Air Forces Central Decimal Files 1917–1938, NARA.

128. Ibid.

129. Ibid.

130. Ibid.

131. Francis H. Dean and Dan Hagedorn, *Curtiss Fighter Aircraft: A Photographic History 1917–1948* (Atglen, PA: Schiffer Military History, 2006), pp. 37–38; Bowers, *Curtiss Aircraft 1907–1947*, pp. 241–42.

132. Annual Report and Condensed Balance Sheet and Income and Deficit Account of the Curtiss Aeroplane & Motor Corporation and Subsidiaries, December 31, 1922, Folder 31 (Curtiss Aeroplane & Motor Co., 1917–1929), Box 25, CMKP, NASM.

133. Ibid.

134. Ibid.

135. C. M. Keys to Frank G. Allen, State Savings Bank & Trust Co., October 23, 1922, Folder 3 (Frank G. Allen, 1922), Box 2, CMKP; C. M. Keys to J. C. Hall, Curtiss Aeroplane & Motor Corporation, February 1, 1923, Folder 7 (Reorganization 1923), Box 11, CMKP, NASM.

136. Frank G. Allen, State Savings Bank & Trust Co., to C.M. Keys, December 30, 1922, Folder 3 (Frank G. Allen, 1922), Box 2, CMKP, NASM; C. M. Keys, "To Holders of the Common and Preferred Stock and to Holders of Voting Trust Certificates for Common and Preferred Stock of Curtiss Aeroplane and Motor Corporation," March 12, 1923, Folder 5 (Curtiss Aeroplane & Motor Co. 1923–1940), CMKP, NASM.

137. C. M. Keys to Frank G. Allen, April 6, 1923, Folder 4 (Frank G. Allen, 1923), Box 2, CMKP, NASM.

138. *The Wall Street Journal*, April 6, 1923, p. 2.

139. *The Wall Street Journal*, April 21, 1923, p. 10.

140. Committee on Stock List, New York Stock Exchange, Curtiss Aeroplane & Motor Company, Inc., Listing Application, October 27, 1927, Folder 9 (Listing Application NYSE legal-size documents 1925–1926), Box 29, CMKP, NASM.

141. C. M. Keys to C. Roy Keys, Curtiss Aeroplane & Motor Corp., Buffalo, June 23, 1923, Folder 13 (Buffalo Office-C. Roy Keys 1918–1930), Box 8, CMKP, NASM.

142. C. M. Keys to Major General Mason M. Patrick, chief of Air Service, July 19, 1923, File 452.1 Curtiss Pursuit Planes, Entry 166, Record Group 18, Records of the Army Air Forces, 1917–1938, NARA.

143. Ibid.

144. Major General Mason M. Patrick, chief of Air Service, to C. M. Keys, July 26, 1923, File 452.1 Curtiss Pursuit Planes, Entry 166, Record Group 18, Records of the Army Air Forces, 1917–1938, NARA.

145. Wagner, *American Combat Aircraft*, p. 58.

146. *Flight*, 15, no. 40 (October 4, 1923): 589.

147. Bowers, *Curtiss Aircraft 1907–1947*, pp. 234–35.

148. *Flight*, 15, no. 45 (November 8, 1923): 679.

Chapter 3

1. Aeronautical Chamber of Commerce, *Aircraft Year Book 1924* (New York, 1924), p. 1.

2. Ibid., p. 1.

3. "Report of the American Aviation Mission on European Tour," *Aerial Age Weekly*, 9, no. 23 (August 18, 1919): 1067.

4. Ibid.

5. "Report of the American Aviation Mission on European Tour," *Aerial Age Weekly*, 9, no. 24 (August 25, 1919): 1103.

6. M. Johnson and V. Houston, *Taking Flight: The Foundations of American Commercial Aviation, 1918–1938* (College Station, TX: Texas A&M University Press, 2019), p. 28.

7. Ibid.

8. National Advisory Committee for Aeronautics: Aeronautics: Seventh Annual Report of the National Advisory Committee for Aeronautics 1921 (Washington, DC: Government Printing Office, 1922), p. 4.

9. Ibid.

10. Keys would articulate this view in his correspondence with key government officials, in his testimony before Congress, and in the press. See Keys's testimony in U.S. Congress, House. Select Committee of Inquiry into the Operations of the United States Air Service. Hearings (Lampert Hearings), 68th Congress, 1st Session, 1924, Part 2 (Washington, DC, 1925), pp. 972–77, 985–91, 1122–70, 1186–89, 1193–94, 1382–1439 (hereinafter referred to as the Lampert Hearings), particularly pp. 1140–42, 1433.

11. Lampert Hearings, p. 1401.

12. Ibid.

13. "Shutting Down the Aviation Industry," *Aviation*, 16, no. 11 (March 17, 1924): 282–83.

14. *Aircraft Year Book 1924*, p. 1.

15. *The New York Times*, March 13, 1923, p. 3.

16. *Aircraft Yearbook 1924*, p. 300.

17. "An Indictment and a Warning," p. 86.

18. *Aircraft Year Book 1924*, p. 298.

19. Ibid., pp. 298, 300.

20. "The United States Naval Air Service, 1922–23," *Aviation*, 16, no. 2 (January 14, 1924): 37.

21. "N.A.A. Annual Banquet a Great Success," *Aviation*, 15, no. 17 (October 22, 1923): 509.

22. Jacob Vander Muelen, *The Politics of Aircraft Procurement: Building an American Military Industry* (Lawrence: University of Kansas Press, 1991), p. 41.

23. Ibid., p. 42.

24. Ibid.

25. Irving B. Holley, *Buying Aircraft: Material Procurement for the Army Air Forces.* United States Army in World War II: Special Studies (Washington, DC: U.S Government Printing Office, 1964), p. 86.

26. C. M. Keys to Major General Mason M. Patrick, November 17, 1921.

27. Ibid.

28. Robert P. White, *Mason Patrick and the Fight for Air Service Independence* (Washington, DC: Smithsonian Institution Press, 2001), p. 95.

29. Major General Mason Patrick, Chief of Air Service, to C. M. Keys, Curtiss Aeroplane & Motor Corporation, November 19, 1921, Office of the Chief of the Air Corps, Major General Mason M. Patrick Correspondence, 1922–1927, I-M, Box 5, Entry 228, Record Group 18, Army Air Forces Central Decimal Files, 1917–1938, NARA.

30. Major General Mason Patrick to C. M. Keys, November 19, 1921.

31. C. M. Keys to Major General John J. Pershing, Chief of Staff, War Department, December 14, 1922, Folder 7 (Aeronautical Chamber of Commerce of America, Inc.: S. S. Bradley (1922–1926), Box 1, CMKP, NASM.

32. C. M. Keys to Major General John J. Pershing, December 14, 1922.

33. Ibid.

34. Major General John J. Pershing, Chief of Staff, War Department, to C. M. Keys, Curtiss Aeroplane & Motor Corporation, January 4, 1923, Folder 7 (Aeronautical Chamber of Commerce of America, Inc.: S. S. Bradley, 1922–1927), Box 1, CMKP, NASM.

35. Edwin Denby, "The Aeronautical Industry and the National Defense," *U.S. Air Service*, 8, no. 1 (January 1923): 9–10.

36. C. M. Keys to Edwin Denby, Secretary of the Navy, January 3, 1923, Records of the Secretary of the Navy 1916–1925, Record Group 80, NARA.

37. Ibid.

38. "The Aircraft Industry," *Aviation*, 14, no. 1 (January 1, 1923): 7.

39. Ibid.

40. Ibid.

41. Vander Meulen, *The Politics of Aircraft*, pp. 45–46.

42. Arthur Stone Dewing, *The Financial Policy of Corporations*, Vol. II (New York: The Ronald Press, 1941), p. 774.

43. Ibid., Vol. I, pp. 50–51, 60–62; David F. Jordon, *Jordon on Investments*, Revised Edition (New York: Prentice Hall, 1930), pp. 4–6.

44. John J. Raskob, Chairman, General Motors Corporation, to C. M. Keys, December 26, 1922, Folder 32 (Raskob, John J. 1922–1924), Box 21, CMKP, NASM.

45. Commander E. S. Land, to Chief of Bureau of Aeronautics, January 13, 1923, Bureau of Aeronautics General Correspondence 1925–1942 (28), Record Group 72, NARA.

46. Ibid.

47. Ibid.

48. Ibid.

49. C. M. Keys, President, Curtiss Aeroplane & Motor Company, to Edwin Denby, Secretary of the Navy, December 4, 1923, copy in Office of the Chief of the Air Corps, Major General Mason Patrick Correspondence, 1922–1927, I-M, Box 5, Entry 228, Record Group 18, Army Air Forces Central Decimal Files, 1917–1938, NARA.

50. C. M. Keys to Edwin Denby, Secretary of the Navy, December 4, 1923, p. 4.

51. Ibid., p. 5.

52. Ibid., p. 6.

53. Robert P. White, *Mason Patrick and the Fight for Air Service Independence*, (Washington, DC: Smithsonian Institution Press, 2001), pp. 94–100; William F. Trimble, *Admiral William A. Moffett: Architect of Naval Aviation* (Washington, DC: Smithsonian Institution Press, 1994), pp. 111–12.

54. Trimble, *Admiral William A. Moffett Architect of Naval Aviation*, p. 113.

55. White, *Mason Patrick and the Fight for Air Service Independence*, pp. 70, 101–06; Trimble, *Admiral William A. Moffett Architect of Naval Aviation*, pp. 113–15, 167–69.

56. *The New York Times*, December 21, 1923, p. 16.

57. Thomas Worth Walterman, "Airpower and Private Enterprise: Federal-Industrial Relations in the Aeronautics Field, 1918–1926," PhD Diss., Washington University, 1970 (U.M.I. Dissertation Service), p. 360.

58. "Current Newspaper Comment," *Aeronautical Digest*, 14, no.1 (January 1924): 41–42.

59. C. M. Keys to John J. Raskob, Chairman of the Board of Directors, Du Pont De Nemeurs E. I. & Co., Inc., November 25, 1924, Folder 34 (Lampert and Hoover Committees 1924–1926), Box 18 CMKP, NASM.

60. Curtiss Aeroplane & Motor Company Annual Report for 1923.

61. Ibid.

62. Ibid.

63. Ibid.

64. Ibid.

65. *The New York Times*, March 10, 1924, p. 24; *The Wall Street Journal*, March 10, 1924, p. 3.

66. *The New York Times*, March 13, 1924, p. 16.

67. Vander Meulen, *The Politics of Aircraft*, p. 46.

68. W. E. Boeing to Major General Mason M. Patrick, Chief of Air Service, March 18, 1924, Office of the Chief of the Air Corps, Major General Mason Patrick Correspondence, 1922–1927, A-H, Entry 228, Record Group 18, Army Air Forces Central Decimal Files, 1917–1938, NARA.

69. Maurer Maurer, *Aviation in the U.S. Army. 1919–1939* (Washington, DC, 1987), p.73.

70. Sean Seyer, *Sovereign Skies: The Origins of American Civil Aviation Policy* (Baltimore, MD: Johns Hopkins University Press, 2021), p. 115.

71. Walterman, *Aviation and Private Enterprise*, p. 451.

72. C. M. Keys to Frank Allen, April 5, 1924, Folder 5 (Frank G. Allen, 1924–26), Box 2, CMKP, NASM.

73. Admiral William A. Moffett, Chief of Bureau of Aeronautics, to C. M. Keys, Curtiss Aeroplane & Motor Co. April 6, 1924, p. 2, Records of Division and Offices within the Bureau of Aeronautics, Office of Services Division, Administrative Services, General Correspondence 1925–1942, Entry 62, QM (26), Vol. 3–5, Box 4364, Record Group 72, Records of the Bureau of Aeronautics, NARA.

74. Admiral William A. Moffett to C. M. Keys, April 6, 1924, p. 2.

75. Exhibit F, Contracts Accepted by United States Flying Services 1923–1924, C. M. Keys to Major General Mason M. Patrick, Chief of Air Service, November 25, 1924, Office of the Chief of the Air Corps, Major General Mason M. Patrick Correspondence 1922–1927, I-M, Box 5, Entry 228, Record Group 18 Army Air Forces Central Decimal Files, 1917–1938, NARA.

76. Wagner, *American Combat Planes*, p. 120.

77. C. M. Keys to Edward G. Wilmer, Goodyear Tire & Rubber Co., October 22, 1924, Folder 21 (Edward G. Wilmer, 1924), Box 25, CMKP, NASM.

78. C. M. Keys to Edward G. Wilmer, Goodyear Tire & Rubber Co., October 14, 1924, Folder 21 (Edward G. Wilmer, 1924), Box 25, CMKP, NASM.

79. C. M. Keys to Admiral William A. Moffett, Chief of Bureau of Aeronautics, October 1, 1924, Bureau of Aeronautics General Correspondence 1925–1942, QM (26), Vol. 3–5, Box 4364, Record Group 72, NARA.

80. Ibid.

81. Ibid.

82. Admiral William A. Moffett to C. M. Keys, October 30, 1924, Bureau of Aeronautics General Correspondence 1925–1942, QM (26), Vol. 3–5, Box 4364, Record Group 72, NARA.

83. Maurer, *Aviation in the U.S. Army. 1919–1939*, pp. 185–86.

84. Foxworth, *The Speed Seekers*, pp. 224–25.

85. Ibid., p. 215.

86. C. M. Keys to Newton D. Baker, December 8, 1924, Folder 6 (Newton D. Baker, 1924), Box 4, CMKP, NASM.

87. *The New York Times*, September 24, 1924, p. 24.

88. The full list of witnesses can be found in U.S. Congress, House, Select Committee of Inquiry into the Operations of the United States Air Services, Inquiry into

the Operations of the United States Air Service, Part 1, Hearing, 68th Congress, 1st Session, 1925 (Washington, DC, 1925), Part 1, pp. 374–76.

89. C. M. Keys testimony to the Navy General Board, November 14, 1924, p. 8. A copy of Keys's testimony to the Navy General Board can be found in Folder 16 (Testimony: Unidentified Investigation), Box 31, CMKP, NASM.

90. C. M. Keys testimony to the Navy General Board, November 14, 1924, p. 9.

91. Ibid., p. 13.

92. Ibid., p. 22.

93. C. M. Keys to Major General Mason M. Patrick, November 25, 1924.

94. *The New York Times*, December 3, 1924, p. 8; "Annual Report of the Chief of Air Service," *Aviation*, 17, no. 24 (December 15, 1924): 1392–93; Tenth Annual Report of the National Advisory Committee for Aeronautics 1924 (Washington, DC: Government Printing Office, 1925), p. 62; *The New York Times*, November 21, 1924, p. 18.

95. "The Investigation of the Lampert Committee," *Aviation*, 18, no. 2 (January 12, 1925): 50.

96. Lampert Hearings, pp. 972–77, 985–91, 1122–70, 1186–89, 1193–94, 1382–1439.

97. Ibid., p. 1134.

98. Ibid., pp. 1134, 1392.

99. Ibid., p. 1391.

100. Ibid., p. 1401.

101. Ibid., p. 1155.

102. Ibid., p. 1132.

103. Ibid., p. 1401.

104. Ibid., p. 1140

105. Ibid., p. 1402.

106. Ibid, p. 1147.

107. Ibid., p. 1401.

108. Ibid., p. 1141.

109. Ibid., pp. 1143, 1404–5.

110. Ibid., p. 1151.

111. *The New York Times*, March 10, 1924, p. 24; *The Wall Street Journal*, March 10. 1924, p. 3.

112. Ibid., pp. 1429–33.

113. Ibid., p. 1150.

114. Frank Burke, President, United States Shipping Board Emergency Fleet Corporation, to C. M. Keys, January 28, 1925, Folder 37 (Frank Burke, 1924–1925), Box 4, CMKP, NASM.

115. Ibid.

116. C. M. Keys to Frank Burke, January 29, 1925, Folder 37 (Frank Burke, 1924–1925), Box 4, CMKP, NASM.

117. C. M. Keys, "Conditions Affecting Aeronautic Progress," *Journal of the Society of Automotive Engineers*, 17, no.5 (November 1925): 448.

118. Ibid.

119. *The New York Times*, February 15, 1925, p. 27.

120. "Report on the Distribution of Aircraft Testimony Books," Folder 35 (Lampert Committee-Testimony Distribution 1925–1926), Box 18, CMKP, NASM.

121. Edgar Gott, Boeing Airplane Company, to C. M. Keys, March 29, 1925; Howard S. Mott to C. M. Keys, March 28, 1925, Folder 35 (Lampert Committee-Testimony Distribution 1925–1926), Box 18, CMKP, NASM.

122. "Food for Thought," *Aero Digest*, 6, no. 4 (April 1925): 210.

123. Curtiss Aeroplane & Motor Company, Annual Report for 1924, March 2, 1925.

124. Ibid.

125. Ibid.

126. Wagner, *American Combat Planes of the 20th Century*, p. 148.

127. Ibid., pp. 62, 85–86; "Curtiss Wins C.O. Award," *Aviation*, 18. no. 13 (March 30, 1925), p. 349; Dean and Hagedorn, *Curtiss Fighter Aircraft*, pp. 47–48.

128. F. H. Russell, Curtiss Aeroplane & Motor Company, to C. M. Keys, March 18, 1925, Folder 11 (Frank H. Russell, 1925), Box 11, CMKP, NASM.

129. C. M. Keys to W. L. Gilmore, Curtiss Aeroplane & Motor Company, April 6, 1925, Folder 12 (W. L. Gilmore, 1922–1925), Box 9, CMKP, NASM.

130. Alfred F. Hurley, *Billy Mitchell: Crusader for Air Power* (Bloomington: Indiana University Press, 1975), p. 101.

131. Walterman, *Airpower and Private Enterprise*, pp. 462–67; Turnbull and Lord, *History of United States Naval Aviation*, pp. 250–51.

132. Rondall R. Rice, *The Politics of Air Power: From Confrontation to Cooperation in Army Aviation-Civil Relations* (Lincoln: University of Nebraska Press, 2004), pp. 46–47.

133. Walterman, *Airpower and Private Enterprise*, pp. 468–69.

134. Report of the President's Aircraft Board, November 30, 1925 (Washington, DC, 1926), pp.2–3.

135. Hearings Before the President's Aircraft Board, October 14 and 15, 1925, Volume 4, (Washington, DC, 1926), pp. 1414–15.

136. Ibid., p. 1414.

137. Foxworth, *Speed Seekers*, pp. 231–32, 459.

138. Ibid., p. 1437.

139. Ibid., pp. 1437–42.

140. Foxworth, *Speed Seekers*, pp. 2314–35, 476.

141. Report of the President's Aircraft Board, p. 29.

142. *The New York Times*, December 3, 1925, p. 1; Report of the President's Aircraft Board (Washington, DC, 1926), pp. 6–14, 19, 21; Walterman, *Airpower and Private Enterprise*, pp. 475–77.

143. *The New York Times*, December 5, 1925, pp. 2, 5.

144. "An "Alarming Situation" in Our Air Service," *The Literary Digest*, 87, no. 13 (December 26, 1925): 5.

145. Report of the Select Committee of Inquiry into Operations of the United States Air Services (Washington, DC, 1925), pp. 8–9.

146. Walterman, *Airpower and Private Enterprise*, pp. 489–92.

147. Trimble, *Admiral William A. Moffet*, p. 167.

148. Annual Report of the Curtiss Aeroplane & Motor Company, Inc., March 4, 1926.

149. Ibid.

150. Trimble, *Admiral William A. Moffet*, p. 177.

151. *Aeronautical Chamber of Commerce: Aircraft Yearbook for 1927* (New York 1927), pp. 3–13; "A Big Military Aviation Program," *The Literary Digest*, 89, no. 12 (June 19, 1926): 9.

152. "The Army Air Bill," *Aviation*, 21, no. 2 (July 12, 1926): 49.

153. *Aircraft Yearbook for 1927*, p. 1.

154. "The Army Air Bill," p. 1.

155. Edwin H. Rutkowski, *The Politics of Military Aviation Procurement, 1926–1934* (Ohio State University Press, 1966), pp. 197–99.

156. Ibid., pp. 199–204.

157. Vander Meulen, *The Politics of Aircraft*, p. 81.

158. *Aircraft Yearbook for 1927*, p. 351; see also Holly, *Buying Aircraft*, pp. 89–93; Rutkowski, *The Politics of Military Aviation Procurement*, pp. 208–13.

159. Clement M. Keys to Hon. Carl Vinson, House of Representatives, January 31, 1927, p. 14, Folder 10 (Carl Vinson, 1926–1927), Box 13, CMKP, NASM.

160. Clement M. Keys to Hon. F. Trubee Davison, Assistant Secretary of War, War Department, August 6, 1926, Folder 7 (U.S. Government-General 1926), Box 13, CMKP, NASM.

161. Curtiss Aeroplane & Motor Company Annual Report for 1926, Annual and Interim Reports, Curtiss Aeroplane and Motor Co., 1917–1929, CMKP, NASM.

162. Ibid.

163. Ibid.

Chapter 4

1. Keys, "The Advance Agent of Prosperity," *World's Work*, 17, no. 3 (January 1909): 11164–70.

2. C. M. Keys to G. C. Moore, *Aeronautical Digest*, November 9, 1923, Box 8, Folder 1, CMKP, NASM.

3. Keys, "The Advance Agent of Prosperity," p. 11165.

4. Ibid.

5. Curtiss Aeroplane & Motor Company, Inc., House of Representatives Select Committee of Inquiry into Operations of the United States Air Services-Testimony of C. M. Keys, President (New York, 1925), p. 3.

6. Seyer, *Sovereign Skies: The Origins of American Civil Aviation Policy*, p. 72.

7. Johnson, *Taking Flight: The Foundations of American Commercial Aviation, 1918–1938*, pp. 30–34; *Aeronautical Chamber of Commerce: Aircraft Yearbook 1922* (New York, 1922), pp. 32–44; David D. Lee, "Herbert Hoover and the Development of Commercial Aviation," *The Business History Review*, 58, no. 1 (Spring 1984): 83–85.

8. Report of American Aviation Mission, p. 16.

9. Ibid., pp. 18–20.

10. Ibid., p. 20.

11. Ibid., p. 12.

12. National Advisory Committee for Aeronautics, *Annual Report for 1920* (Washington, DC: Government Printing Office, 1921), pp. 55–56.

13. Ibid., pp. 10–11.

14. Nick Komons, *Bonfires to Beacons: Federal Civil Aviation Policy under the Air Commerce Act 1926–1938* (Washington, DC: U.S. Department of Transportation, Federal Aviation Administration, 1978), p. 35.

15. Stuart Banner, *Who Owns the Sky?: The Struggle to Control Airspace from the Wright Brothers On* (Cambridge, MA: Harvard University Press, 2008), p. 149.

16. Seyer, *Sovereign Skies*, p. 98; Komons, *Bonfires to Beacons*, p. 23.

17. Ibid., p. 112.

18. Johnson, *Taking Flight*, p. 3.

19. Ibid., pp. 2–3.

20. Lampert Committee Hearings, pp. 1149–50.

21. David D. Lee, "Herbert Hoover and the Development of Commercial Aviation, 1921–1926," *The Business History Review*, 58, no. 1 (Spring 1984): 78.

22. Ibid., p. 79.

23. Ibid., p. 80.

24. National Advisory Committee for Aeronautics, *Annual Report for 1921* (Washington, DC: Government Printing Office, 1922), p. 13.

25. Ibid.

26. Ibid., p. 14.

27. Komons, *Bonfires to Beacons*, p. 46.

28. Ibid.

29. Lee, "Herbert Hoover and the Development of Commercial Aviation, 1921–1926," p. 88.

30. Seyer, *Sovereign Skies*, p. 99.

31. Ibid., p. 103.

32. Ibid., p. 105.

33. Lee, "Herbert Hoover and the Development of Commercial Aviation," p. 89.

34. Ibid.

35. Ibid.

36. Ibid.

37. Seyer, *Sovereign Skies*, pp. 104–105.

38. Lee, "Herbert Hoover and the Development of Commercial Aviation, 1921–1926," p. 90.

39. Seyer, *Sovereign Skies*, p. 105.

40. Lee, "Herbert Hoover and the Development of Commercial Aviation," p. 90; Komons, *Bonfires to Beacons*, p. 57.

41. Komons, *Bonfires to Beacons*, p. 57.

42. See David T. Courtwright, *Sky as Frontier: Adventure, Aviation, and Empire* (College Station, TX: Texas A&M University Press, 2005), pp. 46–52.

43. *Aeronautical Chamber of Commerce: Aircraft Yearbook 1923* (New York, 1923), p. 103.

44. Frank Russell testified that the Curtiss Aeroplane & Motor Corporation believed that over 75 percent of these accidents could have been prevented with effective government regulation. See Hearing Before a Subcommittee of the Committee on Commerce, United States Senate: Civil Aviation in the Department of Commerce, December 19, 1921, p. 24.

45. Aeronautical Chamber of Commerce, *Aircraft Yearbook 1923*, p. 102.

46. Ibid., pp. 91, 301; Aeronautical Chamber of Commerce, *Aircraft Yearbook 1924* (New York, 1924), p. 307.

47. William M. Leary, *Aerial Pioneers: The U.S. Air Mail Service, 1918–1927* (Washington, DC: Smithsonian Institution Press, 1985), pp. 171–77.

48. Ibid., p. 182.

49. Aeronautical Chamber of Commerce, *Aircraft Yearbook 1923*, pp. 361–66.

50. "Advantages of Aerial Mail," *Aerial Age Weekly*, 15, no. 11 (May 22, 1922): 250–51.

51. Komons, *Bonfires to Beacons*, p. 21.

52. Aeronautical Chamber of Commerce, *Aircraft Yearbook 1923*, pp. 343–44.

53. Ibid., pp. 98–99, 343–44.

54. C. M. Keys, Memorandum for Mr. Howard E. Coffin, October 16, 1923, Folder 16 (Paul Henderson, 1923–1924), Box 21, CMKP, NASM.

55. Memorandum to Howard E. Coffin, October 16, 1923.

56. Ibid.

57. C. M. Keys to Paul Henderson, Second Assistant Postmaster General, October 16, 1923, Folder 16 (Paul Henderson, 1923–1924), Box 21, CMKP, NASM.

58. C. M. Keys to Paul Henderson, Second Assistant Postmaster General, November 5, 1923, Folder 16 (Paul Henderson, 1923–1924), Box 21 CMKP, NASM.

59. C. M. Keys to Paul Henderson, November 5, 1923.

60. C. M. Keys to Paul Henderson, Second Assistant Postmaster General, November 7, 1923, Folder 16 (Paul Henderson, 1923–1924), Box 21, CMKP, NASM.

61. C. M. Keys to C. C. Moore, *Aeronautical Digest*, November 9, 1923, Folder 1 (Curtiss Aeroplane and Motor Co., *Aeronautical Digest*), Box 8, CMKP, NASM.

62. Ibid.

63. C. M. Keys to Leonard Kennedy, November 16, 1923, Folder 23 (Leonard Kennedy, 1923–1928), Box 18, CMKP, NASM.

64. Ibid.

65. C. M. Keys to Paul Henderson, Second Assistant Postmaster General, November 24, 1923, Folder 16 (Paul Henderson, 1923–1924), Box 21, CMKP, NASM.

66. Ibid.

67. Paul Henderson, Second Assistant Postmaster General, to C. M. Keys, November 21, 1923, Folder 16 (Paul Henderson, 1923–1924), Box 21, CMKP, NASM.

68. House of Representatives Select Committee of Inquiry into Operations of the United States Air Services: Testimony of C. M. Keys, President, p. 3.

69. Ibid.

70. C. M. Keys interview with Edgar Reinhart, Edgar Reinhart to Lester Gardner, Editor in Chief, Aviation, July 30, 1924, Folder 24 (Aviation, 1924–1928), Box 21, CMKP, NASM.

71. Ibid.

72. Quoted in Komons, *Bonfires to Beacons*, p. 66.

73. Leary, *Aerial Pioneers*, p. 197.

74. Komons, *Bonfires to Beacons*, pp. 195–201; "A New York-Chicago Night Air Mail Service," *Aviation*, 17, no. 22 (December 1, 1924): 1328–29.

75. Leary, *Aerial Pioneers*, p. 203.

76. Aeronautical Chamber of Commerce: *Aircraft Yearbook 1925* (New York, 1925), p. 44.

77. Komons, *Bonfires to Beacons*, p. 66.

78. Ibid.

79. Lampert Committee Hearings, pp. 1430–33.

80. Seyer, *Sovereign Skies*, p. 116.

81. Hearings Before the President's Aircraft Board, Volume 1, Table of Contents.

82. Komons, *Bonfires to Beacons*, p. 80.

83. Ibid., p. 81.

84. Janet R. Daly Bednarek, *America's Airports: Airfield Development, 1918–1947* (College Station, TX: Texas A&M University Press, 2001), p. 41.

85. Komons, *Bonfires to Beacons*, p. 86.

86. Ibid., p. 67.

87. Ibid.

88. See Paul Henderson's testimony to the Lampert Committee, Lampert Committee Hearings, p. 277.

89. C. M. Keys to Harold F. Pitcairn, April 8, 1925, Record Group 46. U.S. Senate, 74th Congress, Special Committee on Investigation of Air-Ocean Mail Contracts, 1934, Entry 17, North American Aviation, Box 146, North American Aviation Correspondence.

90. "Detroit's Ambitious Aeronautical Program," *Aviation*, 18, no. 17 (April 27, 1925): 461.

91. Meyer, Meyer, Austrian, and Platt, *Chicago: Corporate and Legal History of United Airlines and Its Predecessors and Subsidiaries* (Chicago: Twentieth Century Press, 1953), p. 58.
92. Ibid., p. 59.
93. C. M. Keys to Harold F. Pitcairn, April 8, 1925.
94. Meyer et al., *Corporate and Legal History of United Airlines*, p. 59.
95. *The New York Times*, April 13, 1925, p. 1.
96. "America Moves to Take the Lead in Aviation," *The New York Times*, August 23, 1925.
97. Leary, *Aerial Pioneers*, p. 224.
98. *The New York Times*, May 1, 1925, p. 1.
99. "National Air Transport Organized," *Aviation*, 18, no. 22 (June 1, 1925): 598–600.
100. *Aero Digest*, 7, no. 1 (July 1925): 388.
101. "National Air Transport, Incorporated," U.S. Air Services (August 1925), pp. 15–17; *The New York Times*, May 22, 1925; A report titled "National Air Transport, Inc.," prepared for the Air Mail Investigations, contains a listing of the initial stockholders in National Air Transport. Record Group 46, U.S. Senate, 74th Congress. Special Committee on Investigation of Air-Ocean Mail Contracts. United Airlines, File: Boeing Airplane Co., Box 152.
102. "A New York-Chicago Air Service," *The New York Times*, May 2, 1925, p. 14.
103. "National Air Transport Organized," *Aviation*, 18, no.22 (June 1, 1925): 598.
104. Ibid., p. 599.
105. "A Welcome Newcomer," *Aviation*, 16, no. 22 (June 1, 1925): 597.
106. Ibid.
107. Ibid.
108. "Publisher's News Letter," *Aviation*, 18, no. 17 (April 27, 1925): 473.
109. Samuel Taylor Moore, "The Fords Look to the Air," *The Independent*, 115, no. 392 (July 18, 1925): 68.
110. *The New York Times*, July 3, 1925, p. 15; "N.A.T. to Begin Service with Curtiss Planes," *Aviation*, 19, no. 7 (August 17, 1925): 183.
111. "New Air Mail Routes," *Aviation*, 19, no. 4 (July 27, 1925): 92; F. Robert Van der Linden, *Airlines and Air Mail: The Post Office and the Birth of the Commercial Aviation Industry* (Lexington: University Press of Kentucky, 2002), p. 18.
112. Meyer et al., *Corporate and Legal History of United Airlines*, pp. 61–62.
113. "Chicago-Dallas Air Mail Opened," *Aviation*, 20, no. 21 (May 24, 1926):783–84; "Texas-Chicago Air Mail Operates 97 per cent Perfect," *Aviation*, 21, no. 8 (August 23, 1926): 329.
114. National Air Transport, Inc. Minutes: Executive Committee, Box 1, Predecessor Companies and Subsidiaries (National Air Transport), United Airlines Legacy Foundation, Chicago, IL.
115. Meyer et al., *Corporate and Legal History of United Airlines*, p. 62; "Astonishing Air Mail Bids," *Aviation*, 22, No. 4 (January 24, 1927): 170.
116. National Air Transport, Inc., Minutes: Executive Committee, Meeting of March 18, 1926.
117. Ibid., pp. 136–37.
118. Memorandum on National Air Transport, Inc., April 6, 1927, copy in Record Group 46. U.S. Senate, 74th Congress, Special Committee to Investigate Air-Ocean Mail Contracts, Entry 17, North American Aviation, Box 146, North American Aviation Correspondence. See also the Memorandum of March 10, 1927, in the same file which discusses the possibility of capitalizing certain expenses.

119. Meyer et al., *Corporate and Legal History of United Airlines*, pp. 69–71.

120. Memorandum on National Air Transport, Inc., April 6, 1927, copy in Record Group 46. U.S. Senate, 74th Congress, Special Committee to Investigate Air-Ocean Mail Contracts, Entry 17, North American Aviation, Box 146, North American Aviation Correspondence.

121. "N.A.T. Planes Flown over 700,000 Miles during First Year of Operation," *Aviation*, 23, no. 2 (July 11, 1927): 80–82, 102.

122. Ibid.

123. *The New York Times*, February 20, 1927, SM2.

124. Clement M. Keys to H. G. Wells, May 3, 1927, Folder 29, Box 20, CMKP, NASM.

125. Ibid.

126. C. M. Keys to C. Townsend Luddington, March 4, 1931, Folder 7 (Luddington Philadelphia Flying Service 1931), Box 19, CMKP, NASM.

127. Ibid.

128. Memorandum to Howard Coffin, October 16, 1923, Folder 16 (Post Office Department: Paul Henderson, 1923–1924), Box 21, CMKP, NASM.

129. C. M. Keys to Hon. Harry S. New, Postmaster General, August 19, 1926, Folder 17 (Post Office Department: Harry S. New, 1926–1930), Box 21, CMKP, NASM.

130. C. M. Keys to Seward Prosser, Bankers Trust Company, May 10, 1928, Folder 6 (Transcontinental Air Transport: Prosser, Seward 1928), Box 24, CMKP, NASM.

131. van der Linden, *Airlines and Air Mail*, p. 39.

132. D. R. Lane, "How the T.A.T. Line Was Organized," *Airway Age*, 10, no. 7 (July 1929): 1023.

133. C. M. Keys, speech to the Babson Aviation Meeting, September 13, 1927, Folder 5 (Speeches and Articles: by C. M. Keys, 1925–1931), Box 22, CMKP, NASM.

134. I am grateful to Dr. Robert van der Linden, NASM, for bringing this point to my attention.

135. C. M. Keys and Howard Coffin to Col. Charles A. Lindbergh, June 14, 1927, File 251/Box 102, MS 325 Charles Lindbergh Collection, Sterling Memorial Library, Manuscripts and Archives, Yale University, New Haven, CT.

136. Ibid.

137. Chester W. Cuthell to Col. Charles Lindbergh, June 27, 1927, File 251/Box 102, MS 325, Charles Lindbergh Collection, Sterling Memorial Library, Manuscripts and Archives, Yale University, New Haven, CT.

138. Charles A. Lindbergh, *Autobiography of Values* (New York: Harcourt Brace Jovanovich, 1977), pp. 13–14, 80–81; Chester Cuthell to C. M. Keys, July 22, 1927, Folder 7 (Curtiss Aeroplane & Motor Co.: Legal 1923–1928), Box 10, CMKP, NASM.

139. C. M. Keys to J. C. Hunsaker, Federal Aviation Commission, October 3, 1934, Folder 22 (Aircraft Investigations: Howell Committee 1934), Box 27, CMKP, NASM.

140. C. M. Keys, speech to the Babson Aviation Meeting, p. 4.

141. Ibid.

142. Charles A. Lindbergh, *Autobiography of Values* (New York, 1976), pp. 83–96.

143. Ibid.

144. Ibid., pp. 96–101.

145. C. M. Keys to C. Townsend Luddington, March 4, 1931.

146. Lindbergh, *Autobiography of Values*, pp. 101–102.

147. General W. W. Atterbury, *The Railroads Enter Aviation*, Philadelphia: Pennsylvania Railroad Information, Vol. 1, No. 2, 1929, p. 21.

148. The following section on the formation of Transcontinental Air Transport relies
     heavily on Dr. Robert van der Linden's book, *Airlines and Air Mail*, especially
     pp. 39–46, which covers the same subject. I am grateful to Dr. van der Linden for
     sharing with me the results of his own research and for pointing out to me the key
     letters and papers that document Keys's role in the formation of the company.
149. C. M. Keys, "Tentative Plan No. 1," no date, Folder 30 (Transcontinental Air Trans-
     port: Misc. April–June 1928), Box 23; Memo, no date, Folder 7 (Transcontinental
     Air Transport: Publicity-undated), Box 24, CMKP, NASM.
150. Col. Paul Henderson, National Air Transport, Inc., to C. M. Keys, April 26, 1928,
     Folder 16 (Transcontinental Air Transport: Henderson, Paul: April-June 1928),
     Box 23, CMKP, NASM.
151. C. M. Keys, Memorandum of April 27, 1928; Col. Paul Henderson, National Air
     Transport, Inc., to C. M. Keys, April 26, 1928; C. M. Keys to Col. Paul Henderson,
     National Air Transport, April 30, 1928, Folder 16 (Transcontinental Air Transport:
     Paul Henderson: April–June 1928), Box 23, CMKP, NASM.
152. C. M. Keys, Memorandum of April 27, 1928.
153. Undated plan in Keys's papers; see Folder 7 (Transcontinental Air Transport:
     Publicity-undated), Box 24, CMKP, NASM.
154. C. M. Keys to Col. Paul Henderson, National Air Transport, April 30, 1928.
155. Undated plan in Keys's papers.
156. Daniel L. Rust and Alan B. Hoffman, *Come Fly with Me: The Rise and Fall of Trans
     World Airlines* (St. Louis, MO: Missouri Historical Society Press, 2023), p. 18.
157. *The New York Times*, May 16, 1928, p. 1.
158. Ibid.
159. C. M. Keys to Colonel Charles A. Lindbergh, May 23, 1928, Folder 27, Box 44,
     Charles A. Lindbergh Papers, Missouri Historical Society, St. Louis, MO.
160. Ibid.
161. *The New York Times*, May 23, 1928, p. 1.
162. *New York Telegraph*, May 24, 1928.
163. C. M. Keys, "Making the T.A.T.," Folder 5 (Speeches and Articles: by C. M. Keys,
     1925–1931), Box 22, CMKP, NASM,
164. Ibid.
165. Ibid.
166. *The New York Times*, May 29, 1928, p. 27.
167. See C. M. Keys to Fred Harvey, June 25, 1928, Folder 16 (Transcontinental Air Trans-
     port: Paul Henderson, April-June 1928), Box 23; C. M. Keys to Burdette Wright,
     May 29, 1928, Folder 32 (Burdette S. Wright, 1928), Box 24, CMKP, NASM.
168. van der Linden, *Airlines and Air Mail*, p. 44.
169. C. M. Keys to L. H. Piper, Universal Aircraft Corporation, April 19, 1929, Folder 12
     (Universal Aviation Corp. 1928–1929), Box 25, CMKP, NASM.
170. C. M. Keys to Executive Committee, Transcontinental Air Transport, September 13,
     1928, Folder 22 (Transcontinental Air Transport: Reports-Technical Committee
     1928), Box 24, CMKP, NASM.
171. Chester Cuthell to C. M. Keys, August 8, 1928, Folder 2 (Transcontinental Air
     Transport: Cuthell, Hotchkiss, and Mills 1928), Box 23, CMKP, NASM.
172. Ibid.
173. C. M. Keys to Colonel Paul Henderson, October 17, 1928, Folder 18 (Transcontinen-
     tal Air Transport: Paul Henderson, October 1928), Box 23, CMKP, NASM.
174. Ibid.

175. C. H. Mathews Jr., and C. E. McCullough, "Report of Study of Commercial Aviation in Europe," Pennsylvania Railroad, August 10, 1928, copy in File 9: Commercial Aviation in Europe 1928, Box 757, Traffic Department, 1807 Pennsylvania Railroad, Hagley Museum and Library, Wilmington, DE.

176. See C. M. Keys to Executive Committee, Transcontinental Air Transport, September 13, 1928; Charles Lindbergh to C. M. Keys, August 31, 1928, C. S. Jones to C. M. Keys, August 23, 1928, Major E. F. Brainard to C. M. Keys, August 6, 1928, Folder 22 (Transcontinental Air Transport: Reports-Technical Committee 1928), Box 24; C. M. Keys to W. B. Mayo, Ford Motor Company, September 11, 1928. Folder 15 (Transcontinental Air Transport: Ford Co., 1927–1928), Box 23, CMKP, NASM.

177. Charles Lindbergh to C. M. Keys, August 31, 1928, Folder 15 (Transcontinental Air Transport: Ford Co., 1927–1928), Box 23, CMKP, NASM.

178. C. M. Keys to Lewis W. Baldwin, Missouri Pacific Railroad, October 11, 1928, Folder 15 (Transcontinental Air Transport: Missouri Pacific Railway Co. 1928), Box 24, CMKP, NASM.

179. C. M. Keys to Executive Committee, Transcontinental Air Transport, September 13, 1928, Folder 22 (Transcontinental Air Transport: Reports-Technical Committee 1928), Box 24, CMKP, NASM; C. M. Keys to Colonel Charles Lindbergh, September 20, 1928, Folder 21, Box 44, Charles A. Lindbergh Papers, Missouri Historical Society.

180. *The Magazine of Business*, 55 (June 1929): 625.

181. Transcontinental Air Transport Report to Stockholders, February 20, 1929, Folder 4 (Annual and Interim Reports: Transcontinental Air Transport 1929–1932), Box 26, CMKP, NASM.

Chapter 5

1. "Aviation and Its Future," Speech made at Republican Club Luncheon on February 2, 1929, Folder 6 (Curtiss Aeroplane & Motor Co.: Speeches 1929), Box 12, CMKP, NASM. The full text of the speech was printed in the March 1929 edition of the magazine *U.S. Air Services*.

2. C. M. Keys, "Aviation and Its Future," *U.S. Air Services*, 14, no. 3 (March 1929): 46.

3. Keys, "Aviation and Its Future," p. 46.

4. *NAT Bulletin Board*, No. 27 (May 1929), p. 1.

5. National Air Transport Minutes of the Executive Committee, meeting of January 16, 1928.

6. Ibid., meeting of March 1, 1928.

7. *Aviation*, 24, no. 25 (June 18, 1928): 1770.

8. *The Wall Street Journal*, February 27, 1930, p. 7.

9. van der Linden, *Airlines and Air Mail*, pp. 47–48.

10. *NAT Bulletin Board*, No. 19 (September 4, 1928), p. 1.

11. *The Wall Street Journal*, February 27, 1930, p. 7.

12. Ibid.

13. *Barron's*, March 24, 1930, p. 7.

14. *Air Transportation*, January 25, 1930, p. 14.

15. Ibid.

16. Ibid.; Aeronautical Chamber of Commerce: *The Aircraft Yearbook for 1930* (New York: D. Van Nostrand, 1930), p. 22.

17. Ibid.

18. *Air Transportation*, January 25, 1930, p. 14.

19. *Evening Star*, March 5, 1929, p. 4.

20. David D. Lee, "Herbert Hoover and the Golden Age of Aviation," in William M. Leary, *Aviation's Golden Age: Portraits from the 1920s and 1930s* (Iowa City: University of Iowa Press, 1989), p. 138.

21. M. Johnson and V. Houston, *Taking Flight: The Foundations of American Commercial Aviation, 1918–1938* (College Station: Texas A&M University Press, 2019), p. 84.

22. *Air Transportation*, January 25, 1930, p. 14.

23. Ibid.

24. van der Linden, *Airlines and Air Mail*, p. 114.

25. Johnson, *Taking Flight: The Foundations of American Commercial Aviation, 1918–1938*, p. 86.

26. van der Linden, *Airlines and Air Mail*, p. 153.

27. Johnson, *Taking Flight*, p. 86.

28. van der Linden, *Airlines and Air Mail*, p. 114.

29. Ibid., p. 114; *Air Transportation*, January 25, 1930, p. 14.

30. *Air Transportation*, January 25, 1930, p. 14.

31. Ibid., p. 31.

32. Ibid.

33. van der Linden, *Airlines and Air Mail*, p. 66.

34. Johnson, *Taking Flight*, p. 90.

35. Paul T. Dodd, *The Economics of Air Mail Transportation* (Washington, DC: The Brookings Institution, 1934), p. 105.

36. Ibid.

37. van der Linden, *Airlines and Air Mail*, p. 152.

38. Dodd, *The Economics of Air Mail Transportation*, p. 106.

39. Aeronautical Chamber of Commerce, *The Aircraft Yearbook for 1931* (New York: D. Van Nostrand, 1931), p. 19.

40. The section on Transcontinental Air Transport's fortunes later in the chapter discusses the Watres Act and Postmaster General Walter Folger Brown's vision for air commerce in more detail.

41. *Chicago Tribune*, March 25, 1930, p. 15.

42. *The New York Times*, March 25, 1930, p. 5.

43. *Chicago Tribune*, March 27, 1930, p. 23.

44. van der Linden, *Airlines and Air Mail*, p. 144.

45. Ibid., p.144; *The New York Times*, March 27, 1930, p.38; C. M. Keys, Memorandum to the Board of Directors of National Air Transport, Inc., April 2, 1930, U.S. Senate, 74th Congress. Special Committee on Investigation of Air Mail and Ocean Mail Contracts, 1934, Entry 17, North American Aviation, Box 146, North American Aviation Correspondence, RG 46, NARA.

46. National Air Transport Minutes of the Executive Committee, meeting of April 3, 1930.

47. C. M. Keys, Memorandum to the Board of Directors of National Air Transport, Inc., April 2, 1930.

48. van der Linden, *Airlines and Air Mail*, p. 145.

49. Howard Coffin to Frederick Rentschler, United Aircraft & Transport Corporation, April 3, 1930, Record Group 46, U.S. Senate, 74th Congress. Special Committee on Investigation of Air Mail and Ocean Mail Contracts, 1934, Entry 17, United Air Lines, Box 152, National Air Transport, RG 46, NARA; *The New York Times*, April 4, 1930, p. 23.

50. *The New York Times*, April 4, 1930, p. 23.

51. Frederick B. Rentschler, United Aircraft & Transport Corporation, to the Stockholders of National Air Transport, April 4, 1930, Record Group 46, U.S. Senate, 74th Congress. Special Committee on Investigation of Air Mail and Ocean Mail Contracts, 1934, Entry 17, United Air Lines, Box 152, National Air Transport, RG 46, NARA.

52. *The New York Times*, April 4, p. 23; April 8, 1930, p. 43.

53. van der Linden, *Airlines and Air Mail*, p. 146.

54. Ibid.; Charles E. Lawrance and John W. Pattison, National Air Transport, to the Board of Directors, National Air Transport, Inc., April 9, 1930, U.S. Senate. 74th Congress. Special Committee on Investigation of Air Mail and Ocean Mail Contracts, 1934, Entry 17, North American Aviation, Box 146, North American Aviation Correspondence, RG 46, NARA.

55. C. M. Keys press release, April 10, 1930, U.S. Senate, 74th Congress, Special Committee on Investigation of Air Mail and Ocean Mail Contracts, 1934, Entry 17, North American Aviation, Box 146, North American Aviation Correspondence, RG 46, NARA.

56. van der Linden, *Airlines and Air Mail*, p. 148; *The New York Times*, April 11, p. 43; April 12, 1930, p. 12.

57. *The Wall Street Journal*, April 15, 1930, p. 15.

58. van der Linden, *Airlines and Air Mail*, p. 149; *The New York Times*, April 15, p. 48; April 16, p. 20; April 22, 1930, p. 48.

59. *The New York Times*, April 17, 1930, p. 44.

60. C. M. Keys to C. W. Cuthell, April 19, 1930, U.S. Senate, 74th Congress, Special Committee on Investigation of Air Mail and Ocean Mail Contracts, 1934, Entry 17, North American Aviation, Box 146, North American Aviation Correspondence, RG 46, NARA.

61. *The Wall Street Journal*, April 25, 1930, p. 15.

62. *The New York Times*, April 24, p. 21; April 26, 1930, p. 30.

63. *The New York Times*, March 1, 1927, p. 29.

64. *News Wing*, Vol. I, No. 4 (December 1927), p. 1.

65. *News Wing*, Vol. I, No. 9 (May 1928), p. 1.

66. *News Wing*, Vol. I, No. 12 (August 1928), p. 1

67. *News Wings* Vol.2, No.3 (November 1928), pp. 4–5; *The New York Times*, October 28, 1928, p. 148.

68. Frank Kingston Smith, *Legacy of Wings: The Harold F. Pitcairn Story* (Lafayette Hill, PA: T-D Associates, 1981), pp. 137–41, 145–46. Smith's book was based on access to Harold Pitcairn's personal correspondence, but he does not provide a specific source for these discussions between Keys and Pitcairn.

69. Smith, *Legacy of Wings*, p. 147.

70. Smith, *Legacy of Wings*, pp. 147, 150–60; *News Wing*, 2, no. 10 (June 1929): 1; *The New York Times*, June 15, 1929, p. 31, and June 29, 1929, p. 33.

71. *News Wing*, 2, no. 10 (June 1929).

72. North American Aviation, Inc., Minutes of Board Meetings, June 27, July 12, September 18, 1929, Boeing Archives, Seattle, WA.

73. Robert J. Serling, *From the Captain to the Colonel: An Informal History of Eastern Airlines* (New York: Dial Press, 1980), pp. 41–42.

74. *The Wall Street Journal*, January 16, 1930, p. 16.

75. *The Wall Street Journal*, February 17, 1930, p. 8.

76. C. M. Keys to Daniel Sheaffer, Pennsylvania Railroad Company, August 29, 1930, File 077.2: Air Mail Investigation Copies of Correspondence, D. M. Sheaffer, 1929–1931, Box 394, Office of Passenger Transportation, ACC No.1810 Pennsylvania Rail Road, Hagley Museum and Library, Wilmington, Delaware.

77. *The New York Times*, March 13, 1930, p. 13; *Aviation*, 28, no. 13 (March 29, 1930): 662; *News Wing*, Eastern Air Transport, Inc., 3, no. 8, (April 1930).

78. *The New York Times*, August 8, 1930, p. 7.

79. Meetings of the Executive Committee, August 29, October 24, November 21, 1930, North American Aviation, Inc., Minutes of Board Meetings.

80. Meeting of the Executive Committee, November 21, December 19, 1930, North American Aviation, Inc., Minutes of Board Meetings.

81. *Air Transportation*, 11, no.14 (April 19, 1930): 16; 13, no. 3 (August 16, 1930): 6; 13, no. 4 (August 23, 1930): 10; 14, no. 1 (September 13, 1930): 6; 14, no. 8 (November 1, 1930): 1; 14, no. 10 (November 15, 1930): 1; *The New York Times*, August 4, 1930, p. 7; August 16, 1930, p. 28; September 13, 1930, p. 16; December 4, 1930, p. 23; December 7, 1930, p. 148; December 11, 1930, p. 28; *Aviation News*, 1, no. 7 (August 23, 1930), p. 17; 1, no. 15, October 18, 1930, p. 19. The figure for Eastern Air Transport's air mail revenues can be found in a report on the Aviation Corporation, Folder 9 (Aviation Corporation of Delaware-General File 1928–1931), Box 3, CMKP, NASM, and in Financial Construction of North American Aviation, Inc., Subsidiaries and Affiliated Companies—And Other Data, Prepared for the Senate Special Committee to Investigate Foreign and Domestic Ocean and Air Mail Contracts, by E. C. Sauer and S. C. Simon, April 25, 1934, U.S. Senate. 74th Congress. Special Committee on Air-Ocean Mail Contracts, Entry 17, North American Aviation, Box 145, RG 46, NARA.

82. C. M. Keys, "Making the TAT," June 25, 1929, Folder 5 (Speeches and Articles: by C. M. Keys 1925–1931), Box 22, CMKP, NASM.

83. Charles A. Lindbergh to D. M. Sheaffer, Chief, Passenger Transportation, Pennsylvania Railroad, April 30, 1929, Folder 22, Box 44, Charles A. Lindbergh Papers, Missouri Historical Society, St. Louis, Missouri.

84. *TAT Plane Talk*, I, no. 6, June 1929; *Aviation*, 36, no. 19 (May 11, 1929): 1625.

85. R. E. G. Davies, *Airlines of the United States Since 1914* (London: Putnam, 1972), pp. 85–87.

86. *Los Angeles Times*, "Maddux Lines Expanding," December 2, 1928, p. B8.

87. Aviation Corporation of California, Maddux Air Lines Co. Voting Trust Certificates Offering Circular, November 28, 1928, Folder 23 (Circulars[Offering and Special]: Maddux Air Lines Co. 1929), Box 26, CMKP, NASM; Minutes of the Meeting of the Executive Committee of Transcontinental Air Transport, Inc., May 14, 1929, File: Air Mail Investigation (Black Committee) Transcontinental Air Transport 1928–34, Box 394: Office of Passenger Transportation, Acc. No. 1810: Pennsylvania Railroad, Hagley Museum and Library.

88. Minutes of the Meeting of the Executive Committee of Transcontinental Air Transport, Inc., May 14, 1929, File: Air Mail Investigation (Black Committee) Transcontinental Air Transport 1928-34, Box 394: Office of Passenger Transportation, Acc. No. 1810: Pennsylvania Railroad, Hagley Museum and Library; *The New York Times*, June 2, 1929, p. N14.

89. *The New York Times*, July 7, 1929, p. 121.

90. *The New York Times*, June 29, 1929, p. 14.

91. *The New York Times*, July 8, 1929, p. 1; *TAT Plane Talk*, I, no. 7 (July 1929).

92. Keys, "The Business of Aviation: 1929–1930," p. 41.

93. *Western Flying*, 6, no. 4 (October 1929): 72.

94. *The New York Times*, August 13, 1929, p. 3.

95. *TAT Plane Talk*, I, no. 9 (September 1929): 1.

96. *The New York Times*, September 5, 1929, p. 1, September 9, p. 1; *Los Angeles Times*, September 5, 1929, p. 1.

97. C. M. Keys to Directors, October 31, 1929.

98. C. M. Keys to Colonel Charles A. Lindbergh, November 25, 1929, Folder 23, Box 44, Charles A. Lindbergh Papers, Missouri Historical Society, St. Louis, MO.

99. *Air Transportation*, 10, no. 7 (November 30, 1929): 12.

100. *The New York Times*, January 11, 1930, p. 1.

101. Ibid.

102. *The New York Times*, January 20, 1930, p. 1.

103. Charles A. Lindbergh to Colonel Henry Breckenridge, January 24, 1930, Folder 24, Box 44, Charles A. Lindbergh Papers, Missouri Historical Society, St. Louis, MO.

104. Colonel Henry Breckenridge to Charles Lindbergh, January 23, 1930, Folder 24, Box 44, Charles A. Lindbergh Papers, Missouri Historical Society, St. Louis, MO.

105. *The New York Times*, January 27, 1930, p. 35; *Los Angeles Times*, January 27, 1930, p. 3.

106. *Chicago Tribune*, March 18, 1930, p. 23.

107. Annual Report for Transcontinental Air Transport, March 13, 1930, Folder 4 (Annual and Interim Reports: Transcontinental Air Transport 1929–1932), Box 26, CMKP, NASM.

108. Thomas Eastland to Colonel Charles A. Lindbergh, March 27, 1930, File 507, Box 124, Charles Lindbergh Collection, MS 325, Manuscripts and Archives, Sterling Memorial Library, Yale University, New Haven, CT.

109. *The Wall Street Journal*, March 15, 1930, p. 1.

110. Dodd, *The Economics of Air Mail Transportation*, p. 109.

111. van der Linden, *Airlines and Air Mail*, pp. 153–157, 173.

112. Daniel Sheaffer, Memorandum Covering Important Points in the Development of a Consolidated Operation, T.A.T. Inc., and Western Air Express, Inc., As Presented in the Attached Agreement Contemplating the Formation of an Operating Company, July 15, 1930, File 077.2: Air Mail Investigation Copies of Correspondence, W. W. Atterbury, A. J. County, F. J. Fell, W. S. Franklin, J. G. Wilson, 1928–1934, Box 394 Office of Passenger Transportation, ACC No.1810 Pennsylvania Rail Road, Hagley Museum and Library, Wilmington, Delaware.

113. Ibid.

114. van der Linden, *Airlines and Air Mail*, p. 173; Sheaffer memorandum of July 15, 1930.

115. van der Linden, *Airlines and Air Mail*, pp. 176–77.

116. Rust and Hoffman, *Come Fly with Me*, p. 33.

117. van der Linden, *Airlines and Air Mail*, pp. 176–77.

118. *Air Transportation*, 14, no. 11 (November 22, 1930): 5.

119. Daniel Sheaffer to M. W. Clement, October 6, 1930, File-Air Express Service Transcontinental & Western Air-TAT 1929–1937, Box 244-Vice President Finance General Office Files, ACC No. 1810 Pennsylvania Rail Road, Hagley Museum and Library, Wilmington, Delaware.

120. J. Cheever Cowdin to C. M. Keys, D. M. Sheaffer, and Harris Hanshue, January 20, 1931, Folder 13, Box 46, Charles A. Lindbergh Papers, Missouri Historical Society, St. Louis, MO.

121. Ibid.

122. Ibid.

123. Harris M. Hanshue to J. Cheever Cowdin, January 28, 1931, Folder 26, Box 44, Charles A. Lindbergh Papers, Missouri Historical Society, St. Louis, MO.

124. Marc Dierikx, *Fokker: A Transatlantic Biography* (Washington, DC: Smithsonian Institution Press, 1997), pp. 140–43.

125. Ibid., p. 139.

126. *The New York Times*, March 5, 1931, p. 40.

127. *The New York Times*, March 13, 1931, p. 36.

128. C. M. Keys to Charles L. Lawrance, Aeronautical Chamber of Commerce, March 23, 1931, Folder 5 (Aeronautical Chamber of Commerce of America, Inc., 1930–31), Box 1, CMKP, NASM.

129. C. M. Keys to S. S. Fontaine, Benjamin, Hill & Co., March 24, 1931, Folder 18 (S. S. Fontaine, 1929–1931), Box 17, CMKP, NASM.

130. C. M. Keys to General W. W. Atterbury, Pennsylvania Railroad, March 23, 1931, U.S Senate, 74th Congress, Special Committee on Air-Ocean Mail Contracts, Entry 17, North American Aviation Correspondence, Box 146, RG 46, NARA.

131. A. J. County to General W. W. Atterbury, April 24, 1931, U.S. Senate. 74th Congress. Special Committee on Air-Ocean Mail Contracts, Entry 17, North American Aviation Correspondence, Box 146, RG 46, NARA.

132. J. Mel Hickerson, and Ernie Breech, *The Story of His Remarkable Career at General Motors, Ford, and TWA* (New York: Meredith Press, 1968), p. 74.

133. These comments are from a quote from Breech in his biography, unfortunately without supporting documents.

134. Ibid., p. 75.

135. *The New York Times*, July 9, 1931, p. 30.

136. Transcontinental & Western Air, Annual Report for 1931, May 4, 1932, Folder 27, Box 84, Charles A. Lindbergh Papers, Missouri Historical Society, St. Louis, MO.

137. Richard Robbins to Daniel M. Sheaffer, July 14, 1931, Folder 14, Box 46, Charles A. Lindbergh Papers, Missouri Historical Society, St. Louis, MO.

138. C. M. Keys to Richard Robbins, August 21, 1931, Folder 14, Box 46, Charles A. Lindbergh Papers, Missouri Historical Society, St. Louis, MO.

139. Richard Robbins to the TWA Organization, September 22, 1931.

140. Richard Robbins to Colonel Charles A. Lindbergh, October 28, 1931, Folder 15, Box 46, Charles A. Lindbergh Papers, Missouri Historical Society, St. Louis, MO.

141. Memo from Richard Robbins, June 2, 1931, Folder 29, Box 45, Charles A. Lindbergh Papers, Missouri Historical Society, St. Louis, MO.

142. D. W. Tomlinson to Colonel Charles A. Lindbergh, November 24, 1931, Folder 1, Box 46, Charles A. Lindbergh Papers, Missouri Historical Society, St. Louis, MO.

143. *The Wall Street Journal*, April 17, 1931, p. 11.

144. C. M. Keys to Richard Robbins, November 21, 1931, File 077.2 Air Mail Investigation (Black Committee) Copies of Correspondence D. M. Sheaffer 1929–1931, Box 394 Office of Passenger Transportation, Acc. No. 1810 Pennsylvania Railroad, Hagley Museum and Library, Wilmington, DE.

145. C. M. Keys to Richard Robbins, November 21, 1931.

146. Ibid.

147. Ronald Miller and David Sawers, *The Technical Development of Modern Aviation* (London: Praeger Publishers, 1970), pp. 19, 47–49.

148. Clement M. Keys, "The Uplift of the World," *The World's Work*, 14, no. 3 (July 1907): 9094.

149. Clement M. Keys, "The Advance Agent of Prosperity," *The World's Work*, 17, no. 3 (January 1909): 11164–170.

150. See Exhibit A, North American Aviation, February 5, 1929, Intercontinent Aviation, Inc. Minutes of Board Meetings, Book 2, Boeing Historical Archives, The Boeing Company, Seattle, WA.

151. This section on Keys's contribution to building commercial air transport in China is drawn from William M. Leary Jr.'s excellent study, *Dragon's Wings: The China National Aviation Corporation and the Development of Commercial Aviation in China* (Athens, GA: University of Georgia Press, 1976), which contains more details on Keys's efforts. See *Dragon's Wings*, p. 6.

152. C. M. Keys to Clarence Dillon, Dillon, Read & Company, November 26, 1928, Folder 33 (Aviation Exploration, Inc.: General File 1928–1929), Box 3, CMKP, NASM.

153. See Exhibit A, North American Aviation, February 5, 1929, Intercontinent Aviation, Inc.; Minutes of the Executive Committee, March 1, 1929; Minutes of the Board of Directors, March 13, 1929, North American Aviation Minutes of Board Meetings, Book 2, Boeing Historical Archives, The Boeing Company, Seattle, WA.

154. C. M. Keys to Clarence Dillon, Dillon, Read & Company, November 26, 1928, Folder 33 (Aviation Exploration, Inc.: General File 1928–1929), Box 3, CMKP, NASM.

155. *The New York Times*, January 16, 1929, p. 13.

156. Leary, *Dragon's Wings*, p. 11.

157. Ibid.

158. Ibid.

159. Exhibit B, Memorandum for the Executive Committee, North American Aviation, Inc., June 27, 1929, North American Aviation Minutes of Board Meetings, Book 2, Boeing Historical Archives, The Boeing Company; C. W. Webster, "My Early Years in Aviation," pp. 43–50, unpublished manuscript, copy in the collection of the Museum of Flight, Seattle, WA.

160. R. E. G. Davies, *Airlines of Latin America since 1919* (London: Putnam Aeronautical Books, 1983), pp. 289–90.

161. Memorandum of June 27, 1929, and Webster, "My Early Years in Aviation," pp. 48–50.

162. E. C. Sauer and S. C. Simon, "North American Aviation, Inc.," prepared for the Senate Special Committee to Investigate Foreign and Domestic Ocean and Airmail Contracts, p. 31, U.S. Senate, 74th Congress, Special Committee on Air-Ocean Mail Contracts, Entry 17, North American Aviation, Box 142, RG 46 NARA.

163. Leary, *Dragon's Wings*, p. 19.

164. A copy of the original contract can be found at the China National Aviation Corporation website, www.cnac.org/history02. htm accessed February 23, 2022.

165. *The New York Times*, September 20, 1929, p. 20; *Aviation*, 27, no. 13 (September 28, 1929): 678.

166. *Aviation*, 27, no. 13 (September 28, 1929): 678.

167. E. C. Sauer and S. C. Simon, "North American Aviation, Inc.," prepared for the Senate Special Committee to Investigate Foreign and Domestic Ocean and Airmail Contracts, p. 31, Record Group 46, U.S. Senate, 74th Congress, Special Committee, Air-Ocean Mail Contracts, Entry 17, North American Aviation, Box 142, NARA.

168. Memorandum of June 27, 1929, North American Aviation.

169. C. M. Keys, "Memorandum to the Executive Committee, North American Aviation, Inc.," August 30, 1929, North American Aviation Minutes of Board Meetings, Book 2, Boeing Historical Archives, The Boeing Company, Seattle, WA.

170. Ibid.
171. Ibid.
172. Ibid.
173. Ibid.
174. Gene Banning, *Airlines of Pan American Since 1927* (McLean, VA: Paladwr Press, 2001), p. 281.
175. Leary, *The Dragon's Wings*, p. 24.
176. Ibid., p. 29.
177. Ibid., p. 68.
178. Ibid., pp. 33–34.

Chapter 6

1. C. M. Keys to Charles Gray, November 9, 1928, Folder 20 (The Aeroplane 1925–1928), Box 21, CMKP, NASM.
2. James C. Willson & Co., Curtiss and Associated Companies (New York), March 20, 1929, copy in the author's possession.
3. Reeves, *Aviation's Place in Tomorrow's Business*, p. 92.
4. Curtiss and Associated Companies, p. 9.
5. *Aviation's Place in Tomorrow's Business*, pp. 97–98.
6. Ibid., p. 93.
7. Ibid., p. 95.
8. Wagner, *American Combat Planes of the 20th Century*, pp. 48, 62–69, 85–88, 149–52, 154–55.
9. Curtiss Aeroplane & Motor Company Annual Report for 1926, March 1, 1927.
10. Rae, *Climb to Greatness*, p. 49; Aeronautical Chamber of Commerce, *Aircraft Yearbook 1928* (New York, 1928), p. 457.
11. William Robertson, President, Robertson Aircraft Corporation, to C. M. Keys, September 6, 1927, Folder 32 (Curtiss-Robertson Airplane Manufacturing Co.: William B. Robertson, 1927–1928), Box 15, CMKP, NASM; "What an Investor is Entitled to Know," no date, Folder 27 (Curtiss-Robertson Airplane Manufacturing Co.: Misc. 1927–29), Box 15, CMKP, NASM. This was a copy that Robertson sent to Keys after their meeting in New York in August 1927 of a proposal that Robertson had put together for potential investors in St. Louis.
12. C. M. Keys to Charles H. Diefendorf, Vice President, The Marine Trust Company, Buffalo, New York, Folder 27 (C. H. Diefendorf, 1927), Box 8, CMKP, NASM.
13. C. M. Keys to William B. Robertson, Robertson Aircraft Corporation, September 9, 1927, Folder 32 (Curtiss-Robertson Airplane Manufacturing Co.: William B. Robertson, 1927–1928), Box 15, CMKP, NASM.
14. Ibid.
15. "Curtiss and Associated Companies," p. 22.
16. Harold M. Bixby, The State National Bank, to C. M. Keys, Curtiss Aeroplane & Motor Company, October 21, 1927; C. M. Keys to Harold M. Bixby, The State National Bank, October 24, 1927, Folder 27 (Curtiss-Robertson Airplane Manufacturing Co.: Misc. 1927–29), Box 15, CMKP, NASM; C. M. Keys to William B. Robertson, Robertson Aircraft Corporation, November 1, 1927; William B. Robertson, Robertson Aircraft Corporation, November 3, 1927, , Folder 32 (Curtiss-Robertson Airplane Manufacturing Co.: William B. Robertson, 1927–1928), Box 15, CMKP, NASM; Harry H. Knight, Knight, Dysart & Gamble,

to C. M. Keys, November 21, 1927; C. M. Keys to Harry H. Knight, Knight, Dysart & Gamble, November 23, 1927; Harry H. Knight, Knight, Dysart & Gamble, to C. M. Keys, December 16, 1927, Folder 27 (Curtiss-Robertson Airplane Manufacturing Co.: Misc. 1927–29), Box 15, CMKP, NASM.

17. Curtiss and Associated Companies, pp. 25–26; Frank H. Ellis, *Canada's Flying Heritage* (Toronto: University of Toronto Press, 1954), pp. 238, 250; *The New York Times*, December 6, 1928, p. 51; December 8, 1928, p. 31.

18. Curtiss and Associated Companies, pp. 23–24, 27–28; *The Wall Street Journal*, October 6, 1928, p. 7, January 28, 1929, p. 11; *The New York Times*, January 17, 1929, p. 11; G.M-P. Murphy & Co., "Curtiss-Caproni Corporation," January 28, 1929, Folder 15 (Circulars: Curtiss-Caproni Corp. 1929), Box 26, CMKP, NASM; Bill Gunston, *World Encyclopedia of Aircraft Manufacturers*, 2nd ed. (Gloucestershire, UK: Sutton Publishing, 2005), pp. 90–91.

19. Curtiss Aeroplane & Motor Company, Annual Report for 1926 (March 1, 1927).

20. See the untitled memo from March 1928 in Folder 10 (Curtiss Aeroplane Export Co./Curtiss-Wright Export Co. Misc. 1928), Box 14; C. M. Keys to James C. Willson, October 8, 1926, Folder 9 (Curtiss Aeroplane Export Co./Curtiss-Wright Export Co. Misc.1925–1927), Box 14; Listing Bulletin No. 121, Curtiss Aeroplane Export Corporation, March 22, 1928, Folder 8 (Curtiss Aeroplane Export Co./ Curtiss-Wright Export Co. Listing Statement [New York Stock Exchange] 1928), Box 14, CMKP, NASM; Dan Hagedorn, *Conquistadors of the Sky: A History of Aviation in Latin America* (Gainesville: University Press of Florida, 2008), pp. 131–35.

21. See the undated and untitled memos labeled Exhibit A, B, C, and D in Folder 9 (Curtiss Aeroplane Export Co./Curtiss-Wright Export Co. Misc.1925–1927), and C. M. Keys to James C. Willson, December 19, 1927, in the same Folder, CMKP, NASM.

22. C. M. Keys to the Board of Directors, Curtiss Aeroplane & Motor Company, January 16, 1928, Folder 28 (Curtiss Aeroplane & Motor Co.: Directors 1918–1928), Box 8, CMKP, NASM.

23. C. M. Keys to Blair & Company, James C Willson & Company, September 12, 1928, Folder 30 (Curtiss Flying Service/Curtiss Wright Flying Service: Blair and Co. 1928), Box 14, CMKP, NASM.

24. Reeves, *Aviation's Place in Tomorrow's Business*, p. 96.

25. Reeves, p. 96; G.M-P. Murphey & Co., "Curtiss Flying Service, Inc.," Folder 13 (Curtiss Flying Service/Curtiss Wright Flying Service: Stock Trading Account 1928–1930), Box 15, CNKP, NASM; *The New York Times*, September 7, 1928, p. 14.

26. Stockholders Notice, November 30, 1928, Folder 8 (Stockholders Notice: Curtiss Flying Service/Curtiss-Wright Flying Service 1928), Box 27; C. M. Keys to Blair & Company, James C Willson & Company, September 12, 1928, Folder 30 (Curtiss Flying Service/Curtiss Wright Flying Service: Blair and Co., 1928), Box 14, CMKP, NASM; *The New York Times*, December 1, 1928, p. 23; *Air Transportation*, Vol. 5, No. 14 (December 8, 1928), p. 24.

27. Bednarek, *America's Airports: Airfield Development, 1918–1947*, pp. 41–42.

28. *The New York Times*, February 28, 1928, p. 6.

29. *The New York Times*, May 7, 1929, p. 14; May 9, 1929, p. 45; Curtiss Airports Corporation offering circular, May 8, 1929, Folder 16 (Curtiss Airports Co./Curtiss-Wright Airports Co.: Misc. 1929–1930), Box 14, CMKP, NASM.

30. Reeves, *Aviation's Place in Tomorrow's Business*, pp. 94–95.

31. *The Wall Street Journal*, June 6, 1928.

32. C. M. Keys to Commander J. C. Hunsaker, October 10, 1934, Folder 22 (Aircraft Investigations: Howell Committee 1934), Box 27, CMKP, NASM.

33. Ibid.

34. John Francis Fowler Jr., *American Investment Trusts* (New York: Harper and Brothers, 1928), p. 20.

35. Ibid., p. 6; Carosso, *Investment Banking in America*, pp. 281–84, 287.

36. Curtiss and Associated Companies, p. 13.

37. Ibid., pp. 13–14.

38. G.M-P. Murphy & Co., National Aviation Corporation offering circular, June 1928, Folder 25 (Circulars: National Aviation Corporation 1929), Box 26, CMKP, NASM; *The New York Times*, June 26, 1928, p. 34.

39. C. M. Keys to William E. Boeing, the Boeing Company, June 21, 1928, Folder 32 (Transcontinental Air Transport: Boeing Aircraft Company 1928), Box 22, CMKP, NASM.

40. Ibid.

41. Curtiss and Associated Companies, pp. 40–41, 43, 44–45.

42. Earl D. Osborn, "The Industry's Progress during 1928," *Aviation*, 26, no. 1 (January 5, 1929): 24.

43. C. M. Keys to Special Committee to Investigate Foreign and Domestic Ocean and Air Mail Contracts, January 22, 1934, U.S. Senate, 74th Congress. Special Committee on Air-Ocean Mail Contracts, Individuals: Correspondence-Income Tax He-K, Box 126, Entry 14, RG 46, NARA.

44. Ibid.

45. *The New York Times*, December 7, 1928, p. 41.

46. North American Aviation, Inc. offering circular, Folder 34 (North American Aviation: Board of Directors 1928), Box 20, CMKP, NASM.

47. North American Aviation, Inc., Annual Report, February 13, 1929, Folder 1 (Annual and Interim Reports: North American Aviation 1927–1935), Box 26, CMKP, NASM.

48. *The Wall Street Journal*, December 19, 1928, p. 26.

49. North American Aviation, Inc. offering circular, Folder 34 (North American Aviation: Board of Directors 1928), Box 20, CMKP, NASM; *The Wall Street Journal*, December 19, 1928, p. 26.

50. North American Aviation, Inc., Minutes of Board Meetings, Book I: 1928, December 17, 1928, Boeing Historical Archives, The Boeing Company.

51. *The New York Times*, December 24, 1928, p. 30.

52. See Minutes of Board and Executive Committee Meetings for December 17, 1928, January 18, February 7, March 1, and March 13, 1929, North American Aviation, Inc., Minutes of Board Meetings, Book 1, Boeing Historical Archives, The Boeing Company.

53. Aeronautical Chamber of Commerce of America, *Aircraft Yearbook for 1928* (New York, 1928), p. 3.

54. *The New York Times*, New York Curb Market Transactions, May 30, 1927, p. 25; September 12, 1927, p. 41.

55. See, for example, Arthur Leinbach, "Nation's Youngest Industry on Verge of Real Expansion: Business and Financial Aspects of Aviation," *The Magazine of Wall Street*, 40, no. 5 (July 2, 1927): 380–83, 447; Merryle Stanley Rukeyser, "What We May Expect From Commercial Aviation," *Forbes*, 20, no. 1 (July 1927): 9–11, 44–45, 47.

56. *Aircraft Yearbook for 1928*, p. 10.

57. "Report of the Aviation Securities Committee," p. 2.

58. Paul A. Dodd, "Financial Policies of the Aviation Industry," PhD Diss., University of Pennsylvania, 1933, Appendix M, p. 195; *Commercial and Financial Chronicle*, Vol. 126, May 19, 1928, p. 3034; June 16, 1928, p. 3667; Vol. 127, July 14, 1928, p. 176; September 15, 1928, p.1449; October 13, 1928, p. 2008; November 10, 1928, p. 2607; December 15, 1928, p. 3316; Vol. 128, January 19, 1929, p. 320.

59. *The New York Times*, April 21, 1929, p. 37; Dodd, *Financial Policies of the Aviation Industry*, pp. 3–4.

60. Keys and Howard Coffin had met with President Calvin Coolidge as they were organizing National Air Transport in 1925. Coolidge had specifically "laid great emphasis upon the fact that he hoped that in the establishment of the new line we would all undertake to see that the errors of the early railway financing were not repeated." Then Secretary of Commerce Herbert Hoover had given them the same advice. See C. M. Keys to J. C. Hunsaker, Federal Aviation Commission, October 2, 1934, Folder 22 (Aircraft Investigations: Howell Committee 1934), Box 27, CMKP, NASM.

61. Reeves, *Aviation's Place in Tomorrow's Business*, pp. 94–95.

62. Dodd, *Financial Policies of the Aviation Industry* (Philadelphia, 1933), Appendix II.

63. *Aero Analyst*, September–October 1929, p. 12; *National Cyclopedia of American Biography*, Current Vol. C (New York: J. T. White, 1930), p. 329. Dodd, *Financial Policies of the Aviation Industry* (Philadelphia, 1933), Appendix II, lists all the underwriters for the stock issues for Keys's companies.

64. See, for example, C. M. Keys to Blair & Company and James C. Willson & Company, September 12, 1928, concerning the Curtiss Flying Service, Folder 30 (Curtiss Flying Service/Curtiss-Wright Flying Service: Blair & Company, 1928), Box 14; G.M-P. Murphy & Co. , "Curtiss Flying Service," Folder 13 (Curtiss Flying Service/Curtiss-Wright Flying Service: Stock Trading Account, 1928–1930), Box 15, CMKP, NASM.

65. Dodd, *Financial Policies of the Aviation Industry*, p. 8.

66. Alfred D. Chandler Jr., *Strategy and Structure: Chapters in the History of the American Industrial Enterprise* (Cambridge, MA: MIT Press, 1990), pp. 42–45.

67. *The Wall Street Journal*, "Keys Explains Grouping Plan," May 14, 1929, p. 16.

68. Harry W. Hall, "On the Financial Horizon," *Pacific Flyer*, 3, no. 1 (July 1929): 7.

69. C. M. Keys to Messrs. Hall, Spitz, and Rooks, May 16, 1929, Folder 10 (Leo Spitz, 1929), Box 22, CMKP, NASM. This letter, which was apparently never sent, was in response to a complaint from Spitz, a Curtiss Aeroplane shareholder, about use of the Curtiss name to other companies. It provides an interesting discussion of the steps Keys took to expand into new business areas.

70. C. M. Keys to Messrs. Hall, Spitz, and Rooks, May 16, 1929.

71. *The Wall Street Journal*, "Keys Explains Grouping Plan," May 14, 1929, p. 16.

72. See Folders 10–15 (Frank H. Russell), Folders 20–21 (J. A. B. Smith), Box 11, Folders 1–3 (J. A. B. Smith), Box 12, Folder 19 (T. P. Wright), Box 13, CMKP, NASM, for his correspondence to the three key people at Curtiss Aeroplane & Motor Company which includes numerous examples of his requests for information and suggestions for action.

73. C. M. Keys to J. A. B. Smith, Curtiss Aeroplane & Motor Company, October 20, 1928, Folder 3 (J. A. B. Smith, July–December 1928), Box 12, CMKP, NASM.

74. See C. M. Keys to Hon. Hugo Black, United States Senate, January 25, 1934, U.S. Senate. 74th Congress. Special Committee on Air-Ocean Mail Contracts. Individuals, Box 126, Correspondence: Income Tax, He-K, Entry 14, RG 46, NARA.

75. See Blair & Co. to C. M. Keys, December 28, 29, 1928; Hemphill, Noyes & Co. to
C. M. Keys, April 23, 1929, Folder 13 (Curtiss Flying Service/Curtiss-Wright Flying
Service: Stock Trading Account 1928–1930), Box 15, CMKP, NASM; see memo
on Douglas Aircraft Company, Inc., No Par Common Stock in U.S. Senate, 74th
Congress, Special Committee on Air-Ocean Mail Contracts, North American
Aviation, Bancamerica-Blair Corp., Entry 17, Box 142, Entry 17, RG 46, NARA.
76. Summary memo, January 23, 1934, C. M. Keys & Co. and Blair Co. Syndicate Stock
Transactions, U.S. Senate, 74th Congress, Special Committee on Air-Ocean Mail
Contracts, Box 144, Entry 17, RG 46, NARA.
77. *The Wall Street Journal*, May 15, 1928, January 10, 1929; *The New York Times*, May 13,
1928.
78. Col. H. E. Hartney, "A Chance for a Harriman," *The Independent*, 114, no. 390
(March 28, 1925): 355, 364.
79. Earl Reeves, "Why Aviation's Future Is Strictly Business," *Forbes*, April 1, 1929, p. 13.
80. Howard Mingos, "The Harriman of Aviation," *New York Herald Tribune*, June 2,
1929, p. 12.
81. Ibid.
82. Ibid.
83. Forbes, "Aviation Industry Raising Millions of Dollars on Wall Street," April 1, 1929,
p. 82.
84. Rae, *Climb to Greatness*, p. 40. See also van der Linden, *Airlines and Air Mail*,
pp. 48–49.
85. van der Linden, *Airlines and Air Mail*, p. 49.
86. "Three Aircraft Firms Combine in $150,000,000 Deal," *Air Transportation*, 5, no. 17
(December 22, 1928): 1; *The Wall Street Journal*, December 17, 1928, p. 4.
87. Dively, "Investment Status of Aviation Securities," p. 8.
88. van der Linden, *Airlines and Air Mail*, pp. 52–61; "The Aviation Corporation,"
Air Transportation, 12, no. 15 (July 19, 1930): 13–18; *Air Transportation*, 6, no. 10
(March 16, 1929): 1; 7, no. 1 (April 13, 1929): 1.
89. *The New York Times*, May 19, 1929, p. N9.
90. C. M. Keys to Richard F. Hoyt, Hayden, Stone & Company, August 23, 1921, Folder 2
(Hoyt, Richard F. 1921–1929), Box 18, CMKP, NASM.
91. See the correspondence between Keys and Hoyt in Folder 2 (Richard F. Hoyt,
1921–1929), Box 18, CMKP, NASM; "Curtiss and Associated Companies"; *The
Wall Street Journal*, April 9, 1929, p. 5; April 13, 1929, p. 15; April 27, 1929, p. 10.
92. "Business and Industrial Leaders Discuss the Trend of Business," *The Magazine of
Business*, 55 (April 1929): 382; Mingos, "The Harriman of Aviation," p. 18.
93. See The Commercial National Bank and Trust Company, *The American Aviation
Industry* (New York, 1929), and Bowers, *Curtiss Aircraft*, pp. 314–17.
94. Foss, "Merger Groupings in Aviation," p. 424.
95. Dively, "Investment Status of Aviation Securities," p. 8.
96. Ibid.
97. George V. Worthington, "Avoiding Security Air-Pockets: A Critical Analysis of Aviation Investing," *The Magazine of Wall Street*, 43, no. 14 (April 20, 1929): 1098–99.
98. *The New York Times*, June 25, 1929, p. 46.
99. *The Wall Street Journal*, June 28, 1929, p. 6; June 29, 1929, p. 10.
100. *The New York Times*, June 29, 1929, p. 33.
101. C. M. Keys press release, June 28, 1929, Folder 5 (Publicity-Bruno (H.A.)-Blythe
(R.R.) [legal-sized documents 1919]), Box 30, CMKP, NASM.

102. "Curtiss-Wright Corporation," Hayden, Stone & Co., Bancamerica-Blair Corporation, James C. Willson & Co., August 1, 1929, Folder 16 (Circulars (Offering and Special), Curtiss-Wright Corp., 1929–1931), Box 26, CMKP, NASM.

103. Emmett V. Mann, "Aviation Joins Big Business," *New York Motor News* (September 1929): 13.

Chapter 7

1. This brief discussion of the concept of Nemesis is drawn from the website www.theoi.com, the website of the Theoi Project, dedicated to exploring Greek mythology, and from C. Kerenyi's book, *The Religion of the Greeks and Romans* (London: Thames and Hudson, 1962). The quote comes from the reference page on Nemesis.

2. See C. M. Keys to J. Cheever Cowdin, January 19, 1932, Folder 38 (J. Cheever Cowdin, 1932–1934), Box 27, CMKP, NASM; C. M. Keys to Hon. Hugo Black, United States Senate, January 25, 1934; C. M. Keys to A. G. Patterson. Special Senate Committee, February 6, 1934, U.S. Senate, 74th Congress, Special Committee on Air-Ocean Mail Contracts, Individuals, Box 126, He-K, Clement M. Keys, Box 126, Entry 14, RG 46, NARA.

3. Keys's correspondence from 1930–1932 concerning his financial difficulties is missing from his papers. In his correspondence with the Senate Special Committee to Investigate Air-Ocean Mail Contracts, Keys did not specify who these advances were made to, nor did he provide the individual amounts borrowed, but he later did provide a record of the monies still owed to C. M. Keys & Co. by the firm's clients. I am assuming that these were at least some of the clients he lent money to in 1928 and 1929. See Keys to Hon. Hugo Black, January 25, 1934, and C. M. Keys to F. J. Fell Jr., Henry M. Hogan, and Henry G. Hotchkiss, October 5, 1934, Folder 10 (F. J. Fell), Box 28, CMKP, NASM.

4. Carosso, *Investment Banking in America*, pp. 253–54.

5. Keys's papers do not contain a clear description of what exactly transpired with James C. Willson in September and October of 1929. The foregoing paragraph is my own interpretation of events based on the correspondence that has survived in Keys's papers and in the files of the Senate Special Committee to Investigate Air-Ocean Mail Contracts. In Keys's papers see C. M. Keys & Co. to F. H. Hardy, October 24, 1929; James C. Willson & Co. to Messers. Francis H. Hardy & C. M. Keys & Co., May 3, 1930; C. M. Keys to James C. Willson, June 18, 1930; James C. Willson & Co. to C. M. Keys & Co., August 12, 1930; James C. Willson & Co. to Messers. F. H. Hardy and C. M. Keys & Co., August 20, 1930; C. M. Keys to R. J. Haddow, C. M. Keys & Co., August 29, 1930, all in Folder 19 (James C. Willson, 1929–1931), Box 25; Memorandum Re: Morton Securities Corporation, Folder 46 (James C. Willson, 1933–1937), Box 28, CMKP, NASM. Keys's response to an inquiry from A. G. Patterson, an investigator working with the Senate Special Committee, mentions that call loans from North American Aviation were placed with members of the New York Stock Exchange which may refer to the loans to the Willson Holding Company. See C. M. Keys to A. G. Patterson, Special Committee, United States Senate, January 31, 1934, U.S. Senate. 74th Congress. Special Committee on Air-Ocean Mail Contracts, Individuals, Box 126, He-K, Clement M. Keys, Entry 14, RG 46, NARA.

6. *Barron's*, August 19, 1929, p. 10; "Curtiss-Wright Corporation: Survey of Subsidiary Companies," August 15, 1929, p. 128. Copy in the library of the Aviation Hall of Fame and Museum of New Jersey, Teterboro, NJ.

7. *The New York Times*, August 8, 1929, p. 36.

8. *Barron's*, September 9, 1929, p. 19.

9. Ibid.

10. *The New York Times*, October 9, 1929, p. 54.

11. T. Allen Grover, "The Monthly Financial Review," *Aeronautics* 5, no. 3 (September 1929): 75.

12. Aeronautical Chamber of Commerce of America, *Aircraft Yearbook for 1930* (New York: 1930), p. 3.

13. B.C. Forbes and R.W. Schabacker, "Winter Letdown in Air Stocks Foreseen," *Aeronautics*, 5, no. 5 (November 1929): 22.

14. Ibid.

15. *The Wall Street Journal*, October 23, 1929, p. 1.

16. Ibid.

17. Wigmore, *The CRASH and Its Aftermath: A History of Securities Markets in the United States, 1929-1933* (Westport, CT: Greenwood Press, 1985), p. 15; Carosso, *Investment Banking in America*, p. 305.

18. *Air Transportation*, September–October–November issues.

19. *The Wall Street Journal*, November 5, 1929, p. 1.

20. Ibid.

21. Ibid.

22. Ibid.

23. Wigmore, *The CRASH and Its Aftermath*, pp. 5, 12–15.

24. This is my interpretation of what transpired at C. M. Keys & Co. after the October Crash based on the few letters that deal with the situation. Again, there are no records for C. M. Keys & Co. in Keys's papers. See C. M. Keys to Hon. Hugo Black, United States Senate, January 25, 1934, U.S. Senate, 74th Congress, Special Committee on Air-Ocean Mail Contracts, Individuals, Box 126, He-K, Clement M. Keys, Entry 14, RG 46, NARA; C. M. Keys to A. G. Patterson. Special Senate Committee, February 6, 1934, U.S. Senate, 74th Congress, Special Committee on Air-Ocean Mail Contracts, Individuals, Box 126, He-K, Clement M. Keys, Entry 14, RG 46, NARA; North American Aviation, Inc.–: Financial Construction of North American Aviation, Inc., Subsidiaries and Affiliated Companies- and Other Data, Prepared for the Senate Special Committee to Investigate Foreign and Domestic Ocean and Air Mail Contracts, by E. C. Sauer and S. C. Simon, April 25, 1934, U.S. Senate, 74th Congress, Special Committee on Air-Ocean Mail Contracts, North American Aviation, Box 145, Entry 17, RG 46, NARA; C. M. Keys to J. Cheever Cowdin, Bancamerica-Blair Corporation, January 19, 1932, Folder 38 (J. Cheever Cowdin, 1932–1934), Box 27, CMKP, NASM.

25. See C. M. Keys to Hon. Hugo Black, United States Senate, January 25, 1934, U.S. Senate. 74th Congress. Special Committee on Air-Ocean Mail Contracts, Individuals, Box 126, He-K, Clement M. Keys, Entry 14, RG 46, NARA; Memorandum Re Interview with Mr. Talbot in the Late Afternoon of November 11, 1932, U.S. Senate, 74th Congress, Special Committee on Air-Ocean Mail Contracts, Aviation Corporation, Correspondence 1932, Box 139, Entry 17, RG 46, NARA; C. M. Keys to F. J. Fell Jr., Henry M. Hogan, and Henry G. Hotchkiss, October 5, 1934, Folder 10 (F. J. Fell), Box 28, CMKP, NASM.

26. *Los Angeles Times*, December 31, 1929, p. 11.

27. *The Wall Street Journal*, January 4, 1930, p. 7.

28. "What Are the Prospects for 1930?" *Aviation* 28, no. 7 (February 15, 1930): 291.

29. Ibid.

30. Broadus Mitchell, *The Economic History of the United States Volume IX: Depression Decade: From the New Era to the New Deal 1929–1941* (New York: M. E. Sharpe, 1947), pp. 31–32, 83–85; Wigmore, *The Crash and Its Aftermath*, pp. 128–29.

31. Wigmore, *The Crash and Its Aftermath*, pp. 130–31.

32. Ibid., pp. 141–43.

33. Stock price data taken from issues of Air Transportation.

34. Ibid.

35. North American Aviation, Inc., Annual Report for 1929, March 3, 1930.

36. Minutes of the Executive Committee, January 17, February 14, 1930, North American Aviation, Inc., Minutes of the Board of Directors, Book 3, North American Aviation, Inc., Minutes of the Executive Committee, 1928–1934, Boeing Historical Archives, the Boeing Company; Commercial & Financial Chronicle, February 22, 1930.

37. Ibid.

38. Robert H. Montgomry, Ed., *Financial Handbook*, Fifth Printing (New York, 1925), p. 206.

39. *The Wall Street Journal*, September 6, 1930, p. 3.

40. North American Aviation, Inc., Semi-Annual Report, July 15, 1930.

41. See *The Wall Street Journal*, March 22, 1930, p. 1; *Barron's*, April 7, 1930, p. 21.

42. *The Wall Street Journal*, March 22, 1930, p. 1.

43. Curtiss-Wright Corporation, Annual Report for 1929, March 28, 1930.

44. Curtiss-Wright Corporation, Annual Report for 1929, March 28, 1930; Curtiss-Wright Corporation, Interim Report, August 29, 1930.

45. Ibid.

46. Curtiss-Wright Corporation, Annual Report for 1929, March 28, 1930; Curtiss-Wright Corporation, Interim Report, August 29, 1930.

47. United Aircraft & Transport Corporation, Annual Report for the Year Ended December 31, 1929; *The Wall Street Journal*, March 20, 1930, p. 3; *Barron's*, April 7, 1930, p. 21.

48. Curtiss-Wright Corporation, Annual Report for 1929, March 28, 1930.

49. Ibid.

50. *Aviation*, 28, no. 14 (April 5, 1930): 735.

51. C. M. Keys to C. R. Keys, July 7, 1930, Folder 13 (Buffalo Office: C. Roy Keys, 1918–1930), Box 8, CMKP, NASM.

52. C. M. Keys to Charles Wright Jr., Beaumont, Smith & Harris, August 29, 1930, Folder 10 (Howard E. Coffin, General File 1928–1929), Box 7, CMKP, NASM.

53. J. Cheever Cowdin to C. M. Keys, April 29, 1930, Folder 25 (J. Cheever Cowdin, 1928–1932), Box 7, CMKP, NASM.

54. *Barron's*, June 16, 1930, p. 13.

55. *The Wall Street Journal*, September 6, 1930, p. 8.

56. Curtiss-Wright Corporation, Interim Report, August 29, 1930.

57. C. M. Keys to J. Cheever Cowdin, January 1931, Folder 25 (J. Cheever Cowdin, 1928–1932), Box 7, CMKP, NASM.

58. North American Aviation, Inc.: Financial Construction of North American Aviation, Inc., Subsidiaries and Affiliated Companies and Other Data, Prepared for the Senate Special Committee to Investigate Foreign and Domestic Ocean and Air Mail Contracts, by E. C. Sauer and S. C. Simon, April 25, 1934, pp. 11–13, U.S. Senate, 74th Congress, Special Committee on Air-Ocean Mail Contracts, North American Aviation, Box 145, Entry 17, RG 46, NARA.

59. *Barron's*, January 5, 1931, p. 12.

60. Wigmore, *The Crash and Its Aftermath*, p. 144.

61. *Barron's*, January 5, 1931, p. 12.

62. Aeronautical Chamber of Commerce, *Aircraft Yearbook for 1931* (New York, 1931), p. 14.

63. R. R. Doane, "Aeronautical Finance in 1930," *Aviation*, 30, no. 1 (January 1931): 45–47.

64. Ibid., p. 45.

65. "The Industry Comments on 1931 and on Prospects," *Airway Age*, 12, no. 1 (January 1931): 20.

66. Ibid, p. 21.

67. Ibid.

68. Ibid., p. 22.

69. Chandler, *Strategy and Structure: Chapters in the History of the American Industrial Enterprise*, p. 11.

70. Hughes, *American Genesis: A Century of Invention and Technological Enthusiasm, 1870–1970*, pp. 216–17.

71. Peter F. Drucker, *Management: Tasks, Responsibilities, Practices* (New York: Harper and Row, 1974), pp. 380–81.

72. C. M. Keys to T. A. Morgan, April 28, 1950, Folder 39 (Sperry Corp., 1932–1951), Box 28, CMKP, NASM.

73. C. M. Keys to Charles L. Lawrance, Aeronautical Chamber of Commerce, March 23, 1931, Folder 5 (Aeronautical Chamber of Commerce of America, Inc.: 1930–31), Box 1, CMKP, NASM.

74. W. Langhorne Bond (James E. Ellis, Editor), *Wings for an Embattled China* (Bethlehem, PA: Lehigh University Press, 2001), p. 22.

75. North American Aviation, Annual Report for 1930, copy in Folder 1 (Annual and Interim Reports: North American Aviation, 1927–1935), Box 26, CMKP, NASM.

76. North American Aviation, Inc., Minutes of Board Meetings, February 18, 1931, Book 3, Boeing Historical Archives, The Boeing Company, Seattle, WA.

77. *The New York Times*, February 18, 1931, p. 26; March 13, 1931, p. 36; North American Aviation, Inc., Minutes of Board Meetings, February 18, March 19, 1931, Book 3, Boeing Historical Archives, The Boeing Company, Seattle, WA.

78. North American Aviation, Annual Report for 1930, copy in Folder 1 (Annual and Interim Reports: North American Aviation, 1927–1935), Box 26, CMKP, NASM.

79. *The Wall Street Journal*, August 3, 1931.

80. North American Aviation, Annual Report for 1930; *Barron's*, April 13, 1931, p. 17; *The Wall Street Journal*, August 3, 1931.

81. Ibid., p. 17; May 4, 1931, p. 26; *The New York Times*, February 21, 1931, p. 27; North American Aviation, Inc., Minutes of Board Meetings, March 19, 1931, Book 3, Boeing Historical Archives, The Boeing Company, Seattle, WA.

82. *The New York Times*, March 27, 1931, p. 44.

83. *Aviation*, 30, no. 5 (May 1931): 266.

84. Curtiss-Wright Corporation, Annual Report for 1930, copy in Folder 35 (Curtiss-Wright Corporation 1930–1940), Box 25, CMKP, NASM.

85. Curtiss-Wright Corporation, Annual Report for 1930.

86. Curtiss-Wright Corporation, Annual Report for 1930; Aeronautical Chamber of Commerce of America, Inc., *Aircraft Yearbook for 1931* (New York, 1931), pp. 118–19.

87. Curtiss-Wright Corporation, Annual Report for 1930; *The Wall Street Journal*, September 28, 1931, p. 8.

88. Curtiss-Wright Corporation, Annual Report for 1930.

89. Aeronautical Chamber of Commerce of America, Inc., *Aircraft Yearbook for 1931*, p. 496; *The Wall Street Journal*, September 28, 1931, p. 8.

90. Curtiss-Wright Corporation, Annual Report for 1930.

91. Ibid.

92. *The Wall Street Journal*, June 24, 1931, p. 1.

93. Dodd, *Financial Policies in the Aviation Industry*, p. 7.

94. *The Wall Street Journal*, September 28, 1931, p. 8.

95. Ibid.

96. Ibid.

97. *The Wall Street Journal*, June 25, 1931, p. 1.

98. "North American Aviation, Inc.: Financial Construction of North American Aviation, Inc. Subsidiaries and Affiliated Companies, and Other Data" by E. C. Sauer and S. C. Simon contains a statement of C. M. Keys & Co.'s call loans due to North American Aviation for the years 1929–1932 on p. 10. U.S. Senate, 74th Congress, Special Committee on Air-Ocean Mail Contracts, North American Aviation, Box 145, Entry 17, RG 46, NARA.

99. Wigmore, *The CRASH and Its Aftermath*, pp. 235–37, 286–87.

100. Ibid., pp. 237–39; Carosso, *Investment Banking in America*, pp. 307–09.

101. Carosso, *Investment Banking in America*, p. 307.

102. Wigmore, *The CRASH and Its Aftermath*, pp. 217–19.

103. Ibid., pp. 236–37.

104. Quoted in David M. Kennedy, *Freedom from Fear: The American People in Depression and War, 1929–1945* (New York: Oxford University Press, 1999), p. 89.

105. See C. M. Keys to F. J. Fell, Pennsylvania Railroad Company, September 5, 1934; C. M. Keys to F. J. Fell, Henry M. Hogan, and Henry G. Hotchkiss, October 8, 1934, Folder 10 (F. J. Fell, 1934), Box 28, CMKP, NASM.

106. E. C. Sauer and S. C. Simon, "North American Aviation, Inc.: Financial Construction of North American Aviation, Inc. Subsidiaries and Affiliated Companies, and Other Data," p. 10. U.S. Senate, 74th Congress, Special Committee on Air-Ocean Mail Contracts, North American Aviation, Box 145, Entry 17, RG 46, NARA.

107. North American Aviation, Inc., Minutes of Board Meetings, Executive Committee Meeting of January 5, 1932, Book 3, Boeing Historical Archives, The Boeing Company, Seattle, WA.

108. C. M. Keys to J. Cheever Cowdin, January 19, 1932. A note on this letter in Keys's handwriting says that it was never sent. Folder 38 (Cheever Cowdin, 1932–1934), Box 27, CMKP, NASM.

109. *The Wall Street Journal*, January 8, 1932, p. 1.

Chapter 8

1. Richard Robbins, President, Transcontinental & Western Air, Inc. to C. M. Keys, February 19, 1932, Folder 18, Box 46, Charles A. Lindbergh Papers, Missouri Historical Society, St. Louis, MO.

2. Richard Robbins to C. M. Keys, February 19, 1932.

3. Thomas A. Morgan, Memorandum to the Chairman of the Board, Curtiss-Wright Corporation, October 19, 1932, pp. 5–6, copy in the author's possession.

4. North American Aviation, Inc., Minutes of Board Meetings, Book 3, Board of
   Directors, January 6, 1932, Boeing Historical Archives, Seattle, WA.

5. North American Aviation, Inc., Minutes of Board Meetings, Book 3, Board of
   Directors, January 11, February 10, 1932, Boeing Historical Archives, Seattle, WA.

6. North American Aviation, Inc., Minutes of Board Meetings, Book 3, Board of
   Directors, February 10, March 9, 1932, Boeing Historical Archives, Seattle, WA;
   North American Aviation, Inc. Annual Report for 1931, February 26, 1932, copy
   in Folder 1 (Annual and Interim Reports: North American Aviation 1927–1935),
   Box 26, CMKP, NASM.

7. *The Wall Street Journal*, March 12, 1932, p. 1.

8. Banning, *Airlines of Pan American since 1927*, p. 280.

9. Leary, *Dragon's Wings*, pp. 70–73.

10. *The New York Times*, January 22, 1933, p. N7.

11. *The New York Times*, January 27, p. 28; March 19, p. N5; April 27, p. 25; April 29, p. 19;
    May 25, 1933, p. 29; *The Wall Street Journal*, April 6, 1933, p. 7; *The Los Angeles Times*,
    June 15, 1933, p. 14.

12. C. M. Keys to Hon. Hugo Black, U.S. Senate, January 25, 1934, U.S. Senate, 74th
    Congress, Special Committee on Air-Ocean Mail Contracts, Individuals Income
    Tax: He-K, Clement M. Keys, Entry 14, Box 126, RG 46, NARA.

13. Report "North American Aviation, Inc.: Financial Construction of North American
    Aviation, Inc. Subsidiaries and Affiliated Companies, and Other Data" by E. C. Sauer
    and S. C. Simon contains a statement for the amounts C. M. Keys & Co. owed to the
    North American subsidiaries on p.10. U.S. Senate, 74th Congress, Special Committee
    on Air-Ocean Mail Contracts, North American Aviation, Entry 17, Box 145, RG 46,
    NARA; for the amount owed to Transcontinental Air Transport, see C. M. Keys
    to Hon. Hugo Black, U.S. Senate, January 25, 1934, U.S. Senate. 74th Congress.
    Special Committee on Air-Ocean Mail Contracts, Individuals Income Tax: He-K,
    Clement M. Keys, Entry 14, Box 126, RG 46, NARA.

14. North American Aviation, Inc., Minutes of Board Meetings, Book 3, Executive
    Committee, February 10, 1932, Boeing Historical Archives, Seattle, WA.

15. C. M. Keys to F. J. Fell, Henry M. Hogan, Henry G. Hotchkiss, October 5, 1934,
    Folder 10 (Fell, F. J. 1934), Box 28, CMKP, NASM.

16. Certificate of Continuation of Partnership Name Pursuant to Sections 80 and
    81 of the Partnership Law, June 16, 1932, copy in the Office of the County Clerk,
    New York County, New York City, New York.

17. See C. M. Keys's letter to Col. Charles A. Lindbergh, December 8, 1938, File 517,
    Box 18, Charles A. Lindbergh Collection, MS 325, Manuscripts and Archives,
    Sterling Memorial Library, Yale University, New Haven, CT.

18. Interview with Gordon Keys, February 21, 2005.

19. Interview with Gordon Keys, August 27, 2005.

20. C. M. Keys to C. N. Griffis, Lima, Peru, October 3, 1947, Folder 15 (C. N. Griffis,
    1947), Box 28, CMKP, NASM; "Eagle Hatched," *Time*, November 25, 1947, p.56;
    The New York Times, "Highways and Byways of Finance," October 31, 1947.

21. *The New York Times*, November 12, 1946, p. 2; December 6, 1946, p. 37; December 29,
    1946, p. 28; March 1, 1949, p. 50; Davies, *A History of the World's Airlines*, p. 349.

22. *The New York Times*, July 7, 1948, p. 33; March 1, 1949, p. 50; Davies, *A History of the
    World's Airlines*, p. 349.

23. *The New York Times*, January 13, 1952, p. 89.

24. Howard Mingos, "The Harriman of Aviation," *New York Herald Tribune*, June 2, 1929, pp. 12–13, 18.
25. Joseph Schumpeter, *The Theory of Economic Development: An Inquiry into Profits, Capital, Credit, Interest, and the Business Cycle* (Cambridge, MA: Harvard University Press, 1934), pp. 93–94.
26. C. M. Keys, "The Business End of Aviation, May 8, 1929, Folder 5 (Speeches and Articles by C. M. Keys 1925–1931), Box 22, CMKP, NASM.
27. Schumpeter, *The Theory of Economic Development*, p. 88.
28. Hughes, "The Evolution of Large Technological Systems," pp. 50–51.
29. Schumpeter, *The Theory of Economic Development*, p. 89.
30. Ibid.
31. Ibid., p. 70.
32. Byttebier, *The Curtiss D-12 Engine*, p. 109.
33. Ian Lloyd, *Rolls-Royce: The Years of Endeavour* (London: Macmillan Press, 1978), pp. 96–97.
34. Ibid., pp. 160–61.
35. My thanks to Dr. Robert van der Linden, Curator in the Aeronautics Department at the National Air and Space Museum for developing this idea.
36. Leary, *Dragon's Wings*, p. 69.
37. Thomas A. Morgan, Memorandum to the Chairman of the Board, Curtiss-Wright Corporation, October 19, 1932, p. 2, copy in the author's possession.
38. Schumpeter, *The Theory of Economic Development*, p. 78.
39. Mayo and Nohria, *In Their Time*, p. xxiii; Chandler, *The Visible Hand*, p. 7.
40. C. M. Keys to Thomas Morgan, April 28, 1950, Folder 39, Box 28, CMKP, NASM.
41. Thomas A. Morgan, Memorandum to the Chairman of the Board, Curtiss-Wright Corporation, October 19, 1932.
42. "Clement Melville Keys Dies," U.S. Air Service (February 1952), p. 24.

Appendix

1. Folder 25, Box 18, Clement Melville Keys Papers, NASM.

BIBLIOGRAPHY

Archives

Boeing Historic Archives

North American Aviation, Inc., Minutes of the Executive Committee, 1928–1934

Hagley Museum and Library, Wilmington, Delaware.

Accession Number 1810: Records of the Pennsylvania Railroad: Office of Passenger Transportation.

Accession Number 1807: Records of the Pennsylvania Railroad, Board of Directors Minutes

Houghton Library, Harvard University, Cambridge, Massachusetts.

The Papers of Walter Hines Page

The Library of Congress

The Papers of Glenn L. Martin

Missouri Historical Society, St. Louis, Missouri

Charles A. Lindbergh Collection

Museum of Flight, Seattle, Washington

Curtiss-Wright Collection

National Air and Space Museum, Washington, DC

Clement Melville Keys Papers, Accession Number XXXX.0091

National Archives and Records Administration, Washington, DC

Record Group 18, Records of the Army Air Forces, 1917–1938, Entry 22.

Record Group 18, Records of the Army Air Forces, 1917–1938, Entry 87, Records of the Aircraft Production Board.

Record Group 18, Records of the Army Air Forces, 1917–1938, Entry 110, Office of the Chief of the Air Service, Correspondence, Reports, Studies, and other Records Relating to the Functions of the Bureau of Aircraft Production and the Division of Military Aeronautics, during World War I, 1917–1921.

Record Group 18, Records of the Army Air Forces, 1917–1938, Entry 111, General Records of the Chief of the Air Service.

Record Group 18, Records of the Army Air Forces, 1917–1938. Entry 166, File 452.1 Curtiss Airplanes.

Record Group 18, Records of the Army Air Forces, 1917–1938. Entry 228, Office of the Chief of the Air Corps, Correspondence of Maj. Gen. Mason Patrick 1922–1927.

Record Group 18, Army Air Forces Central Decimal Files, 1917–1938. Entry 230, Correspondence of Maj. Gen. James Fechet.

Record Group 18, Records of the Army Air Forces, 1917–1938. Entry 166, File 452.1 Sale of Planes Abroad.

Record Group 45, Naval Records, Collection of the Office of Naval Records and Library, Subject File 1911–1927 GC: Aircraft Construction.

Record Group 46. U.S. Senate. Special Committee on Investigation of Air Mail and Ocean Mail Contracts, 1934.

Record Group 72, Bureau of Aeronautics General Correspondence 1925–1942, QM (26), Curtiss Aeroplane and Motor Company.

Record Group 72, Bureau of Aeronautics, Entry 5, General Correspondence Initiated in the Office of the Chief of Naval Operations, 1917–1925.

Record Group 80, Records of the Secretary of the Navy, 1916–1925.

Record Group 151, Records of the Bureau of Foreign and Domestic Commerce.

United Airlines Legacy Foundation, Chicago, Illinois.

Predecessor Companies & Subsidiaries (National Air Transport, Inc.).

Yale University, New Haven, Connecticut.

Charles A. Lindbergh Collection.

Government Documents

America's Munitions 1917–1918: Report of Benedict Crowell, Assistant Secretary of War, Director of Munitions. Washington, DC, 1919.

Report of American Aviation Mission (Aircraft Journal, August 23, 1919).

Interstate Commerce Commission. *Nineteenth Annual Report on the Statistics of Railways in the United States for the Year Ending June 30, 1906.* Washington, DC: Government Printing Office, 1907.

President's Aircraft Board (Morrow Board) *Report.* Washington, DC: Government Printing Office, 1925.

United States Army Aircraft Production Facts, Col. G. W. Mixter and Lt. H. H. Emmons. Washington, DC: Government Printing Office, 1919.

U.S. Congress. House. Committee on Military Affairs. *The Curtiss Elmwood Airplane Plant, Buffalo, N.Y., Hearing Before a Subcommittee of the Committee on Military Affairs,* 66th Congress, 1st Session, November 10, 1919. Washington, DC: Government Printing Office, 1920.

U.S. Congress. House. Committee on Military Affairs. *Reorganization of the Army.* Hearings on H.R. 7925, 66th Congress, 2nd Session, August 29, 1919. Washington, DC: Government Printing Office, 1920.

U.S. Congress. House. Committee on Ways and Means. *Importation of Surplus Airplanes.* Hearings, 66th Congress, 3rd Session, Part I, May 28, 1920; Part 2, June 1, 1920. Washington, DC: Government Printing Office, 1920.

U.S. Congress. House. Select Committee of Inquiry Into Operations of the United States Air Services. Hearings, 68th Congress, 1st Session. Washington, DC: Government Printing Office, 1925.

U.S. Congress. House. Subcommittee No. 1 (Aviation) of the Select Committee on Expenditures in the War Department. *Hearings on War Expenditures* Volume 3, 66th Congress, 1st Session, November 14, 1919. Washington, DC: Government Printing Office, 1920.

U.S. Congress. House. Subcommittee of the Military Affairs Committee. *A United Air Service.* Hearing on H.R. 10380 and H.R. 9804, 66th Congress, 2nd Session, December 4, 1919. Washington, DC: Government Printing Office, 1920.

U.S. Congress. Senate. Committee on Military Affairs. *Aircraft Production: Hearings Before the Subcommittee of the Committee on Military Affairs,* 65th Congress, 2nd

Session, May 29–August 15, 1918, Volumes I and II. Washington, DC: Government Printing Office, 1918.

U.S. Congress. Senate. Subcommittee of the Committee on Military Affairs. *Reorganization of the Army*, 66th Congress, 1st Session, August 7–November 5, 1919. Washington, DC: Government Printing Office, 1919.

U.S. Congress. Senate. Special Committee to Investigate Air and DC, Ocean Mail Contracts. Hearings. *Investigation of Air Mail and Ocean Mail Contracts*, 73rd Congress, 2nd Session, September 26, 1933, to May 25, 1934. Washington, DC: Government Printing Office, 1934.

War Office. *America's Munitions 1917–1918: Report of Benedict Crowell the Assistant Secretary of War, Director of Munitions*. Washington, DC: Government Printing Office, 1919.

Magazines and Newspapers

Aerial Age Weekly
Aero Analyst
Aero Digest
Aeronautics
Air Transportation
Aircraft Journal
Aircraft Yearbook
Airway Age
Automotive Industries
Aviation
Aviation and Aeronautical Engineering
Barron's
Commercial and Financial Chronical
Forbes
Literary Digest
NAT Bulletin Board
News Wing
Railway Age
Review of Reviews
The Curtiss Flyleaf
The Curtiss Fuselage
The Los Angeles Times
The Magazine of Wall Street
The New York Times
The Slipstream
The Wall Street Journal
The World's Work
U.S. Air Services
Western Flying

Primary Sources

Allen, C. B. "Curtiss Moves to Buffalo." *Aero Digest*, Vol. 18, No. 5 (May 1931), pp. 54–55.

"Air Mail Bins Opened by Postmaster Show United Company Low Bidder." *Air Transportation*, Vol. 13, No. 5 (August 30, 1930), pp. 1, 24.

"Airline Merger Reports Definitely Denied by Western Air and T.A.T." *Air Transportation*, Vol. 12, No. 10 (June 14, 1930), p. 1.

"All Aboard the Lindbergh Limited!" *Literary Digest* (March 2, 1929), pp. 55–58.

"An Indictment and a Warning." *Aviation*, Vol. 16, No. 4 (January 28, 1924), p. 86.

"Astonishing Air Mail Bids." *Aviation*, Vol. 22, No. 4 (January 24, 1927), p. 170.

Atterbury, W. W. (General). "Linking Rail and Air Transport." *Aero Digest*, Vol. 13, No. 6 (December 1928), pp. 1113–14, 1120.

———. "Linking Rail and Air Transport: The 'Iron Horse' Shares Honors with the 'All-Metal Bird.'" *The American Aviator: Airplanes and Airports*, Vol. 2, No. 2 (May 1929), pp. 32–39.

———. *The Railroads Enter Aviation*. Philadelphia: Pennsylvania Railroad Information, Vol. 1, No. 2, 1929.

"Aviation Corporation Buys Seven Fairchild Companies." *Air Transportation*, Vol. 7, No. 1 (April 13, 1929), pp. 1, 41.

"Aviation, Inc., $200,000,000 Aircraft Holding Corporation." *Air Transportation*, Vol. 6, No. 10 (March 16, 1929), p. 15.

Black, Archibald. "Air Transport Investment and Operating Cost." *Automotive Industries*, Vol. 53, No. 13 (September 17, 1925), pp. 456–59, 65.

Bliss, Harry. "Clement M. Keys: He Bought a Company Nearly Insolvent." *Air Transportation* (July 27, 1929), pp. 51, 55.

Bowen, R. Sidney, Jr. "Trends of the Industry during 1930." *Aviation*, Vol. 30, No. 1 (January 1931), pp. 15–17.

Campbell, E. G. *The Reorganization of the American Railroad System 1893–1900*. New York: Columbia University Press, 1938.

Carson, William E. "Profits in Aeroplanes." *The Magazine of Wall Street*, Vol. 20, No. 9 (August 4, 1917), pp. 582–86.

Chamberlain, Lawrence. *The Principles of Bond Investment*. New York: Henry Holt and Company, 1911.

———. *The Work of the Bond House*. New York: Moody's Magazine Book Department, 1912.

"Chicago-Dallas Air Mail Line." *Aviation*, Vol. 20, No. 14 (April 5, 1926), pp. 490–91.

"Chicago-Dallas Air Mail Opened." *Aviation*, Vol. 20, No. 21 (May 24, 1926), pp. 783–85.

"Chicago-Dallas Air Mail Service." *Aviation*, Vol. 20, No. 19 (May 10, 1926), pp. 706–7.

"Clement Melville Keys Dies." *U.S. Air Service* (February 1952).

Cleveland, Frederick A., and Fred Wilbur Powell. *Railroad Finance*. New York: D. Appleton and Company, 1912.

The Commercial National Bank and Trust Company of New York. *Financial Handbook of the American Aviation Industry*. New York, 1929.

Conant, Charles A. "The Concentration of Capital in New York, and Those Who Manage It." *The Bankers' Magazine*, Vol. 75, No. 5 (November 1907), pp. 659–66.

"Conference of Air Transport Officials." *Aviation*, Vol. 20, No. 10 (March 8, 1926), pp. 333–34.

"Curtiss-Keys Group Buys Pitcairn Aviation Inc." *Air Transportation*, Vol. 8, No. 4 (June 22, 1929), pp. 1, 20.

"Curtiss-Wright Assets Approximate $220,000,000." *Air Transportation*, Vol. 8, No. 6 (July 6, 1929), pp. 1, 12.

"Curtiss-Wright Corporation Unites Large Aviation Groups." *Aviation*, Vol. 27, No. 1 (July 6, 1929), pp. 45, 50.

Crowell, Benedict, and Robert Forrest Wilson. *How America Went to War: The Armies of Industry II*. New Haven, CT: Yale University Press, 1921.

———. *How America Went to War: Demobilization*. New Haven, CT: Yale University Press, 1921.

Daggett, Stuart. *Railroad Reorganization*. Boston: Houghton, Mifflin and Company, 1908.

"Dallas Reports Pennsylvania Railroad Plans Rail and Air Combination Journeys." *Air Transportation*, Vol. 2, No. 11 (March 31, 1928), pp. 1–2.

Darling, Velva G. "Across the Continent in Forty-Eight Hours." *The World's Work*, Vol. LVIII, No. 9 (September 1929), pp. 52–56.

Denby, Edwin. "The Aeronautical Industry and the National Defense." *U.S. Air Service*, Vol. 8, No. 1 (January 1923), pp. 9–10.

Dewing, Arthur Stone. *The Financial Policy of Corporations*, Vols. I and II. New York: Ronald Press, 1941.

Dively, George S. "Investment Status of Aviation Securities." *Barron's*, Vol. 9, No. 28 (July 15, 1929), p. 8.

Doane, R. R. "Aeronautical Finance in 1930." *Aviation*, Vol. 30, No. 1 (January 1931), pp. 45–49.

Durand, Dr. W. F. "An Analysis of the Need for Civil Aviation." *Automotive Industries*, Vol. 41, No. 18 (October 30, 1919), pp. 872–75.

———. "The Need for Civil Air Transport: Part II." *Automotive Industries*, Vol. 41, No. 19 (November 6, 1919), pp. 922–24.

———. "The Need for Civil Air Transport: Part III." *Automotive Industries*, Vol. 41, No. 20 (November 13, 1919), pp. 968–71.

———. "The Need for Civil Air Transport." *Automotive Industries*, Vol. 41, No. 21 (November 20, 1919), pp. 1020–24.

Eaton, J. Shirley. "Railroad Operations: How to Know Them from a Study of the Accounts and Statistics." *The Railroad Gazette*. New York, 1900.

Forbes, B. C. *Automotive Giants of America: Men Who Are Making Our Motor Industry*. New York: B. C. Forbes Publishing Company, 1926.

Forbes, B. C., and R. W. Schabacker. "Winter Letdown in Air Stocks Foreseen." *Aeronautics*, 5, issue 5 (November 1929), pp. 22, 98.

Foss, Charles W. "Merger Groupings in Aviation." *Airway Age*, Vol. 10, No. 4 (April 1929), pp. 424–27.

Fowler, John Francis, Jr. *American Investment Trusts*. New York: Harper and Brothers, 1928.

Froelich, Michael H. "Bright 1930 Ahead Is Consensus of Industry's Executives." *Air Transportation*, Vol. 10, No. 11 (December 28, 1929), pp. 1–2.

Guggenheim, Harry F. "Air Transportation and the Railroads." *The Slipstream*, Vol. 9, No. 3 (March 1928), pp. 7–9.

Gwinn, Sherman. "Aviation Is about to Come of Age." *American Magazine*, Vol. 108 (September 1929), pp. 51–53, 99–100.

Hall, Harry W. "On the Financial Horizon." *Pacific Flyer*, Vol. 3, No. 1 (July 1929), pp. 7–8.

Harrington, John Walker. "He Put Wings on American Aviation." *Management* (October 1929), pp. 36–39, 72–73.

Hartney, Col. H. E. "A Chance for a Harriman: Dropping the War Jinx in Commercial Aviation." *The Independent*, Vol. 114, No. 390 (March 28, 1925), pp. 355–56, 364.

Henderson, Col. Paul. "More Flying-Less Talking." *Western Flying*, Vol. 1, No. 4 (April 1926), pp. 8–9.

Hendrick, Burton J. *The Life and Letters of Walter H. Page*. New York: Doubleday, Page and Company, 1926.

"Hoyle." *The Game in Wall Street, and How to Play It Successfully*. New York: J. S. Ogilvie Publishing Company, 1898.

Johnson, Emory R. *American Railway Transportation*, Revised Edition. New York: D. Appleton and Company, 1908.

Jones, Eliot. *Principles of Railroad Transportation*. New York: Macmillan Company, 1931.

Jones, H. A. *The War in the Air: Being the Story of the Part Played in the Great War by the Royal Air Force*, Vol. IV. Oxford: Clarendon Press, 1937.

Jordon, David F. *Jordon on Investments*, Revised Edition. New York: Prentice Hall, 1930.

Keys, Clement M. "The Advance Agent of Prosperity." *The World's Work*, Vol. 17, No. 3 (January 1909), pp. 11164–67.

———. "Air Transport and the Railroads." *Air Transportation*, Vol. 15, No. 17 (December 22, 1928), p. 21.

———. "Air Transport Grows Up." *Pacific Flyer*, Vol. 2, No. 1 (January 1929), pp. 9–10.

———. "Air Transportation—When Will It Pay?" *The World's Work*, Vol. 52, No. 3 (July 1926), pp. 239–40.

———. "The Airplane in Commerce." *The Financial Diary*, Vol. I, No. 3 (April 1929).

———. "Airplanes Grow Useful." *Air Transportation* (August 28, 1929), p. 6.

———. "An Era of Better Railroads." *The World's Work*, Vol. 17, No. 4 (February 1909), p. 11238243.

———. "As Many Railroad Methods as Railroad Kings." *The World's Work*, Vol. 10, No. 5 (September 1905), pp. 6652–59.

———. "Aviation and Its Future." *U.S. Air Service*, Vol. 14, No. 3 (March 1929), pp. 43, 46–48.

———. "The Business of Aviation—1929–30." *Airway Age*, Vol. 11, No. 1 (January 1930), pp. 40–42.

———. "Cassatt and His Vision." *The World's Work*, Vol. 20, No. 3 (July 1910), pp. 13187–204.

———. "Conditions Affecting Aeronautic Progress." *Journal of the Society of Automotive Engineers*, Vol. 17, No. 5 (November 1925), pp. 446, 449–51.

———. "The Contest for Pacific Traffic." *The World's Work*, Vol. 10, No. 4 (August 1905), pp. 6503–509.

———. "A 'Corner' in Pacific Railroads." *The World's Work*, Vol. 9, No. 4 (February 1905), pp. 5816–22.

———. "The Curtiss Aeroplane and Motor Corporation after the War." *Aerial Age Weekly* (June 6, 1921), pp. 295–96.

———. "The Gambler's Chance—and the Penalty." *The World's Work*, Vol. 18, No. 2 (June 1908), pp. 11648–650.

———. "Harriman I: The Man in the Making: His Early Life and Start." *The World's Work*, Vol. 13, No. 3 (January 1907), pp. 8455–464.

———. "Harriman II: The Building of His Empire." *The World's Work*, Vol. 13, No. 4 (February 1907), pp. 8537–552.

———. "Harriman III: The Spinner of Golden Webs." *The World's Work*, Vol. 13, No. 5 (March 1907), pp. 8651–664.

———. "Harriman IV: The Salvage of the Two Pacifics." *The World's Work*, Vol. 13, No. 6 (April 1907), pp. 8791–803.

———. "How Men Get Rich Now." *The World's Work*, Vol. 11, No. 3 (January 1906).

———. "The Large Corporation: The Standard Oil Company." *The World's Work*, Vol. 16, No. 4 (August 1908), pp. 10571–590.

———. "The Money Kings I: The Realm of Credit and Its Rulers," *The World's Work*, Vol. 14, No. 6 (October 1907), pp. 9475–481.

———. "The Money Kings II: Wall Street and the Banks." *The World's Work*, Vol. 15, No. 1 (November 1907, pp. 9520–534.

———. "The Money Kings III: Regulating Banks by Vigilance Committee." *The World's Work*, Vol. 15, No. 2 (December 1907), pp. 9705–711.

———. "The Money Kings IV: Safeguarding the Trust Companies." *The World's Work*, Vol. 15, No. 4 (February 1907), pp. 9907–912.

———. "My Impressions of European Aviation." *The Slipstream*, Vol. 19, No. 1 (January 1928), pp. 9–11.

———. "The New Morals of Business." *The World's Work*, Vol. 27, No. 6 (April 1914), pp 620–25.

———. "The Newest Railroad Power." *The World's Work*, Vol. 10, No. 2 (June 1905), pp. 6302–13.

———. "The Outlook for Commercial Aviation." *Machinery* (February 1929), pp. 419–21.

———. "The Overlords of Railroad Traffic." *The World's Work*, Vol. 13, No. 3 (January 1907), pp. 8437–445.

———. "The Railroad Fights for Life." *The World's Work*, Vol. 20, No. 5 (September 1910), pp. 13419–431.

———. "The Shifting Railroad Control." *The World's Work*, Vol. 20, No. 2 (January 1910), pp. 13045–56.

———. "The Shipper's Fight for Life." *The World's Work*, Vol. 20, No. 6 (October 1910), pp. 13555–564.

———. "A Society Mystery." *The Canadian Magazine*, Vol. 15 (June 1900).

———. "The Swindling Promoter at Work." *The World's Work*, Vol. 18, No. 5 (September 1909), pp. 11987–990.

———. "Trends in Military Airplanes." *Journal of the Society of Automotive Engineers*, Vol. 21, No. 5 (November 1927), pp. 452, 596.

———. "Value of Racing Planes." *Aero Digest*, Vol. 7, No. 4 (October 1925), pp. 535–36, 568.

———. "What the Rail-Air Business Means." *Pennsylvania Railroad Information* (January 1929), pp. 1–12.

Knappen, Theodore M. "Consolidation for Mastery of the Air." *The Magazine of Wall Street*, Vol. 46, No. 8 (August 9, 1930), pp. 598–600, 628–29.

———. *Wings of War: An Account of the Important Contribution of the United States to Aircraft Invention, Engineering, Development and Production during the World War.* New York: G. P. Putnam, 1920.

Lane, D. R. "How the T.A.T. Line Was Organized." *Airway Age*, Vol. 10, No. 7 (July 1929), pp. 1021–24.

Leinbach, Arthur M. "Nation's Youngest Industry on Verge of Real Expansion." *The Magazine of Wall Street*, Vol. 40, No. 5 (July 2, 1927), pp. 380–83, 447.

Lewis, Frederick. "Curtiss Aeroplane-Its Great Business." *The Magazine of Wall Street*, Vol. 22, No. 7 (July 6, 1918), pp. 528–30.

"Lindbergh's Boss." *New Yorker* (June 16, 1928), pp. 16–17.

Lloyd, Ian. *Rolls-Royce: The Years of Endeavour*. London: Macmillan Press, 1978, pp. 96–97.

Marache, Paul J. "Unscrambling the Aviation Eggs." *Western Flyer*, Vol. 15, No. 3 (March 1935), pp. 10–12.

Marsh, E. L. *The History of the County of Gray*. Owen Sound, Ontario: Fleming Publishing Company, 1931.

Marshall, Edward. "The Job of Getting Uncle Sam into the Air." *Forbes*, Vol. 20, No. 6 (September 15, 1927), pp. 12–14, 18.

———. What the Lindbergh Flight Means to Business." *Forbes*, Vol. 19, No. 12 (June 15, 1927).

Meiklejohn, Bernard. "The Conquest of the Air." *The World's Work*, Vol. 13, No. 2 (December 1906), pp. 8283–96.

Mingos, Howard. "Flying Speeds the Pace of Transportation." *The Magazine of Wall Street*, Vol. 48, No. 2 (May 16, 1931), pp. 83–85, 124.

Mingos, Howard."The Harriman of Aviation." *New York Herald Tribune* (June 2, 1929), pp. 12–13.

Mixter, Col. G. W., and Lt. H. H. Emmons. *United States Army Aircraft Production Facts* Washington, DC: Government Printing Office, 1919, pp. 5–6.

Montgomery, Robert H., Ed. *Financial Handbook*. New York: Ronald Press, 1925.

Moody's Investors Service. *Moody's Manual of Investments*.

Morrow, John. "Outlook for Three War Brides." *The Magazine of Wall Street*, Vol. 24, No. 6 (July 5, 1919), pp. 524–28.

Mosher, Clinton. L. "Money Miracle Saved Aviation." *Brooklyn Eagle* (March 4, 1928).

"National Air Transport, Incorporated." *U.S. Air Services* (August 1925), pp. 15–17.

"National Air Transport Organized." *Aviation*, Vol. 18, No. 22 (June 1, 1925), pp. 598–600.

"N.A.T. Awarded New York–Chicago Contract," *Aviation*, Vol. 22, No. 15 (April 11, 1927), p. 731.

"N.A.T. Contracts with American Express." *Aviation*, Vol. 21, No. 20 (November 15, 1926), p. 850.

"N.A.T. Planes Flown over 700,000 Miles during First Year of Operation." *Aviation*, Vol. 23, No. 2 (July 11, 1927), p. 80.

"N.A.T. Votes Million Dollar Increase." *Air Transportation*, Vol. 3, No. 9 (June 16, 1928), p. 2.

Newcomb, H. T. *Railway Economics*. Philadelphia: Railway World Publishing Company, 1898.

"New Mail Routes Bring about Realignment of Lines: T.A.T.–Western Air Name New Company for Line Operation." *Air Transportation*, Vol. 14, No. 5 (October 11, 1930), pp. 1, 4.

Nixon, L. A. "Aeronautical Industry Unhurt by Panic on Stock Exchange Belief of Major Executives in the Trade." *Air Transportation*, Vol. 10, No. 4 (November 9, 1929), pp. 1–2.

Nixon, L. A. "Railroads Posted on Planes But Not Ready to Buy Reports of Plane Makers Say." *Air Transportation*, Vol. 2, No. 5 (February 18, 1928), pp. 1–2.

"North American Aviation Would Acquire Assets of Baltimore Firm." *Air Transportation* (June 21, 1930), p. 2.

Noyes, Alexander, D. *Forty Years of American Finance*. New York: G. P. Putnam's Sons, 1909.

Noyes, Alexander D. *The War Period of American Finance: 1908–1925*. New York: G. P. Putnam's Sons, 1926.

Page, Walter Hines. "On a Tenth Anniversary." *The World's Work*, Vol. 21, No. 3 (January 1911), pp. 13903–917.

Pratt, Serano S. *The Work of Wall Street: An Account of the Functions, Methods, and History of the New York Money and Stock Markets*. New York: D. Appleton and Company, 1903.

Pratt, Serano S. *The Work of Wall Street: An Account of the Functions, Methods, and History of the New York Money and Stock Markets*. New York: D. Appleton and Company, 1912.

Pynchon & Co. *The Aviation Industry*. New York, 1928.
"Railroad Executives Visit Europe to Study Air Line Services." *Air Transportation*, Vol. 13, No. 9 (June 16, 1928), p. 11.
Reeves, Earl. *Aviation's Place in Tomorrow's Business*. New York: B. C. Forbes Publishing Company, 1930.
Reeves, Earl. "Why Aviation's Future Is Strictly Business." *Forbes*, Vol. 13, No. 7 (April 1, 1929), pp. 13, 48, 50–51.
Rentschler, Frederick B. "Is Aviation Over Expanded?" *The Magazine of Business*, Vol. 56 (July 1929), pp. 37–38.
"Reorganization of the Curtiss Corporation." *Aviation*, Vol. 14, No. 13 (March 26, 1923), p. 342.
"Review of Curtiss Aeroplane & Motor Co., Inc." *The Slipstream*, Vol. 18, No. 4 (April 1927), p. 20.
Rukeyser, Merryle S. "What We May Expect from Commercial Aviation." *Forbes*, Vol. 20, No. 1 (July 1, 1927), pp. 9–11, 44–45, 47,
Schumpeter, Joseph A. *The Theory of Economic Development: An Inquiry into Profits, Capital, Credit, Interest, and the Business Cycle*. Harvard Economic Studies, Vol. 46. Cambridge, MA: Harvard University Press, 1934.
"Shutting Down the Aviation Industry." *Aviation*, Vol. 16, No. 11 (March 17, 1924), pp. 282–83.
Slocum, Arthur N. "Curtiss Aeroplane Status and Prospects." *The Magazine of Wall Street*, Vol. 21, No. 3 (November 10, 1917), pp. 217–220.
Snyder, Carl. *American Railroads as Investments*. New York: G. P. Putnam's Sons, 1907.
Sprague, F. Desmond. "Air Transport Making Rapid Progress." *Railway Age* (February 26, 1927), pp. 577–79.
Sterwart, Andrew S. "Curtiss: The Romance of One of the Oldest Names in Aviation." *Air Transportation*, Vol. 9, No. 11 (September 28, 1929), pp. 26–34.
Sweetser, Arthur. *The American Air Service: A Record of Its Problems, Its Difficulties, Its Failures, and Its Final Achievements*. New York: Doubleday and Company, 1919.
"Texas-Chicago Air Mail Operates 97 per cent Perfect." *Aviation*, Vol. 21, No. 8 (August 23, 1926), p. 320.
"The Big Merger," *Aviation*, Vol. 27, No. 1 (July 6, 1929), pp. 24–25.
"The Curtiss-Wright Corporation: Its Historical Development, Scope of Activities, and Executive Personnel." *Aero Digest*, Vol. 24, No. 4 (April 1934), pp. 28–35, 60.
"The Engineering and Business Genius Back of the Curtiss Aeroplane and Motor Company." *Aerial Age Weekly*, October 21, 1918, Vol. 8, No. 6.
"The T.A.T. Air-Rail Service." *Railway Age*, Vol. 87, No. 1 (July 6, 1929), pp. 12–15.
"The United States Naval Air Service, 1922–23." *Aviation*, Vol. 16, No. 2 (January 14, 1924), p. 37.
Thompson, Slason: *A Short History of American Railroads*. Chicago: Bureau of Railway News and Statistics, 1925.
University of Toronto. *The University of Toronto and Its Colleges, 1827–1906*. Toronto: The University Library, 1906.
———. *Honors Classics in the University of Toronto, by a group of classical graduates, with a foreword by Sir Robert Falconer*. Toronto: University of Toronto Press, 1929.
Van Antwerp, W. C. *The Stock Exchange from Within*. New York: Doubleday, Page and Company, 1913.
Van Oss, S. F. *American Railroads as Investments*. New York: G. P. Putnam & Sons, 1893.

Wagner, Ted. "The Curtiss Robin." *Air Transportation*, Vol. 4, No. 11 (September 29, 1928).

Warner, Edward P. "Commercial Aviation in 1923." *The Journal of the Society of Automotive Engineers*, Vol. 14, No. 4 (April 1924), pp. 457–62.

Westervelt, Cdr. G. C., Cdr. H. C. Richardson, and Lt. Cdr. A. C. Read: *The Triumph of the NCs*. New York: Doubleday, Page & Company, 1920.

"What Are the Prospects for 1930?" *Aviation* (February 15, 1930), pp. 290–305.

Willys, John North. "Flying's Commercial Future." *Collier's, The National Weekly*, Vol. 65, No. 11 (March 13, 1920), pp. 10, 32.

Wines, James P. "The 48 HR. Coast to Coast Air-Rail Service of Transcontinental Air Transport." *Aviation*, Vol. 27, No. 1 (July 6, 1929), pp. 26–29.

"Winning Profits from Aviation." *Review of Reviews*, Vol. 79 (June 1929), pp. 110, 112, 114, 116.

Woodlock, Thomas F. *The Anatomy of a Railroad Report and Ton Mile Cost*. New York: S. A. Nelson, 1900.

Worthington, Gerard V. "Avoiding Security Air-Pockets: A Critical Analysis of Aviation Investing." *The Magazine of Wall Street*, Vol. 43, No. 14 (April 20, 1929), pp. 1098–100, 1158.

Secondary Sources

Abramson, Rudy. *Spanning the Century: The Life of W. Averell Harriman 1891–1986*. New York: William Morrow and Company, 1992.

Appleby, Joyce. *The Relentless Revolution: A History of Capitalism*. New York: W. W. Norton, 2010.

Banner, Stuart. *Who Owns the Sky?: The Struggle to Control Airspace from the Wright Brothers On*. Cambridge, MA: Harvard University Press, 2008.

Banning, Gene. *Airlines of Pan American since 1927*. McClean, VA: Paladwr Press, 2001.

Baruch, Bernard M. *Baruch: My Own Story*. New York: Henry Holt and Co., 1957.

Bauer, Eugene. *Boeing: The First Century*. Enumclaw, WA: TABA Publishing, 2000.

Beattie, Kim. *Ridley: The Story of A School*, Volume One and Two. St. Catherines, Ontario: Ridley College, 1963.

Bednarek, Janet R. Daly. *America's Airports: Airfield Development 1918–1947*. College Station, TX: Texas A&M University Press, 2001.

Berg, A. Scott. *Lindbergh*. New York: Putnam, 1998.

Bijker, Weibe, Thomas P. Hughes, and Trevor Pinch. *The Social Construction of Large Technological Systems: New Directions in the Sociology and History of Technology*. Cambridge, MA: MIT Press, 2021,

Bilstein, Roger E. *Flight in America: From the Wrights to the Astronauts*, Rev. Ed. Baltimore, MD: Johns Hopkins University Press, 1994.

———. *Flight Patterns: Trends in Aeronautical Development in the United States, 1918–1929*. Athens, GA: University of Georgia Press, 1983.

———. *The American Aerospace Industry*. New York: Twayne Publishers, 1996.

Bissell, Claude T., Ed. *University College: A Portrait 1953–1953*. Toronto: University of Toronto Press, 1953.

Bond, W. Langhorne (James E. Ellis, Editor). *Wings for an Embattled China*. Bethlehem, PA: Lehigh University Press, 2001.

Bowers, Peter. *Curtiss Aircraft 1907–1947*. London: Putnam, 1979.

Brebner, John Bartlett. *Canada: A Modern History*, Rev. Ed. Ann Arbor: University of Michigan Press, 1970.

Burgess, George H., and Miles C. Kennedy. *Centennial History of the Pennsylvania Railroad 1846–1946*. Philadelphia: Pennsylvania Railroad Company, 1949.

Byttebier, Hugh T. *The Curtiss D-12 Engine*. Smithsonian Annals of Flight, Number 7. Washington, DC: Smithsonian Institution Press, 1972.

Carosso, Vincent P. *Investment Banking in America: A History*. Cambridge, MA: Harvard University Press, 1970.

Carter, Susan B., Ed. *Historical Statistics of the United States: Earliest Times to the Present, Millennial Edition, Volume 4, Part D: Economic Section*. New York: Cambridge University Press, 2006.

Casari, Robert B. *Encyclopedia of U.S. Military Aircraft: The World War I Production Program*, Volume 3: *The Curtiss Jennies*. Chillicothe, OH: Military Aircraft Publications, 1975.

Casey, Louis S. *Curtiss: The Hammondsport Era 1907–1915*. New York: Crown Publishers, 1981.

Chambers, John Whiteclay. II. *The Tyranny of Change: America in the Progressive Era, 1890-1920*, 2nd edition. New Brunswick, NJ: Rutgers University Press, 2000.

Chandler, Alfred D., Jr. *Strategy and Structure: Chapters in the History of the American Industrial Enterprise*. Cambridge, MA: MIT Press, 1990.

———. Ed. *The Railroads: The Nation's First Big Business: Sources and Readings*. New York: Arno Press, 1965.

———. *The Visible Hand: The Managerial Revolution in American Business*. Cambridge, MA: Belknap Press, 1977.

Chernow, Ron. *Titan: The Life of John D. Rockefeller, Sr.* New York: Random House, 1998.

Churella, Albert J. *The Pennsylvania Railroad*, Volume I: *Building an Empire, 1846–1917*, Philadelphia: University of Pennsylvania Press, 2013.

Courtwright, David T. *Sky as Frontier: Adventure, Aviation, and Empire*. College Station: Texas A&M University Press, 2005.

Curcio, Vincent, *Chrysler: The Life and Times of an Automotive Genius*. New York: Oxford University Press, 2000.

Davilla, James J., and Arthur M. Soltan. *French Aircraft of the First World War*. Stratford, CT, 1997.

Davies, R. E. G. *A History of the World's Airlines*. London: Oxford University Press, 1964.

———. *Airlines of Asia since 1920*. London: Putnam Aeronautical Books, 1997.

———. *Airlines of Latin America since 1919*. London: Putnam Aeronautical Books, 1983.

———. *Airlines of the United States Since 1914*. London: Putnam & Company, 1972.

———. *Eastern: An Airline and Its Aircraft*. McClean, VA: Paladwr Press, 2003.

———. *TWA: An Airline and Its Aircraft*. McClean, VA: Paladwr Press, 2000.

Davilla, James J., and Arthur M. Soltan. *French Aircraft of the First World War*. Stratford, CT: Flying Machine Press, 1997.

Davis, Patricia. *The End of the Line: Alexander J. Cassatt and the Pennsylvania Railroad*. New York: Neale Watson Academic Publications, 1978.

Dean, Francis H., and Dan Hagedorn. *Curtiss Fighter Aircraft: A Photographic History 1917 1948* Atglen, PA: Schiffer Military History, 2006.

Dierikx, Marc. *Fokker: A Transatlantic Biography*. Washington, DC: Smithsonian Institution Press, 1997.

———. *Anthony Fokker: The Flying Dutchman Who Shaped American Aviation*. Washington, DC: Smithsonian Books, 2018.

Dodd, Paul A. "Financial Policies of the Aviation Industry." PhD Diss., University of Pennsylvania, 1933.

Doolittle, General James H. "Jimmy," with Carroll V. Glines. *I Could Never Be So Lucky Again*. New York: Bantam Books, 1991.

Ellis, Frank H. *Canada's Flying Heritage*. Toronto: University of Toronto Press, 1954.

Eltscher, Louis R., and Edward M. Young. *Curtiss-Wright: Greatness and Decline*. New York: Twayne Publishers, 1998.

Faulkner, Harold U. *The Decline of Laissez Faire 1897–1917*. The Economic History of the United States Volume VII. New York: Rinehart, 1951.

Fowler, John Francis, Jr. *American Investment Trusts*. New York: Harper and Brothers, 1928.

Foxworth, Thomas G. *The Speed Seeker*. New York: Doubleday and Company, 1975.

Frey, Robert (Ed.). *The Encyclopedia of Business History and Biography: Railroads in the Nineteenth Century*. New York: Bruccoli Clark Layman, 1988.

Geisst, Charles R. *Wall Street: A History from Its Beginnings to the Fall of Enron*. New York: Oxford University Press, 2004.

Graham, Margaret B. W. "Entrepreneurship in the United States, 1920–2000," in David S. Landes, Joel Mokyr, & William Baumol, eds., *The Invention of Enterprise: Entrepreneurship from Ancient Mesopotamia to Modern Times*. Princeton, NJ: Princeton University Press, 2010, pp. 401–42.

Grant, John Webster. *A Profusion of Spires: Religion in Nineteenth-Century Ontario*. Toronto: University of Toronto Press, 1988.

Gregory, Ross. *Walter Hines Page: Ambassador to the Court of St. James*. Lexington: University Press of Kentucky, 1970.

Griese, Noel L. *Arthur W. Page: Publisher, Public Relations Pioneer, Patriot*. Atlanta, GA: Anvil Publishing, 2001.

Hagedorn, Dan. *Conquistadors of the Sky: A History of Aviation in Latin America*. Gainesville: University Press of Florida, 2008.

Hawley, Ellis W. *The Great War and the Search for a Modern Order*. Prospect Heights, IL: St. Martin's Press, 1997.

Hennessy, Juliet A. *The United States Army Air Arm: April 1861 to April 1917*. Washington, DC: Office of Air Force History, 1985.

Heppenheimer, T. A. *Turbulent Skies: The History of Commercial Aviation*. New York: John Wiley & Sons, 1995.

Hickerson, J. Mel. *Ernie Breech: The Story of His Remarkable Career with General Motors, Ford, and TWA*. New York: Meredith Press, 1968.

Hidy, Ralph W., and Muriel E. *History of Standard Oil (New Jersey): Pioneering in Big Business 1882–1911*. New York: Harper & Row, 1955.

Higham, Robin. *Britain's Imperial Air Routes 1918 to 1939: The Story of Britain's Overseas Airlines*. Hamden, CT: Shoe String Press, 1961.

Holley, Irving B. *Buying Aircraft: Material Procurement for the Army Air Forces*. United States Army in World War II: Special Studies. Washington, DC: U.S. Government Printing Office, 1964.

Horgan, James J. *City of Flight: The History of Aviation in St. Louis*. Gerald, MO: Patrice Press, 1984.

Hughes, Thomas P. *American Genesis: A Century of Invention and Technological Enthusiasm, 1870–1970*. Chicago: University of Chicago Press, 2004.

———. *Networks of Power: Electrification in Western Society, 1880–1930*. Baltimore, MD: Johns Hopkins University Press, 1988.

———. "The Evolution of Large Technological Systems." In Weibe Bijker, Thomas P. Hughes, and Trevor Pinch, *The Social Construction of Large Technological Systems:*

New Directions in the Sociology and History of Technology. Cambridge, MA: MIT Press, 2021, pp. 45–76.

Hurley, Alfred F. *Billy Mitchell: Crusader for Air Power*. Bloomington: Indiana University Press, 1975.

Johnson, W., and Houston, V. *Taking Flight: The Foundations of American Commercial Aviation, 1918–1938*. College Station: Texas A&M University Press, 2019.

Kennedy, David M., and Kennedy, David M. *Over Here: The First World War and American Society*, 25th Anniversary Edition. New York: Oxford University Press, 2004.

———. *The Oxford History of the United States Volume IX: Freedom from Fear: The American People in Depression and War, 1929–1945*. New York: Oxford University Press, 1999.

Klein, Maury. *The Genesis of Industrial America, 1870–1930*. Cambridge, UK, 2007.

———. *The Life and Legend of E. H. Harriman*. Chapel Hill: University of North Carolina Press, 2000.

———. *Rainbow's End: The Crash of 1929*. New York: Oxford University Press, 2001.

Komons, Nick A. *Bonfires to Beacons: Federal Civil Aviation Policy under the Air Commerce Act 1926–1938*. Washington, DC: Department of Transportation, Federal Aviation Administration, 1978.

Lamoreaux, Naomi R. *The Great Merger Movement in American Business, 1895–1904*. Cambridge: Cambridge University Press, 1985.

Lamoreaux, Naomi R., and Kenneth Sokoloff. *Financing Innovation in the United States 1870 to the Present*. Cambridge, MA: MIT Press, 2007.

Landes, David S., Joel Mokyr, and William Baumol, eds. *The Invention of Enterprise: Entrepreneurship from Ancient Mesopotamia to Modern Times*. Princeton, NJ: Princeton University Press, 2010.

Launius, Roger D., and Janet R. Daly Bednarek, eds. *Reconsidering a Century of Flight*. Chapel Hill: University of North Carolina Press, 2003.

Leary, William M. *Aerial Pioneers: The U.S. Air Mail Service, 1918–1927*. Washington, DC: Smithsonian Institution Press, 1985.

———. ed. *Aviation's Golden Age: Portraits from the 1920s and 1930s*. Iowa City: University of Iowa Press, 1989.

———. *The Dragon's Wings: The China National Aviation Corporation and the Development of Aviation in China*. Athens, GA: University of Georgia Press, 1976.

Lee, David. "Herbert Hoover and the Development of Commercial Aviation, 1921–1926." *The Business History Review*, Vol. 58, No. 1, Transportation (Spring, 1984), pp. 78–102.

Levy, Jonathan. *Ages of American Capitalism: A History of the United States*. New York: Random House, 2021.

Lindbergh, Charles A. *Autobiography of Values*. New York: Harcourt Brace Jovanovich, 1977.

Lloyd, Ian. *Rolls-Royce: The Years of Endeavour*. London: Macmillan Press, 1978.

Locklin, D. Philip. *Economics of Transportation*, Revised Edition. Chicago: R. D. Irwin, 1946.

Lloyd, Ian. *Rolls-Royce: The Years of Endeavour*. London: Macmillan Press, 1978.

Markham, Jerry W. *A Financial History of the United States, Volume II: From J.P. Morgan to the Institutional Investor (1900–1970)*. Armonk, NY: M. E. Sharpe, 2002.

Marks, Lynne. *Revivals and Roller Skates: Religion, Leisure, and Identity in Late Nineteenth Century Small Town Ontario*. Toronto: University of Toronto Press, 1996.

Martin, Albro. *Enterprise Denied: Origins of the Decline of American Railroads 1897–1917*. New York: Columbia University Press, 1971.

———. *James J. Hill and the Opening of the Northwest*. New York: Oxford University Press, 1976.

———. *Railroads Triumphant: The Growth, Rejection, and Rebirth of a Vital American Force*. New York: Oxford University Press, 1992.

Maurer, Maurer. *Aviation in the U.S. Army 1919–1939*. Office of Air Force History. Washington, DC: U.S. Government Printing Office, 1987.

Mayo, Anthony J., and Nitin Nohria. *In Their Time: The Greatest Business Leaders of the Twentieth Century*. Boston: Harvard Business School Press, 2005.

McGerr, Michael. *A Fierce Discontent: The Rise and Fall of the Progressive Movement in America*. New York: Oxford University Press, 2003.

McKillop, A. B. *Matters of Mind: The University in Ontario 1791–1951*. Toronto: University of Toronto Press, 1994.

Mehtidis, Alexis. *Italian Military Aviation in World War I 1914–1918*. Tiger Lily Publications LLC, 2005.

Meyer, Meyer, Austrian, and Platt, Chicago: *Corporate and Legal History of United Air Lines and its Predecessors and Subsidiaries*. Chicago: Twentieth Century Press, 1953.

Miller, Ronald, and David Sawers. *The Technical Development of Modern Aviation*. London: Praeger Publishers, 1970.

Mitchell, Broadus. *The Economic History of the United States Volume IX: Depression Decade: From the New Era to the New Deal 1929–1941*. New York: M. E. Sharpe, 1947.

Morris, Edmund. *Theodore Rex*. New York: Random House, 2001.

Morrow, John H., Jr. *The Great War in the Air: Military Aviation from 1909 to 1921*. Washington, DC: Smithsonian Institution Press, 1993.

Mott, Frank Luther. *A History of American Magazines*, Volume IV: 1885–1905. Cambridge, MA: Harvard University Press, 1957.

Murray, Robert K. *The Harding Era: Warren G. Harding and His Administration*. Minneapolis: University of Minnesota Press, 1969.

Navin, Thomas R., and Marian V. Sears. "The Rise of the Market for Industrial Securities, 1987–1902." *The Business History Review*, Vol. 29, No. 2 (June 1955), pp. 105–38.

Nevins, Allen, and Frank Ernest Hill. *Ford: Expansion and Challenge 1915–1933*. New York: Scribner's, 1957.

O'Sullivan, Mary A. *Dividends of Development: Securities Markets in the History of US Capitalism*. Oxford: Oxford University Press, 2016.

Ott, Julia C. *When Wall Street Met Main Street: The Quest for an Investors' Democracy*. Cambridge, MA: Harvard University Press, 2011.

Pattillo, Donald M. *Pushing the Envelope: The American Aircraft Industry*. Ann Arbor: University of Michigan Press, 1998.

Pigott, Peter. *Flying Colors: A History of Commercial Aviation in Canada*. Toronto: Douglas & McIntyre, 1997.

Porter, Glenn. *The Rise of Big Business: 1860–1920*, 2nd Edition, Harlan Davidson. Arlington Heights, IL: Crowell, 1992.

Putnam, Donald, and Donald Kerr. *A Regional Geography of Canada*. Toronto: J. M. Dent, 1965.

Rae, John B. *Climb to Greatness: The American Aviation Industry 1920–1960*. Cambridge, MA: MIT Press, 1968.

Rondall, R. Rice. *The Politics of Air Power: From Confrontation to Cooperation in Army Aviation-Civil Relations*. Lincoln: University of Nebraska Press, 2004.

Roseberry, C. R. *Glenn Curtiss: Pioneer of Flight*. New York: Doubleday and Company, 1972.

Rosenberg, Jerry M. *Inside the Wall Street Journal: The History and Power of Dow Jones & Company and America's Most Influential Newspaper*. New York: Macmillan, 1982.

Rusnak, Robert J. *Walter Hines Page and The World's Work 1900–1913*. Washington, DC: University of America Press, 1982.

Rust, Daniel L., and Alan B. Hoffman. *Come Fly with Me: The Rise and Fall of Trans World Airlines*. St. Louis: Missouri Historical Society Press, 2023.

Rutherford, Paul. *A Victorian Authority: The Daily Press in Late Nineteenth-century Ontario*. Toronto: University of Toronto Press, 1982.

Rutkowski, Edwin H. *The Politics of Military Aviation Procurement, 1926–1934: A Study in the Political Assertion of Consensual Values*. Columbus: Ohio State University Press, 1966).

Salsbury, Stephen. "The emergence of an early large-scale technical system: The American railroad network," in Mayntz, Renata, and Thomas P. Hughes, *The Development of Large Technical Systems*. New York: Routledge, 2019, pp. 37–68.

Schleit, Philip: *Shelton's Barefoot Airlines*. Cortland, NY: Max Graphics, No Date.

Schlesinger, Arthur M., Jr. *The Age of Roosevelt: The Crisis of the Old Order, 1919–1933*. Boston: Houghton Mifflin, 1957.

———. *The Age of Roosevelt: The Coming of the New Deal, 1934–1935*. Boston: Houghton Mifflin, 1958.

Schull, Joseph. *Ontario Since 1867*. Toronto: McClelland and Stewart, 1978.

Serling, Robert. *From the Captain to the Colonel: An Informal History of Eastern Airlines*. New York: Dial Press, 1980.

———. *Howard Hughes' Airline: An Informal History of TWA*. New York: St. Martin's Press, 1983.

Seyer, Sean. *Sovereign Skies: The Origins of American Civil Aviation Policy*. Baltimore, MD: Johns Hopkins University Press, 2021.

Simonson, G. R. "Demand for Aircraft and the Aircraft Industry." *Journal of Economic History*, Vol. 20, No. 3 (September 1960), pp. 361–82.

Sloan, Alfred P., Jr. *My Years with General Motors*. New York: Doubleday, 1964.

Smiley, Gene. "The Expansion of the New York Securities Market at the Turn of the Century." *The Business History Review*, Vol. 55, No. 1 (Spring 1981), pp. 75–85.

Smith, Frank Kingston. *Legacy of Wings: The Harold F. Pitcairn Story* T-D Associates. Lafayette Hill, PA: T-D Associates, 1981.

Smith, Henry Ladd. *Airways: The History of Commercial Aviation in the United States*. New York: Alfred K. Knopf, 1942.

Sobel, Robert. *The Great Bull Market: Wall Street in the 1920's*. New York: Norton, 1968.

Soule, George. *The Economic History of the United States*. Volume VIII: *Prosperity Decade: From War to Depression 1917–1929*. New York: Rinehart, 1947.

Stover, John. *The Life and Decline of the American Railroad*. New York: Oxford University Press, 1970.

Swanborough, Gordon, and Peter M. Bowers. *United States Military Aircraft Since 1909*. New Edition. London: Putnam, 1989.

Thetford, Owen. *British Naval Aircraft Since 1912*, Fifth Revised Edition. London: Putnam, 1982.

Trimble, William F. *Wings for the Navy: A History of the Naval Aircraft Factory*. Washington, DC: Naval Institute Press, 1990.

———. *Admiral William A. Moffett: Architect of Naval Aviation*. Washington, DC: Smithsonian Institution Press, 1994.

Turnbull, Archibald D., and Clifton L. Lord. *History of United States Naval Aviation*. New Haven, CT: Yale University Press, 1949.

Turner, HyB. *When Giants Ruled: The Story of Park Row, New York's Great Newspaper Street*. New York: Fordham University Press, 1999.

United Air Lines. *Corporate and Legal History of United Air Lines and Its Predecessors and Subsidiaries, 1925–45*. Chicago: The Twentieth Century Press, 1953.

Van der Linden, F. Robert. *Airlines and Air Mail: The Post Office and the Birth of the Commercial Aviation Industry*. Lexington: University Press of Kentucky, 2002.

Vander Muelen, Jacob. *The Politics of Aircraft Procurement: Building an American Military Industry*. Lawrence: University of Kansas Press, 1991.

Van Deurs, George. *Wings for the Fleet: A Narrative of Naval Aviation's Early Development, 1910-1916*. Annapolis, MD: United States Naval Institute, 1966.

Wagner, Ray. *American Combat Planes of the 20th Century*. Reno, NV: Jack Bacon & Company 2004.

Walterman, Thomas Worth. "Airpower and Private Enterprise: Federal-Industrial Relations in the Aeronautics Field, 1918–1926." PhD Diss., Washington University, 1970, (U.M.I. Dissertation Service).

Weiss, Richard. *The American Myth of Success: From Horatio Alger to Norman Vincent Peale* Basic Books. New York: Basic Books, 1969.

Wendt, Lloyd. *The Wall Street Journal: The Story of Dow Jones and the Nation's Business Newspaper*. Chicago: Rand McNally, 1982.

Westfall, William. *Two Worlds: The Protestant Culture of Nineteenth Century Ontario*. Kingston, Ontario: McGill-Queen's University Press, 1989.

White, Robert P. *Mason Patrick and the Fight for Air Service Independence*. Washington, DC: Smithsonian Institution Press, 2001.

Wigmore, Barrie A. *The CRASH and Its Aftermath: A History of Securities Markets in the United States, 1929–1933*, Westport, CT: Greenwood Press, 1985.

Wilson, Timothy. "Broken Wings: The Curtiss Aeroplane Company, K Boats, and the Russian Navy, 1914–1916." *Journal of Military History*, Vol. 66, No. 4 (October 2002), pp. 1061–1063.

Wyllie, Irvin G. *The Self-Made Man in America: The Myth of Rags to Riches*. New Brunswick, NJ: Rutgers University Press, 1954.

Unpublished Documents and Dissertations

Curtiss Aeroplane & Motor Corporation. *Before the Commissioner of Internal Revenue and the Committee on Appeal and Review, In the Matter of the War and Excess Profits and Income Taxes of Curtiss Aeroplane & Motor Corporation for the Years 1917, 1918, and 1919, Memorandum Submitted by the Corporation*, January 31, 1922. Copy in Black 3-ring binder, Box TY5 Curtiss-Wright Tax and Legal Documents, 1916 to 1946, Wright Collection, Museum of Flight, Seattle, Washington.

Curtiss Aeroplane & Motor Company, Inc. *House of Representatives Select Committee of Inquiry into Operations of the United States Air Services: Testimony of C. M. Keys, President*, No Date. Copy in the author's possession.

Curtiss-Wright Corporation. "Record Dates in the History of Curtiss-Wright Corporation," 1940, copy Box 6, Curtiss-Wright Corporation Histories and Publications Files, Wright Collection, Museum of Flight, Seattle, Washington.

Dodd, Paul A. "Financial Policies in the Aviation Industry," A Thesis in Economics Presented to the Faculty of the Graduate School in Partial Fulfillment of the Requirements for the Degree of Doctor of Philosophy, University of Pennsylvania, 1933.

Gregg, A. B. "Bristol Fighter," September 1919, Record Group 18, General Records of the Chief of the Air Service, Entry 111, Box 4, National Archives.

———. "History of the SE-5 Single Seat Pursuit Plane," No date, Records Group 18, General Records of the Chief of the Air Service, Entry 110, File 452.1, Box 37, National Archives.

———. "SPAD—"Single Seat Pursuit Plane," October 1919, Records Group 18, General Records of the Chief of the Air Service, Entry 111, Box 4, National Archives.

McFarland, R. M. "Bureau of Aircraft Production: History of the Bolling Aeronautical Mission," October 29, 1919, Records Group 18, General Records of the Chief of the Air Service, Entry 111, Box 4, National Archives.

North American Aviation, Inc. "Brief History of North American Aviation, Inc. Date of Incorporation Through December 31, 1934," no date. Box 290, Hatfield Collection, Museum of Flight, Seattle, Washington.

Seymour, Lester D. *Annual Report to the Executive Committee of National Air Transport, Inc. for the Year 1929,* U.S. Senate.74th Congress. Special Committee on Investigation of Air Mail and Ocean Mail Contracts, 1934. Entry 17: United Airlines, Box 152, Record Group 46, NARA.

———. *Summary of National Air Transport Activity as of March 1, 1929.* Copy in the author's possession.

"Supply of Aircraft." Record Group 45, Naval Records, Collection of the Office of Naval Records and Library, Subject File 1911–1927 GC: Aircraft Construction. Box 135, Folder 2, National Archives and Records Service, Washington, DC.

Transcontinental Air Transport, Inc. "Transcontinental Air Transport, Inc." 1929 copy in the author's possession

Walterman, Thomas Worth. "Airpower and Private Enterprise: Federal-Industrial Relations in the Aeronautics Field, 1918–1926." PhD Diss., Washington University, 1970.

James C. Willson & Co. "Curtiss Aeroplane & Motor Company, Inc. and Associated Companies," March 20, 1929, copy in the author's possession.

INDEX

horizontal expansion plans, 177–83;
Keys as Harriman of aviation, 200–
202; merging of Curtiss and Wright
groups, 203–8; North American
Aviation, 189–91; stock market, raising
capital in, 191–94; vertical expansion plans, 183–86. *See also* North
American Aviation; United Aircraft &
Transport Corporation (UATC)
aviation industry, 1–6. *See also* commercial aviation; military aviation
aviation records, 59, 63–65, 70–71, 89–90

Baker, Newton, 24, 45, 46, 56–57, 71, 90
Baldwin, Frederick W. "Casey," 9
Bancamerica-Blair Corporation, 207
bank failures (1931), 226, 234–35
barnstorming era, 111
BAT (Boeing Air Transport), 133, 138, 144,
147, 148, 158, 202–3, 222
Bell, Alexander Graham, 18, *19*
Bendix Aviation Corporation, 204
Bilstein, Roger, xiii
Bingham, Hiram, 117
Bixby, Harold, 131–32, 180
Blair, Lt. Col. James, 46, 190
Blair & Company, 190, 191
Boeing, William E., 4–5, 86, 133, 188, 200,
202–3
Boeing Airplane and Transport
Corporation, 202–3
Boeing Airplane Company, 97, 132–33,
188, 202
Boeing Air Transport (BAT), 133, 138, 144,
147, 148, 158, 202–3, 222
Boeing Model 80, 141
Bolling Mission, 26
bond broker, Keys as, 17, 18
Bradley, Samuel, *46*, 109, 110
Breech, Ernest, 166
Bristol F2B two-seat fighter, 26, 27–28,
33, 37
Brown, Walter Folger, 144–47, 153–54,
160–61, 166, 168
Buffalo, New York factory, 20, 25–27, 34,
37, 70, 89, 231
Bureau of Aeronautics, 82, 87, 109
Bureau of Aircraft Production, 33–34
Burke, Frank, 95–96

C. M. Keys & Co.: Depression's burdens
on, 234–36; establishment of, 17–18;
as financer of aviation projects, 176;
and Great Crash of 1929, 215–17; great
expectations and losses (1931), 224–
28; growth of, 198–99; Keys's financial
recovery process, 241–42; and Keys's
misuse of funds, 209–12, 215–17, 236;
and North American Aviation, 191;
underwriting of Curtiss-Robertson
Airplane, 181; use of to finance loan
for North American Aviation, 211–12,
216, 225
C-12 (formerly K-12) engine, 59–60
Canadian Pacific Railway Company, 15
Caproni, Gianni, 183
Caproni tri-plane bomber, 26
Cassatt, Alexander, 11, 14, 15
CAT (Colonial Air Transport), 124, 144
CD-12 (formerly C-12) engine, 63
Chance Vought Company, 202, 203
Chandler, Alfred, 228
Chapin, Roy, 120
Chicago, Rock Island & Pacific Railway, 15
Chicago–Dallas air mail service route,
123, 125
Chile, Intercontinent investment in, 172
China, international airline development,
170–71, 173, 174–75
China Airways Federal, 173, 174
China National Aviation Corporation
(CNAC), 171, 174–75
Cia De Aviacion Faucett S.A., 172
civil aviation. *See* commercial aviation
Coffin, Howard: on Aircraft Production
Board, 24, 26–27, 34; and commer-
cial aviation framework, 117–18; and
federal regulation drive for aviation,
96, 109–10, 112–13; and initial offer
to Lindbergh, 128; and loss of NAT
to Boeing Air Transport, 148–51;
as Montauk Beach Development
Company director, 199; as NAT president, 120–21, 122, 124, 126–27; photos,
46, *120*
Colonial Air Transport (CAT), 124, 144
Commerce Department, as center for
commercial aviation regulation, 108,
109–10, 117

236–37; Keys's use of to finance loan from C. M. Keys & Co., 211–12, 216, 225; post-Crash adjustments at, 218–21, 227, 228–31; and purchase of Pitcairn Aviation, 152; recapitalization, 240; and sale of NAT to UATC, 149; as separate from Curtiss-Wright, 206
North Elmwood factory, Buffalo, NY, 26–27, 30–31, 33, 50–51, 60
Northern Pacific Railroad, 15
Northrup Alpha airplanes, 167–68
Northrup Corporation, 167–68
NYSE (New York Stock Exchange), 192, 219

Ordnance Engineering Corporation, 52
Orenco D pursuit aircraft, 52

Pacific Air Transport, 203, 222
Page, Walter Hines, 14–16
PAIC (Pittsburgh Aviation Industries Corporation), 161–62
Pan American Airways, 203, 204, 242
Parker, James, 117
passenger air transport: air-rail services, 132–34, 140, 155–56, 239; challenges of, 144; and creating TAT, xiv, 126–41; European acceptance of flying, 130; National Air Transport Flying Service, 142–43; overcoming individual reluctance to fly, 42–43; post-WWI visions for, 41. *See also* airline systems
Patrick, Maj. Gen. Mason, 66–67, 70, 77–79, 82–84, 86, 89, 98
Pennsylvania Railroad, 14, 15, 131–33, 156, *162*, 165
Pershing, Maj. Gen. John, 79–80
Peru, Intercontinent investment, 172
Peruvian International Airways (PIA), 242–43
Pitcairn, Harold, 151–53
Pitcairn Aviation, 151–53, 216
Pittsburgh Aviation Industries Corporation (PAIC), 161–62
Plesman, Albert, 129
Post Office. *See* airmail service
Pratt & Whitney Company, 4, 202
Prince, J. F., 31–32

private aviation: Keys and Hoyt's misguided belief in demand for, 205–6, 226–27, 249; Lindbergh's inspiration of, 179, 181, 185; planning for, 40, 183; post-Crash decline of, 222
proprietary design protection issue with military contracts, 75, 76, 87, 90–91, 104–5
Pulitzer, Ralph, 59
Pulitzer Trophy Races, 59, 71, 83

racing planes and races, 59, 63–65, 70–71, 89–90
Radio Corporation of America, 141
railroad industry: air-rail services, 132–34, 155–56, 239; challenges of transcontinental connection, 128; comparison of Keys to scions of, 200–201; cornering markets in, 15; growth compared to aviation, 1; as inspiration for Keys's aviation ambitions, 55–56, 106; Keys's education on, 10–14, 192–93; Pennsylvania Railroad, 14, 15, 131–33, 156, *162*, 165
Raskob, John J., 81–82
Reeves, Earl, 200
registration of aircraft, 40, 108
regulation: Air Mail Act passage as initiation for, 116–17; air mail's contribution to, 111–12; drive to get government to adopt, 107–10, 112–13; establishment of, 2; lack of progress in aviation without, 111; NACA, 23–24, 74, 108–10
Reid Aircraft Company, 182
Reid Rambler, 182
Rentschler, Frederick, 4, 147–48, 150–51, 202–3, 227
research and development, importance of business incentive to, 76–77, 80
Reynolds, Earle, 147, 150
Ridley College, 9
Rittenhouse, Lt. David, 70–71, 90
Robbins, Richard, 166–68
Robertson, William B., 131, 170–72, 179–81
Robertson Aircraft Corporation, 179
Rockefeller, William A., 120
Rock Island Line, 14
Rolls-Royce Kestrel engine, 247